Lecture Notes in Physics

Edited by J. Ehlers, München, K. Hepp, Zürich
R. Kippenhahn, München, H. A. Weidenmüller, Heidelberg
and J. Zittartz, Köln

154

Macroscopic Properties of Disordered Media

Proceedings of a Conference
Held at the Courant Institute
June 1–3, 1981

Edited by R. Burridge, S. Childress, and G. Papanicolaou

Springer-Verlag
Berlin Heidelberg New York 1982

Editors

R. Burridge
S. Childress
G. Papanicolaou
Courant Institute, New York University
New York, NY 10012, USA

ISBN 3-540-11202-2 Springer-Verlag Berlin Heidelberg New York
ISBN 0-387-11202-2 Springer-Verlag New York Heidelberg Berlin

Printed in Germany

Printing and binding: Beltz Offsetdruck, Hemsbach/Bergstr.
2153/3140-543210

TABLE OF CONTENTS

CONFERENCE ON THE MACROSCOPIC PROPERTIES OF DISORDERED MEDIA

June 1-2-3, 1981

COURANT INSTITUTE, NEW YORK UNIVERSITY, 251 MERCER STREET,
NEW YORK, N.Y. 10012

MONDAY JUNE 1

9:00 - 9:40	R. Burridge, Courant Institute, Poroelastic equations from microstructure
9:40 - 10:20	P. Saffman, Cal. Inst. Tech., Fingering in porous media
10:20 - 10:40	BREAK
10:40 - 11:20	J. Berryman, Bell Laboratories, Elastic waves in fluid-saturated porous media
11:20 - 12:00	S. Childress, Courant Institute, Approximations of Brinkman type
12:00 - 1:30	LUNCH
1:30 - 2:10	R. O'Connell, Harvard Univ., Porous media, self-consistent methods
2:10 - 2:50	D. Johnson, Schlumberger-Doll, Elastodynamics in porous fluid-saturated solids
2:50 - 3:10	BREAK
3:10 - 3:50	V. Twersky, Univ. of Illinois, Chicago Circle, Propagation and attenuation in composite media
3:50 - 4:30	J. Lebowitz, Rutgers Univ., Kinetics of cluster growth in quenched alloys
4:30 - 5:10	L. Tartar, Univ. of Paris Orsay, Optimal bounds in homogenization and applications
5:10 - 5:50	D. J. Bergman, Tel Aviv University, Resonances in the bulk properties of composite media -- theory and applications

TUESDAY JUNE 2

9:00 - 9:40	N. Ashcroft, Cornell University, Electromagnetic propagation in mixed media
9:40 - 10:20	Ping Sheng, Exxon Res., 'Structural units' and the effective medium calculation of composite dielectric constants
10:20 - 10:40	BREAK
10:40 - 11:20	G. Papanicolaou, Courant Institute, Upper and lower bounds for conductivities in random media
11:20 - 12:00	I. Webman, Courant Institute, Diffusion on random networks
12:00 - 1:30	LUNCH
1:30 - 2:10	M. Cohen, University of Chicago, Topology, geometry and physical properties of porous rocks
2:10 - 2:50	M. Lax, City University of N.Y., Coherent medium approach to hopping conductivity

CONFERENCE PROGRAM, continued

Tuesday, continued

2:50 - 3:10 BREAK

3:10 - 3:50 P. Sen, Schlumberger Res., A self-similar model for acoustic and electrical response in rods

3:50 - 4:30 T. Odagaki, City University of N.Y., Hopping conduction in one dimensional chains

4:30 - 5:10 M. Vogelius, Courant Institute, A projection method applied to diffusion in a periodic structure

WEDNESDAY JUNE 3

9:00 - 9:40 A. Acrivos, Stanford University, Slow flow and heat transfer past isotropic arrays of spheres and cylinders

9:40 - 10:20 J. Percus, Courant Institute, Fluctuation corrections to the mean field description of a nonuniform fluid

10:20 - 10:40 BREAK

10:40 - 11:20 J. B. Keller, Stanford University, Reflection and scattering from rough surfaces

11:20 - 12:00 R. V. Kohn, Courant Institute, Homogenization as a tool for engineering design optimization

12:00 - 1:30 LUNCH

1:30 - 2:10 D. Drew, Rensselaer Poly. Inst., Wave propagation in a bubbly liquid

2:10 - 2:50 W. Kohler, Virginia Poly. Inst. and State Univ., Upper and lower bounds for effective parameters

2:50 - 3:10 BREAK

3:10 - 3:50 V.V. and V.K. Varadan, Ohio State Univ., Computation of the frequency dependence of the average dielectric constant in discrete random media

3:50 - 4:30 D. McLaughlin, Courant Inst. and Univ. of Arizona, Self consistent convection of flows with microstructure

4:30 - 5:10 G. W. Milton, Cornell University, A comparison of two methods for deriving bounds on the effective conductivity of composites

The conference is sponsored by a grant from the Exxon Research Corp.

EDITORS' PREFACE

During the first three days of June, 1981, a conference was held at the Courant Institute on the macroscopic behavior of disordered media. We are grateful to the Exxon Research Corporation for the grant which made this conference possible.

The present volume contains most of the papers presented at this conference.

Research work on the macroscopic behavior of disordered media is very broad, ranging from applied engineering and experimental studies to more basic physics. It covers such diverse areas as numerical simulations, the numerical solution of phenomenological macroscopic equations, and mathematical investigations of the validity of macroscopic equations obtained from microscopic laws. In order to facilitate communication between the participants and to limit the length of the conference we decided to focus attention on the passage from microscopic to macroscopic laws and associated mathematical problems. We therefore excluded experimental and numerical studies. However, even within the areas selected, we left many problems such as localization, surface effects, fluid mixtures, etc., unrepresented.

During the course of the conference it became clear that there is a significant concentration of scientists in the New York metropolitan area actively engaged in research on macroscopic effects. Professor Morrel Cohen suggested that perhaps an informal conference among specialists take place annually (or more frequently) in order to exchange ideas and identify problems of theoretical and applied interest.

We expect to hold such an informal gathering at the Courant Institute in early summer 1982.

R. Burridge, S. Childress, G. Papanicolaou

ELECTROMAGNETIC PROPAGATION IN MIXED MEDIA

N. W. Ashcroft
Laboratory of Atomic and Solid State Physics
and
Materials Science Center
Cornell University, Clark Hall, Ithaca, NY 14853

I. Introduction

Colloids and other dilute aggregates of fine metal particles are frequently characterized by rather striking optical properties in the visible range of the spectrum. These and other interesting physical curiosities of such systems gave historical impetus to the study of an important form of condensed matter that, as we would now say, possesses microscopic order but it macroscopically disordered. By fine we imply a scale of length L that is small compared to the wavelength λ of the electromagnetic radiation probing the system. Actually in a mixed medium consisting of two constituents, each of which is characterized by a bulk (scalar) dielectric constant ϵ_i and magnetic permittivity μ_i, there are two intrinsic wave vectors $k_i = (\epsilon_i\mu_i)^{\frac{1}{2}}2\pi/\lambda$, and five possible long wavelength conditions.[1] These are: $|k_1L| << 1$, $|k_2L| << 1$, $|kL| << 1$, $|k_1b| << 1$, and $|kb| << 1$. Here k is the wave vector for the composite medium and is generally complex. The last two conditions are introduced to reflect the common experimental circumstance that in composite media physical clustering or other local ordering often takes place. The length scale characterizing such ordering is taken to be b. An important example is the periodic dielectric in which identical structures of constituent 2 (say) are embedded in crystalline fashion in a host of constituent 1. Here b will be taken as the magnitude of the largest primitive lattice vector. When the physical circumstances are such that all 5 conditions are met, the system is said to be in the quasistatic or infinite wavelength limit, and in this limit it can often be described by a simple average dielectric constant (or average permittivity). Otherwise the quantity of direct physical importance is the average product $(\epsilon\mu)_{av}$, as will be seen below.

II. Effective Medium Theories

The spatial arrangement of the constituents in the mixture will certainly be reflected in the manner in which boundary charges and local fields are established. In particular the connectedness of the composite has a great deal to say about the structure of the approximate theory used to describe the system. This can be demonstrated most easily for the two-component case by considering the mean field description of systems conforming to two physically important geometries. We shall first let the composite be divided into cells, each of which we imagine to be embedded in an effective medium of dielectric constant $\bar{\epsilon}$ (we take $\mu_1 = \mu_2 = 1$ for the present). Then in the quasi-static limit just described, there is an implicit averaging whose effect is to render the composite to be both translationally and rotationally invariant. Accordingly as a first approximation it is plausible to replace each irregularly shaped individual cell by spherical average cells. The question of interest is now this: how shall $\bar{\epsilon}$ of the effective medium be chosen to ensure that scattering of radiation from the average cell actually vanishes? The attributes of connectedness now enter in dictating the nature of those average cells. Consider first a medium in which particles of constituent 2 are always surrounded by constituent 1 and that the fractional occupancy of constituent 2 is η. In this example there is a single average cell. It is a coated sphere (cs) which, though properly to be regarded in its own right as a composite body (constituent 2 coated with constituent 1), can nevertheless be replaced in the long wavelength limit by a uniform sphere with dielectric constant

$$\bar{\epsilon}_{cs} = \epsilon_1 \left[\frac{2\epsilon_1 + \epsilon_2 + 2\eta(\epsilon_2 - \epsilon_1)}{2\epsilon_1 + \epsilon_2 - \eta(\epsilon_2 - \epsilon_1)} \right]$$

It follows that if the coated sphere is embedded in an effective medium with dielectric constant $\bar{\epsilon}$ chosen to be

$$\bar{\epsilon} = \epsilon_1 \left[\frac{2\epsilon_1 + \epsilon_2 + 2\eta(\epsilon_2 - \epsilon_1)}{2\epsilon_1 + \epsilon_2 - \eta(\epsilon_2 - \epsilon_1)} \right] \qquad (1)$$

then the no-scattering condition is trivially satisfied. This result can also be derived by standard Lorentz local field arguments.[2] Note that (1) can also be rewritten as

$$\bar{\epsilon} = \epsilon_1 \left[\frac{\epsilon_1 + (P(1-\eta) + \eta)(\epsilon_2 - \epsilon_1)}{\epsilon_1 + P(1-\eta)(\epsilon_2 - \epsilon_1)} \right] \qquad (2)$$

where P is the depolarizing factor which, for this spherical geometry, is 1/3. Equation (1) is the well known Maxwell-Garnett (MG) result[2,3] and is appropriate to systems conforming to the cermet geometry.

Apart from the basic topological statement requiring regions of 2 always to be surrounded by regions of 1, it contains no reference to microscopic features of the structure of the system. It is a particularly interesting result because irrespective of such details, and indeed, irrespective of connectedness, it can be shown[4] that (1) together with the complement formed by interchange of 1 and 2, form a pair of bounds on $\bar{\epsilon}$ when ϵ_1 and ϵ_2 are real. The generalization of such bounds to complex ϵ_i has been given by Milton[5] and Bergman.[6]

A quite different arrangement of the constituents, and one that is also physically realizable, leads to a very different choice of average cell. In the aggregate topology the components are arranged in such a way that in terms of connectedness each occurs on a completely equal footing, so that in contrast to the cermet topology neither can be considered host. Then with the same implicit averaging procedures in mind, there must be two types of average cell to be embedded in the effective medium. Each cell is again a sphere: we therefore have two spheres of each constituent and the no-scattering condition is very easily implemented by relating the fields inside and outside of them according to

$$E_{in}/E_{out} = 3/(2 + \epsilon_{in}/\epsilon_{out}).$$

A volume average then leads to the equation

$$\frac{3(1 - \eta)}{2 + \epsilon_1/\bar{\epsilon}} + \frac{3\eta}{2 + \epsilon_2/\bar{\epsilon}} = 1 \qquad (3)$$

whose solution gives $\bar{\epsilon}$. This result is originally due to Bruggeman[7]: it is also widely known as the effective medium result, though both arguments just given are effective medium constructs. It will be noted that the symmetry present in the specification of the spatial arrangement of the aggregate topology is reflected in the symmetric occurrence of ϵ_1 and ϵ_2 in (3). Otherwise, however, there is no further reference to the microscopic arrangement of material in the system.

A real composite system must ultimately be described statistically through the specifications of correlation functions that specify the average microscopic arrangement of the constituents. There are correlations in such systems that go far beyond the gross connectedness properties just described. For example, in a dense composite we surely expect clustering effects in which regions of component 2 will be wholly surrounded by regions of component 1, <u>and vice versa</u>. It is possible to partially amalgamate (2) and (3) to reflect these correlations by appealing to an assumption of statistical independence in the occurrence of such regions. For a region in which 2 is enveloped by 1 we have

$$\bar{\epsilon}_1(\eta) = \epsilon_1\left[\frac{2\epsilon_1 + \epsilon_2 + 2\eta(\epsilon_2 - \epsilon_1)}{2\epsilon_1 + \epsilon_2 - \eta(\epsilon_2 - \epsilon_1)}\right]$$

while for the opposite case

$$\epsilon_2(\eta) = \epsilon_2\left[\frac{2\epsilon_2 + \epsilon_1 + 2(1 - \quad)(\epsilon_1 - \epsilon_2)}{2\epsilon_2 + \epsilon_1 - \quad(\epsilon_1 - \epsilon_2)}\right]$$

If the composite is regarded as an aggregate of such regions with relative occupancy preserved, then the average dielectric function is given by the solution to

$$\frac{3(1 - \bar{\eta})}{2 + \bar{\epsilon}_1(\eta)/\epsilon} + \frac{3\bar{\eta}}{2 + \bar{\epsilon}_2(\eta)/\epsilon} = 1 \tag{4}$$

The proper determination of $\bar{\epsilon}_1$ and $\bar{\epsilon}_2$ is dictated by microscopic considerations reflecting the structure of the system. An approach that takes a point of view similar to this has recently been given by Sheng.[8] The point to note is that the analytic structure of the expressions determining ϵ can be changed substantially by the further imposition of physical constraints. In mean field arguments such as those just given many of the details of the structure are obliterated by the averaging procedures. Such effects as are preserved are mainly the manifestations of dipolar fields. To incorporate structural multipole effects it is necessary to develop a theory that includes a more detailed treatment of correlations in such systems.

III. <u>Multiple Scattering Methods</u>

An example of a highly correlated mixed medium is a periodic array of identical inclusions (type 2, say) in a host of type 1. This crystalline arrangement falls within the definition of the cermet topology, as described above. If the inclusions have radius a, and a mean number density $\rho = N/V$, then Maxwell's equations lead to a wave equation for, say $\vec{H}(\vec{r})$, namely

$$\vec{\nabla} \times (\vec{\nabla} \times \vec{H}(\vec{r})) - (\omega^2/c^2)\epsilon(\vec{r})\mu(\vec{r})\vec{H}(\vec{r}) - \epsilon^{-1}(\vec{r})(\vec{\nabla}\epsilon(\vec{r})) \times (\vec{\nabla} \times H(\vec{r})) = 0 \tag{5}$$

where $\epsilon(\omega,\vec{r}) = \epsilon_1(\vec{r})$ for $\vec{r}$ in the host, and ϵ_2 otherwise (similarly for μ). There is a complementary equation for $\vec{E}$. Equation (5) can be regarded as a vector-field analog of the corresponding scalar field problem familiar from the context of one-electron band structure in crystalline systems. In the latter we are required to solve a Schrödinger equation in which the potential is periodic. The existence of a group of translations allows the solutions for the energy E to be indexed by a wave-vector $\vec{k}$, the Bloch wave-vector, which is restricted to a primitive cell of reciprocal space (the first Brillouin zone). The energy is then a function of the quasi continuous $\vec{k}$; the corresponding bands are bounded below and near their minima are quadratic in $\vec{k}$. Indeed for systems with sufficiently high symmetry, the energy for small k mimics that expected of a free particle but with an effective mass (rather than the free particle mass) which is the sole reminder of interactions that the electron has with the system. There is an immediate parallel with the periodic dielectric: again the translation symmetry allows us to index the solutions for $\vec{H}$ by a Bloch wave-vector. For a given frequency ω we shall seek solutions for wave-vectors of the form $\vec{k} = \vec{k}(\omega)$. In the long wavelength limit we shall then define an effective product $(\epsilon\mu)_{av}$ by the relation

$$(\epsilon\mu)_{av} = k^2(\omega)c^2/\omega^2 \qquad (k \to 0) \tag{6}$$

as appropriate to plane waves.

The solution of equation (5) (or its complement for $\vec{E}$) is obtained from a variant[1] on the Green's function or Korringa-Kohn-Rostoker method.[9-11] A synopsis of this variant is now described: full details may be found in Ref. 1. The equation for $\vec{H}$ is rewritten as an integral equation

$$\vec{H}(\vec{r}\,') = -\int_S ds\{[\hat{n}\cdot\vec{\nabla}G_{\vec{k},\kappa}(\vec{r}\,',\vec{r})]\vec{H}(\vec{r}) - G_{\vec{k},\kappa}(\vec{r}\,',\vec{r})\hat{n}\cdot\vec{\nabla}\vec{H}(\vec{r})\} \tag{7}$$

with $\kappa = (\epsilon\mu_1)^{\frac{1}{2}}\omega/c$, and the surface S is taken to be the primitive cell boundary and a surface infinitesimally outside of the spherical inclusion. Further, $\hat{n}$ is a normal that is outward and inward on cell and inclusion, respectively. The quantity G is defined by

$$G_{\vec{k},\kappa}(\vec{r}\,',\vec{r}) = \frac{1}{4\pi}\sum_{\vec{R}} \frac{\exp i\kappa|\vec{r}-\vec{r}\,'-\vec{R}|\exp i\vec{k}\cdot\vec{R}}{|\vec{r}\,'-\vec{r}-\vec{R}|} \tag{8}$$

and is the Green's function (the centers of the spherical inclusions being located by the set of lattice vectors $\vec{R}$). Equation (7) is further reduced by performing an expansion of $\vec{H}(\vec{r})$ in electric and magnetic multipoles:

$$H(\vec{r}) = \vec{H}_E(\vec{r}) - \frac{ic}{\omega\mu} \vec{\nabla} \times \vec{E}_H(\vec{r}) \tag{9}$$

where

$$\vec{H}_E(\vec{r}) = \sum_{\ell m} A^E_{\ell m} f^E_\ell(r)\vec{L}Y_{\ell m}(\hat{r}) \tag{10}$$

with

$$\vec{L} = - i\vec{r} \times \vec{\nabla}$$

The terms in (10) are the electric multipoles. Equation (9) (and its complement) automatically satisfies the Maxwell divergence equations: the other two are satisfied by insisting that (10) also satisfies the wave equation. As noted in Ref. 1, the radial function $f^E_\ell(r)$ then satisfies a radial equation whose solutions for the spherical inclusion problem can be stated most succinctly in terms of a set of phase shifts. The details of the arrangement of these inclusions actually reside in the specification of the set of vectors $\{\vec{R}\}$, and hence in the Green's function G which has itself a standard expansion when $r' < r$. The expansion coefficients $A^E_{\ell m}$ can therefore be determined from the set of equations that emerge when the expanded quantities ($\vec{H}$ and G) are substituted into equation (7) and the result integrated over the surface of the spherical inclusion.

The solution of this secular problem leads via (6) in the long-wavelength limit to an effective product $(\epsilon\mu)_{av}$ which in general does not further separate. However, if the expansion for $\vec{H}$ (or $\vec{E}$) is truncated at $\ell = 1$ (dipoles) then for small ka it can be shown[1] that

$$(\epsilon\mu)_{av} = \epsilon_{MG}\mu_{MG} \tag{11}$$

i.e., the product of Maxwell-Garnett results[12] of the form (1). This verifies that the effective medium argument as applied to the cermet topology includes only dipolar effects: the result is obtained from a multiple scattering theory, and it is clear that structural effects will be included by retaining terms in (10) beyond $\ell = 1$. This is done by using matrix folding techniques on the secular equation determining the relation between k and ω. (The convergence of this procedure is generally assured providing $Ka << 1$ and $k_2a << 1$. That the latter can be easily violated in metallic systems simply means that eddy currents, i.e., magnetic multipoles, can certainly be important.) The result is an expression for the average product that is again separable (see equation (11)) in which the dielectric constant E has the form given in equation (2). In this case, however, the depolarizing factor is no longer that appropriate to a sphere: it is modified by the presence of structurally induced multipoles. The latter now depend quite explicitly on the structure of the system and can give rise to multipole resonances in physical observables whenever the condition

$$\frac{\epsilon_2(\omega) - \epsilon_1}{(\ell + 1)\epsilon_1 + \ell\epsilon_2(\omega)} \approx 0 \qquad (\ell > 1)$$

is satisfied quasistatically. Physically these resonances relate to the higher excited modes of the free dielectric inclusion.

Though derived from a theory devised to describe a periodic array of dielectric inclusions, it is clear that (11) makes no reference to either the arrangement or the size of the constituents. It therefore may be anticipated that a straightforward modification of the theory can be constructed that will also hold true as the lowest order solution for an aperiodic system with the cermet topology. As shown in Ref. 1, this is in fact the case. The required modification is to take a macroscopic sample of aperiodic cermet, to periodically extend it, and to regard the resulting <u>grand crystal</u> as a Bravais lattice with an exceedingly large basis. Though notationally very cumbersome, the method just described is equally applicable to a system with a basis: it resembles closely the associated band structure problem for electrons moving in a crystal described as a lattice with a basis. Accordingly, in demanding a non-trivial solution for the multipole expansion coefficients, a secular equation again results, the principal difference from the previous case being that the matrix elements are distinguished according to site. Again, truncation at $\ell = 1$ gives the simplest results. The effective product $(\epsilon\mu)_{av}$ is separable, and the average $\bar{\epsilon}$ has the Maxwell-Garnett form. Multipole effects can again be included by matrix folding techniques and the result is that provided that $Ka << 1$ and $k_2a << 1$ there is a product rule[12] and an expression for $\bar{\epsilon}$ similar to (2) in which the depolarizing factor is modified by structural multipoles which now reveal the disorder in the system. Indeed, the determination of these corrections requires structural information normally provided by the pair and higher correlation functions for the inclusions. As shown in Ref. 1, the multiple scattering approach can be extended still further to include the fact that the sizes of particles in the system may also be dispersed. To handle this case it is only necessary to assume that the size of a given inclusion is uncorrelated with its spatial position, an assertion that is physically plausible. For this situation also, the results just described for the monodispersed case continue to hold with minor modifications (not surprisingly the form of the structural multipole modifications are somewhat more complex).

Thus, at least for mixed media distributed with the cermet topology, it is possible systematically to improve the mean field treatments of their electromagnetic properties. To do so, however, it is necessary to have access to structural information on the two-point, three-point, and higher correlation functions of the inclusions. In principle, it is possible to obtain such information on pair distributions from scattering experiments. However, for triplet functions and beyond it is necessary at present to resort to theoretical models. Ideally this structural information would also reveal the degree to which clustering and other local aberrations from average disorder are present. The occurrence of such regions leads, as noted earlier, to other length scales and hence to other conditions that must be satisfied in order to physically approach the long-wavelength and quasi-static limits.

Finally it should be noted that while the methods just described are suitable for the cermet topology they are clearly not applicable to the aggregate topology. In the electronic analog--the disordered alloy problem--the Bruggeman approach is related to the coherent potential approximation (the correspondence in the case of Maxwell-Garnett is to the average t-matrix approximation). In multiple scattering approaches to the alloy problem it is known[13] that some correlation effects can be included, albeit approximately, which suggests that provided there is adequate structural information, analogous procedures may be used here. For the periodic version of the aggregate topology it is certainly possible to use standard plane wave expansions in an attempt to solve (5). The intent is again to obtain the dispersion $\omega = \omega(\vec{k})$ and thence via (6) to an effective product $(\epsilon\mu)_{av}$. But on physical grounds and from experience with the electronic case, the convergence of such a procedure can only be expected to be satisfactory when the difference $\epsilon_2 - \epsilon_1$ is small compared with ϵ_1 and ϵ_2.

Acknowledgments

This work has been supported in part by the Solar Energy Research Institute (Grant #XH-9-8158-1) and in part by the National Science Foundation through the facilities of the Materials Science Center at Cornell University (Grant #DMR70-24008-A02). It is largely based on Ref. 1 toward which my co-authors Drs. Lamb and Wood made major contributions.

References and Footnotes

1. W. Lamb, D. M. Wood and N. W. Ashcroft, Phys. Rev. B21, 2248 (1980).
2. R. Landauer, Proceedings of the First Conference on the Electrical and Optical Properties of Inhomogeneous Media, Ohio State University, 1977, edited by J. C. Garland and D. B. Tanner (AIP, New York, 1978), No. 40, p. 2.
3. J. C. Maxwell-Garnett, Philos. Trans. R. Soc. London 203, 385 (1904).
4. W. Brown, Jr., J. Chem. Phys. 23, 1514 (1955); Z. Hashin and S. Shtrikman, J. Appl. Phys. 33, 3125 (1962).
5. G. W. Milton, Appl. Phys. Letts. 37, 300 (1980).
6. D. Bergman, Phys. Rev. Letts. 44, 1285 (1980).
7. D.A.G. Bruggeman, Ann. Phys. (Leipzig) 24, 636 (1935).
8. P. Sheng, Phys. Rev. Letts. 45, 60 (1980).
9. J. Korringa, Physica (Utrecht) 13, 392 (1947).
10. W. Kohn and N. Rostoker, Phys. Rev. 94, 1111 (1954).
11. P. M. Morse, Proc. Nat. Acad. Sci. (USA) 42, 276 (1956).
12. C. G. Granqvist, Z. Phys. B30, 29 (1978).
13. R. J. Elliot, J. A. Krumhansl and P. L. Leath, Rev. Mod. Phys. 46, 465 (1974).

RESONANCES IN THE BULK PROPERTIES OF

COMPOSITE MEDIA - THEORY AND APPLICATIONS*

by

David J. Bergman
Department of Physics and Astronomy
Tel-Aviv University
69978 Tel-Aviv, Israel†

and

Department of Physics
The Ohio State University
Columbus, Ohio 43210

Contents

I. INTRODUCTION

A composite material is an inhomogeneous medium where the scale of the inhomogeneities is macroscopic. There are small regions in which the properties are uniform and equal to the bulk properties of one of the components. On larger scales, the composite can usually be described by a set of effective bulk properties, as though it too were uniform. A central problem in the theory of composites is the calculation of such bulk properties. The main difficulty arises when the bulk property depends on the detailed microgeometry of the composite. For a complete solution, one needs to calculate the physical fields inside the composite, including all their distortions caused by the inhomogeneous structure. Even when the microgeometry is known precisely, this is a very difficult task that has usually been approached by attempting to solve numerically the partial differential equations for the local fields, including continuity conditions at the interfaces between the different pure components.

We note here that even when the composite is made of separate, simply shaped inclusions (e.g., spheres) of one pure material embedded in another pure material, a

*Research supported in part by the United States - Israel Binational Science Foundation under Grant No. 2006/79.

†Permanent address

calculation along these lines does not simplify. That is because, unless the density of inclusions is very low, or they differ very little from the host material in their material property, the distortive effects of different inclusions are not additive. Consequently the solution of the problem for an isolated spherical inclusion does not help us, and we must consider all of them simultaneously. This is similar to the case of a multiple scattering problem, where the field impinging on any scatterer includes the scattered field from all the other scatterers.

An entirely different approach to such problems, which we will advocate here, is to focus attention on the resonance states of the problem. These resonances are eigenfunction solutions of the same equation that determines the local field in the composite in the presence of an external field, but satisfy different boundary conditions on the exterior surface of the system (i.e., no external field is present). It is a very natural idea to try to expand the local field of the real problem (i.e., the one where there is an external field present) in these eigenfunctions. In this way, a very convenient representation is obtained both for the field and for other interesting quantities, in particular for the effective bulk material properties of the composite.

Some advantages of this type of representation are: a) The resonances depend on the microgeometry, but they are independent of the specific material properties of the inclusions, and in some cases (e.g., the dielectric constant) they are also independent of the host properties. Once they are known, they provide us with a solution for any type of inclusion that has the same microgeometry. b) When the microgeometry is only partially known, the positions and strengths of the resonances can be treated as free parameters that are varied in order to find rigorous upper and lower bounds on the effective bulk properties. This turns out to be a very efficient procedure for obtaining such bounds. c) When the microgeometry is precisely known, the resonances can often be calculated in stages. If the inclusions can be described as made up of simply shaped grains (e.g., spheres that do or do not overlap), then we can first find the resonances of each grain by treating it as an isolated inclusion. The totality of these eigenstates from all the grains are then used as a basis in the function space on which to set up the problem for the entire composite. The resonances of the entire composite are then the eigenvectors of an infinite matrix whose diagonal elements are the individual grain eigenvalues, while the off-diagonal elements are overlap integrals between eigenstates of two different grains. This means that instead of having to solve a multiple scattering problem for the actual local field, we only have to solve a problem of two-body interactions to find the resonances. This approach enables us to take the utmost advantage of any symmetries of the individual grain (e.g., if the grains are spherical the isolated grain eigenfunctions and even the two-body overlap integrals are very easy to obtain) even when the total symmetry of the composite is very low. If the composite has a periodic structure, a Bloch theorem can be shown to hold for the eigenfunctions and a further simplification results. d) Since the resonances are the singular points of the effective bulk property as a function of the material properties of the pure components, they completely determine the analytical properties of that function. By considering the resonance spectrum, it is possible to make some rather general statements about particular singularities of that function. For example, by considering a system or an ensemble of systems that has a conductivity threshold at some value of a continuously varying parameter, we can predict singular behavior at that point for a variety of other physical properties of the system.

To date, this approach has been applied to electrostatic problems (Bergman, 1978 a,b, 1979 a,b), to finite frequency, quasi-electrostatic problems (Bergman, 1980, 1981, a,b), to random-resistor-networks (Bergman and Kantor, 1981), to elastostatic problems (Kantor and Bergman, 1981 a,c), and to the scattering of electromagnetic waves (Bergman and Stroud, 1980). An application to quantum mechanical scattering and to other classical problems in composite systems are under consideration. Since the method was first proposed in the context of electrostatic problems, and since that is the field where the most applications have been made, we will mostly discuss those problems here, pointing out where differences arise in the discussion of other types of problems.

II. ELECTROSTATIC PROBLEMS - GENERAL THEORY

The basic properties that are crucial to all later developments are: a) For an $(m+1)$-component composite, with pure component dielectric constants $\epsilon_1 \dots \epsilon_{m+1}$, the effective dielectric constant $\epsilon_e(\epsilon_1 \dots \epsilon_{m+1})$ is an analytic function in most of the complex space of $\epsilon_1 \dots \epsilon_{m+1}$. b) The function $\epsilon_e(\epsilon_1 \dots \epsilon_{m+1})$ is homogeneous of order one, and consequently we can eliminate its dependence on any single component (called the "host material"), e.g., ϵ_{m+1}.

These properties can be demonstrated in different ways. We will choose a method that is useful also for further practical developments. The problem of calculating the electrostatic (e.s.) field ϕ in a composite that lies between the plates of a large parallel plate condenser is set up as an integral equation (compare with Bergman, 1979 a,b)

$$\phi = z + \sum_{i=1}^{m} u_i \Gamma_i \phi$$

$$\Gamma_i \phi \equiv \int dV' \; \theta_i(\vec{r}') \frac{\partial G(\vec{r},\vec{r}')}{\partial \vec{r}'} \cdot \frac{\partial \phi(\vec{r}')}{\partial \vec{r}'}$$

$$\theta_i(\vec{r}) = \begin{Bmatrix} 1 & \text{for } \vec{r} \text{ inside } \epsilon_i \text{ material} \\ 0 & \text{otherwise} \end{Bmatrix}$$

$$u_i \equiv \frac{1}{s^{(i)}} \equiv 1 - \frac{\epsilon_i}{\epsilon_{m+1}} \quad . \qquad \text{(II.1)}$$

Here $G(\vec{r},\vec{r}')$ is Green's function for the parallel plate condenser

$$\nabla^2 G(\vec{r},\vec{r}\,') = -\delta^3(\vec{r}-\vec{r}\,')$$

$$G = 0 \quad \text{for } \vec{r} \text{ on the plates}$$

$$\frac{\partial G}{\partial n} = 0 \quad \text{for } \vec{r} \text{ on the walls,} \qquad \text{(II.2)}$$

and for an infinitely large condenser it is equal to the Coulomb potential of a point charge

$$G(\vec{r},\vec{r}\,') = \frac{1}{4\pi|\vec{r}-\vec{r}\,'|} \quad . \qquad \text{(II.3)}$$

The inhomogeneous term z in (II.1) takes care of the boundary condition on ϕ that arises from assuming that the condenser plates are perpendicular to the z-axis, and that a potential difference is applied that is equal (numerically) to the distance between them. (This means that if the condenser were filled with a homogeneous dielectric, the electric field would be 1.) The linear integral operator Γ_i, which uses values of ϕ only inside ϵ_i material, is hermitian under the following definition of a scalar product

$$(\psi,\phi)_i \equiv \int dV\, \theta_i\, \vec{\nabla}\psi^* \cdot \vec{\nabla}\phi \quad . \qquad \text{(II.4)}$$

Restricting ourselves, for the time being, to a two-component composite, it is easy to convince ourselves that the eigenvalues s_n of $\Gamma(\equiv\Gamma_1)$ must all lie on the real axis between 0 and 1. Any other (real) value for an eigenvalue s_n would mean that there exists no solution for the electrostatic potential in a charged capacitor filled with a composite made of two materials with certain real, positive values of ϵ_1,ϵ_2 - clearly an impossibility.

The (differentiable) functions in the ϵ_1-volume form a Hilbert space under the scalar product of (II.4), and we can use the eigenfunctions of Γ as a basis set, but only inside the ϵ_1-volume. Using the usual expansion for the unity-operator, and introducing a convenient bra-ket notation, we get the following equation for the expansion coefficients of the potential function ϕ

$$\langle\phi_n|\phi\rangle = \langle\phi_n|z\rangle + \sum_m \langle\phi_n|u\Gamma|\phi_m\rangle\langle\phi_m|\phi\rangle \quad . \qquad \text{(II.5)}$$

From this we get

$$\langle\phi_n|\phi\rangle = \frac{\langle\phi_n|z\rangle}{1-u\,s_n} \qquad \text{(II.6)}$$

and hence, inside the ϵ_1-volume, the following expansion for ϕ

$$|\phi\rangle = \sum_n |\phi_n\rangle \frac{\langle\phi_n|z\rangle}{1-u\,s_n} \quad . \qquad \text{(II.7)}$$

In order to get an expansion for ϕ that is valid also outside the ϵ_1-volume, we substitute (II.7) on the right hand side of the integral equation for ϕ, thus getting

$$|\phi\rangle = |z\rangle + u\,\Gamma|\phi\rangle = |z\rangle + \sum_n |\phi_n\rangle \frac{s_n\langle\phi_n|z\rangle}{s - s_n} \quad , \tag{II.8}$$

where we have introduced the new variable

$$s \equiv \frac{1}{u} = \frac{\varepsilon_2}{\varepsilon_2 - \varepsilon_1} \quad . \tag{II.9}$$

The bulk effective dielectric constant of the composite ε_e can be defined by requiring that if the capacitor were filled by a fictitious homogeneous dielectric ε_e, the (uniform) displacement field would be equal to the z-component of the volume averaged displacement field in the inhomogeneous sample

$$\varepsilon_e \equiv \frac{1}{V} \int \varepsilon(\vec{r}) \frac{\partial\phi}{\partial z}\, dV \quad . \tag{II.10}$$

Alternatively, one could require the electrostatic energy stored in the condenser to be reproduced correctly - the two definitions can easily be shown to be equivalent. Noting that we can represent the local dielectric constant $\varepsilon(\vec{r})$ with the help of the θ-functions as follows

$$\varepsilon(\vec{r}) = \varepsilon_2(1 - u\,\theta_1(\vec{r})) \quad , \tag{II.11}$$

we can transform (II.10) into

$$F(s) \equiv 1 - \frac{\varepsilon_e}{\varepsilon_2} = \frac{u}{V} \int dV\, \theta_1 \frac{\partial\theta}{\partial z} = \frac{u}{V}\langle z|\phi\rangle = \sum_n \frac{F_n}{s - s_n} \tag{II.12a}$$

$$F_n \equiv \frac{1}{V}|\langle z|\phi_n\rangle|^2 \quad , \tag{II.12b}$$

where we have now introduced and defined one of the "characteristic geometric functions" of the composite, F(s). This is the basic representation for ε_e of a two-component composite in terms of its resonances.

As we noted previously, the poles s_n (i.e., eigenvalues of Γ) are all between 0 and 1. Due to completeness of the set ϕ_n, the weights F_n obey the following sum rule

$$\sum_n F_n = \frac{1}{V} \sum_n \langle z|\phi_n\rangle\langle\phi_n|z\rangle = \frac{1}{V}\langle z|z\rangle = p_1 \quad , \tag{II.13}$$

where p_1 is the volume fraction of ε_1 material. Since the weights are all positive, it obviously follows that they must all lie between 0 and p_1. It can be shown that, for a composite with isotropic or cubic point symmetry, another sum rule holds, namely,

$$\sum_n F_n s_n = \frac{1}{3}\, p_1 p_2 \tag{II.14}$$

where $p_2 = 1 - p_1$ is the volume fraction of ε_2 (see Bergman, 1978 a,b). The higher moments of the pole spectrum

$$\sum_n F_n s_n^r \quad , \qquad r \geq 2 \tag{II.15}$$

depend on (r+1)-order correlation functions of the form

$$\theta_1(\vec{r}_1)\ \theta_1(\vec{r}_2)\ldots\theta(\vec{r}_{r+1}) \quad . \tag{II.16}$$

A knowledge of these moments can be used to generate an expansion for F(s) in powers of u = 1/s:

$$\begin{aligned} F(s) &= \sum_{r=0}^{\infty} \frac{\sum_n F_n s_n^r}{s^{r+1}} \\ &= p_1 u + \frac{1}{3} p_1 p_2 u^2 + \sum_n F_n s_n^2 \cdot u^3 + \ldots \quad . \end{aligned} \tag{II.17}$$

A further property of the resonance expansion (II.12) that is worth emphasizing is that the poles s_n and the weights F_n are entirely determined by the microgeometry of the composite, and are totally independent of the physical properties of the pure components, i.e., ε_1 and ε_2. The values of the two dielectric constants enter only through the variable s. Consequently, a knowledge of the parameters s_n, F_n enables us to calculate ε_e for any two-component composite with the same microgeometry, and even other bulk properties, such as conductivity, magnetic permeability, diffusivity - since all these physical processes are described by the same mathematical structure as the electrostatic potential field in a dielectric.

Before concluding this discussion of the two-component composite system, I would like to point out that there have been published some incorrect statements about the distribution of singularities of the function F(s) (see, e.g., Bergman 1978 b, p. 384). A careful analysis reveals that the qualitative features of this distribution depend on the singularities in the shape of the $\varepsilon_1,\varepsilon_2$ interface. In particular, for a finite system with a smooth interface (i.e., no corners and no isolated points or lines of contact), the spectrum of s_n is a discrete spectrum, with a single accumulation point at $s = \frac{1}{2}$ (Bergman 1981 c).

In the case of a multicomponent composite, (i.e., more than two pure components,) we can still discuss $\varepsilon_e(\varepsilon_1 \ldots \varepsilon_{m+1})$ in terms of a characteristic geometric function, defined by

$$F(s^{(1)} \ldots s^{(m)}) \equiv 1 - \frac{\varepsilon_e}{\varepsilon_{m+1}} \quad , \qquad m > 1 \quad . \tag{II.18}$$

The analytical properties of F have not been worked out completely for this case. If the interfaces are all smooth and the system is finite, the singularities of F hopefully lie upon a set of (m - 1)-dimensional hypersurfaces in the complex space of $s^{(1)} \ldots s^{(m)}$. It can be shown that, at any real singular point, at least one of the $s^{(i)}$ must always lie between 0 and 1 (see Bergman, 1978 b). In practice, the multicomponent system has always been dealt with, until now, by choosing an appropriate trajectory in the complex space of $s^{(1)} \ldots s^{(m)}$. E.g., in attempting to derive rigorous bounds for multicomponent systems, the useful approach to take is to find a trajectory

$s^{(i)}(s)$ that passes through all the important points in that space, such that $F(s^{(i)}(s))$ has a representation similar to (II.12a), together with appropriate sum rules (see Bergman 1978 b).

III. RIGOROUS BOUNDS FOR REAL AND COMPLEX ϵ_e IN A TWO-COMPONENT COMPOSITE

In principle, for a two-component composite with real values of ϵ_1,ϵ_2 one should start from (II.12a), treat the F_n and s_n as free parameters, and attempt to maximize or minimize F(s) subject to various constraints - equalities such as the sum rules (II.13),(II.14), and inequalities such as

$$0 \le s_n < 1$$

$$0 < F_n \le p_1 \quad . \tag{III.1}$$

In practice, it is often better to define new functions, with properties similar to those of F(s), that incorporate some of these constraints (i.e., the sum rules) automatically.

The first bound is obtained by noting that, when both ϵ_1 and ϵ_2 are positive (i.e., true physical dielectrics) then s is real and lies outside the segment (0,1). In that case we can write

$$F(s) = \sum_n \frac{F_n}{s - s_n} > \frac{\sum F_n}{s} = \frac{p_1}{s} \quad , \quad \text{for} \quad s \notin (0,1). \tag{III.2}$$

Translated into $\epsilon_1,\epsilon_2,\epsilon_e$ language this inequality becomes

$$\epsilon_e < p_1\epsilon_1 + p_2\epsilon_2 \quad , \tag{III.3}$$

i.e., a well known inequality for composites (Wiener, 1912).

A similarly well known inequality involving reciprocal ϵ's, namely (Wiener, 1912),

$$\frac{1}{\epsilon_e} < \frac{p_1}{\epsilon_1} + \frac{p_2}{\epsilon_2} \tag{III.4}$$

can be obtained by considering the function

$$H(t) \equiv 1 - \frac{\epsilon_2}{\epsilon_e} = \frac{F(s)}{F(s) - 1}$$

$$t \equiv \frac{\epsilon_1}{\epsilon_1 - \epsilon_2} = 1 - s \quad . \tag{III.5}$$

Starting from the properties of F(s), it can be shown that H(t) has very similar properties, namely (see Bergman 1981 a,b),

$$H(t) = \sum_n \frac{H_n}{t - t_n}$$

$$\sum H_n = p_1 \tag{III.6a}$$

$$\sum H_n t_n = \frac{2}{3} p_1 p_2$$

$$0 \leq t_n < 1$$

$$0 < H_n \leq p_1 \qquad \text{(III.6b)}$$

(the only difference is that in the second sum rule, the factor 1/3 has been replaced by 2/3). Obviously, H(t) contains the same information as F(s), and therefore, it too is a "characteristic geometric function" of the composite. Repeating the procedure of (III.2) for H(t) we obtain

$$H(t) > \frac{p_1}{t} \quad , \quad \text{for } t \notin (0,1) \quad , \qquad \text{(III.7)}$$

and when this is translated into $\varepsilon_1, \varepsilon_2, \varepsilon_e$ language, the result is (III.4).

A better pair of bounds for real ε_e can be obtained if we know not only the volume fractions p_1, p_2, but also that the system is isotropic or that it is cubic. To achieve this, we must include the sum rule (II.14) in our considerations. This is most easily done by defining a new, auxiliary geometric function,

$$C(s) \equiv \frac{1}{p_1} - \frac{1}{s\,F(s)} \quad . \qquad \text{(III.8)}$$

It is a straightforward matter to show that C(s) has a structure similar to F(s) (see Bergman 1981 b), namely

$$C(s) = \sum_n \frac{C_n}{s - s'_n}$$

$$0 \leq s'_n < 1$$

$$0 < C_n \leq \frac{1}{3} \frac{p_2}{p_1} \quad . \qquad \text{(III.9)}$$

Furthermore, the zero moment sum rule for F(s), (II.13), has gotten translated into the requirement that $C(s) \to 0$ as $s \to \infty$, while the first moment sum rule for F(s) (II.14), has gotten translated into a zero moment sum rule for C(s)

$$\sum_n C_n = \frac{1}{3} \frac{p_2}{p_1} \quad . \qquad \text{(III.10)}$$

If we now apply the procedure of (III.2) to C(s), we get

$$C(s) > \frac{p_2}{3 p_1 s} \quad . \qquad \text{(III.11)}$$

Translated back to F(s) and ε_e, this becomes

$$F(s) \gtrless \frac{p_1}{s - \frac{1}{3} p_2} \quad \text{for} \quad s \gtrless {}^{1}_{0} \qquad \text{(III.12a)}$$

$$\varepsilon_e \lesseqgtr \varepsilon_2 + \cfrac{p_1}{\cfrac{1}{\varepsilon_1 - \varepsilon_2} + \cfrac{p_2}{3\varepsilon_2}} \quad \text{for} \quad \varepsilon_1 \lesseqgtr \varepsilon_2 \quad . \tag{III.12b}$$

The last form is nothing but the well known pair of bounds discovered by Hashin and Shtrikman (1962).

In order to improve the bounds still further, more information is necessary. Such information could come in the form of higher moments of the pole spectrum s_n. Those are in general very difficult to obtain for systems with arbitrary microgeometries. It is however possible to obtain them for some simple, exactly known geometries, e.g., cubic arrays of spherical inclusions. This approach has been used by McPhedran and Milton (1981) to obtain a series of successively narrower pairs of bounds on ε_e for such systems as an alternative to the direct calculation of ε_e, such as we will describe in Sec. IV below.

A different way to include additional information in our considerations arises when we know the value of F at some other value of its argument - say s^+. This can happen if someone has measured a different physical property in the same composite, or in a composite with the same microgeometry. It can even happen if someone has measured the same property, e.g., ε_e, but at a different temperature or frequency, where $\varepsilon_1, \varepsilon_2$ and hence s have different values (see Prager 1969).

A knowledge of $F(s^+)$ leads to a further constraint on the variation of s_n, F_n of the form

$$F(s^+) = \sum_n \frac{F_n}{s^+ - s_n} \quad . \tag{III.13}$$

The best way to treat this case is again to define a new auxiliary geometric function

$$B(s) \equiv \frac{3}{p_1 p_2} - \frac{1}{s^2 F(s) - s\, p_1} \quad , \tag{III.14}$$

which can be shown to have a form similar to that of F(s), namely (see Bergman 1981 b)

$$B(s) = \sum_n \frac{B_n}{s - s''_n}$$

$$s''_o = 0, \quad 0 < s''_n < 1 \quad \text{for} \quad n \neq o$$

$$0 < B_n \quad . \tag{III.15}$$

In this case, both the zero moment and the first moment sum rules of F(s) are automatically incorporated in the definition of B(s), and we only have to worry about the new constraint (III.13). A straightforward calculation leads to

$$\frac{s\,B(s) - s^+ B(s^+)}{s - s^+} = - \sum_{n \neq o} \frac{B_n s_n}{(s - s_n)(s^+ - s_n)} \lesseqgtr 0 \quad , \tag{III.16a}$$

$$\text{for} \quad s,s^{+} \quad \text{on} \quad \begin{Bmatrix} \text{the same side} \\ \text{opposite sides} \end{Bmatrix} \quad \text{of} \quad (0,1) \quad . \tag{III.16b}$$

This gives us a bound for B(s), and hence for F(s) and ε_e.

To get the other bound, we start from another characteristic geometric function

$$E(s) \equiv 1 - \frac{\varepsilon_1}{\varepsilon_e} = \frac{1 - s\,F(s)}{s(1 - F(s))} \quad , \tag{III.17}$$

which is similar to H(t) but with the roles of $\varepsilon_1, \varepsilon_2$ reversed, and from it define an auxiliary geometric function similar to B(s)

$$B_E(s) = \frac{3}{2p_1p_2} - \frac{1}{s^2E(s) - s\,p_2} \quad . \tag{III.18}$$

This function can again be shown to have properties similar to B(s). Applying to it the procedure of (III.16) we then obtain the other bound for F(s) and ε_e.

Translated into $\varepsilon_1, \varepsilon_2, \varepsilon_e$ language, the two bounds can be found by solving the following two equations for ε_e

$$\frac{\varepsilon_2^{+} - \varepsilon_1^{+}}{\varepsilon_e^{+} - \langle\varepsilon^{+}\rangle} - \frac{\varepsilon_2 - \varepsilon_1}{\varepsilon_e - \langle\varepsilon\rangle} = \frac{3}{p_1p_2}\,\frac{\varepsilon_2^{+}\varepsilon_1 - \varepsilon_1^{+}\varepsilon_2}{(\varepsilon_2^{+} - \varepsilon_1^{+})(\varepsilon_2 - \varepsilon_1)}$$

$$\frac{\tilde{\varepsilon}_2^{+} - \tilde{\varepsilon}_1^{+}}{\tilde{\varepsilon}_e^{+} - \langle\tilde{\varepsilon}^{+}\rangle} - \frac{\tilde{\varepsilon}_2 - \tilde{\varepsilon}_1}{\tilde{\varepsilon}_e - \langle\tilde{\varepsilon}\rangle} = \frac{3}{2p_1p_2}\,\frac{\tilde{\varepsilon}_2^{+}\tilde{\varepsilon}_1 - \tilde{\varepsilon}_1^{+}\tilde{\varepsilon}_2}{(\tilde{\varepsilon}_2^{+} - \tilde{\varepsilon}_1^{+})(\tilde{\varepsilon}_2 - \tilde{\varepsilon}_1)} \quad ,$$

$$\tilde{\varepsilon} \equiv 1/\varepsilon, \qquad \langle\varepsilon\rangle \equiv p_1\varepsilon_1 + p_2\varepsilon_2 \quad . \tag{III.19a}$$

The first equation yields the upper (lower) bound and the second equation yields the lower (upper) bound on ε_e when the sign of

$$\frac{\varepsilon_2^{+}\varepsilon_1 - \varepsilon_1^{+}\varepsilon_2}{\varepsilon_1 - \varepsilon_2} \tag{III.19b}$$

is positive (negative). Obviously, we could go on and include more pieces of information of this type, and attempt to get even better bounds.

A further piece of information that has not been included until now, is an inequality that must hold for $\varepsilon_e(\varepsilon_1,\varepsilon_2)$ and $\varepsilon_e(\varepsilon_2,\varepsilon_1)$. The latter quantity represents a composite with the same microgeometry, but with the two components interchanged. For an isotropic or cubic composite, it can be shown that the following inequality must hold (see Schulgasser 1976, Bergman 1981 b)

$$\varepsilon_e(\varepsilon_1,\varepsilon_2)\ \varepsilon_e(\varepsilon_2,\varepsilon_1) \geq \varepsilon_1\varepsilon_2, \qquad \text{for} \quad \varepsilon_1,\varepsilon_2 > 0 \quad . \tag{III.20}$$

In order to include this too in our considerations, we define a new auxiliary function

$$A(s) \equiv \frac{s - \frac{1}{3} p_1}{p_1} - \frac{1}{F(s)} \quad . \tag{III.21}$$

This can again be shown to have the usual type of structure

$$A(s) = \sum \frac{A_n}{s - s_n'''}$$

$$0 < s_n''' < 1$$

$$0 < A_n \quad , \tag{III.22}$$

and the two sum rules on F(s) are both automatically incorporated. Note that the poles of A(s) are the same as the poles of C(s) (i.e., zeros of F(s)), except for s = 0 which may be a pole of C but may not be a pole of A.

In terms of A(s), (III.20) becomes

$$A(s) + A(1 - s) \leq \frac{1}{3} \frac{p_2}{p_1} \quad , \qquad \text{for} \quad s \notin (0,1) \quad . \tag{III.23}$$

Further inequalities that A(s) must satisfy for particular values of s are

$$A(0) + \frac{1}{3} \frac{p_2}{p_1} = - \frac{1}{F(0)} \geq 0 \tag{III.24}$$

$$A(1) - \frac{2}{3} \frac{p_2}{p_1} = 1 - \frac{1}{F(1)} \leq 0 \quad . \tag{III.25}$$

We note that (III.25) is also a consequence of (III.24) and of (III.23) taken at s = 0 or at s = 1.

Because (III.23) must hold for any s, it is not immediately clear how many independent constraints it is in fact equivalent to. For the time being we therefore restrict ourselves to considering only (III.23) at s = 0 (or at s = 1, which leads to the same inequality), and (III.24), thus

$$- A(0) \leq \frac{1}{3} \frac{p_2}{p_1} \tag{III.26}$$

$$A(0) + A(1) \leq \frac{1}{3} \frac{p_2}{p_1} \quad . \tag{III.27}$$

In addition to these inequalities, we must include the information about $A(s^+)$. In terms of the parameters A_n, s_n''' these three constraints become

$$\sum_n \frac{A_n}{s_n'''} \leq \frac{1}{3} \frac{p_2}{p_1} \tag{III.28}$$

$$\sum_n \frac{A_n(2s_n - 1)}{s_n(1 - s_n)} \leq \frac{1}{3} \frac{p_2}{p_1} \tag{III.29}$$

$$\sum_n \frac{A_n}{s^+ - s_n} = A(s^+) \quad . \qquad \text{(III.30)}$$

Note that these constraints, as well as the quantity we would like to maximize or minimize, $A(s)$, are all linear in the weights A_n. To these constraints we should also add the usual requirement that all $A_n \geq 0$. Each of these linear constraints defines a hyperplane in the space of the A_n.

According to linear programming theory (see, e.g., Dantzig 1963), an extremum point of $A(s)$ with respect to the A_n (the s_n''' are kept fixed for the time being) can only occur at a vertex of these hyperplanes. What this means is that most of the hyperplanes defining that vertex must be the trivial coordinate hyperplanes $A_n = 0$. Some of these trivial hyperplanes (at least one, at most three) can be replaced by the non-trivial hyperplanes of (III.28)-(III.30). Therefore we conclude that the bounding expression for $A(s)$ consists of at least one and at most three poles.

We now introduce another characteristic geometric function

$$G(t) \equiv 1 - \frac{\varepsilon_e}{\varepsilon_1} = \frac{1 - s\,F(s)}{1 - s}$$

$$t \equiv \left(1 - \frac{\varepsilon_2}{\varepsilon_1}\right)^{-1} = 1 - s \quad , \qquad \text{(III.31)}$$

which obviously is very similar to $F(s)$ except that the roles of ε_1 and ε_2 have been interchanged. By analogy with (III.21), we can define the auxiliary function

$$A_G(t) \equiv \frac{t - \frac{1}{3}\,p_1}{p_2} - \frac{1}{G(t)} \quad , \qquad \text{(III.32)}$$

and show that the following relation holds between $A(s)$ and $A_G(t)$

$$\left(\frac{p_2}{p_1}(s - \tfrac{1}{3}) - A(s)\right)\left(\frac{p_1}{p_2}(t - \tfrac{1}{3}) - A_G(t)\right) = ts \quad . \qquad \text{(III.33)}$$

We now consider the case where the bounding expression for $A(s)$ has three poles, and both (III.24) and (III.25) (as well as (III.27)) are satisfied as equalities. It is easy to see from (III.33) that in this case, $A_G(t)$ will have only two poles, and that it satisfies the analogues of (III.24) and (III.25) as strict inequalities

$$- A_G(0) < \frac{1}{3}\,\frac{p_1}{p_2} \qquad \text{(III.34)}$$

$$A_G(1) < \frac{2}{3}\,\frac{p_1}{p_2} \quad . \qquad \text{(III.35)}$$

From (III.33) we can also derive the following equation for $A(s) + A(t)$

$$\frac{1}{3}\frac{p_2}{p_1} - \left(A(s) + A(t)\right) = ts\left[\frac{1}{3}\frac{p_1}{p_2} - \left(A_G(s) + A_G(t)\right)\right]\left[\frac{p_1}{p_2}(t - \tfrac{1}{3}) - A_G(t)\right]^{-1}\left[\frac{p_1}{p_2}(s - \tfrac{1}{3}) - A_G(s)\right]^{-1} . \qquad \text{(III.36)}$$

From this is follows that the analogue of (III.27) is satisfied as an equality

$$A_G(0) + A_G(1) = \frac{1}{3}\frac{p_1}{p_2} \quad . \tag{III.37}$$

Thus, since $A_G(t)$ is as good a tool as $A(s)$ for constructing bounds on $F(s)$ or ε_e, we can always restrict ourselves to at most a two-pole bound for $A(s)$ or $A_G(t)$.

Returning now to consider $A(s)$, we assume a bounding expression that has two poles

$$\frac{A_0}{s - s_0} + \frac{A_1}{s - s_1} \quad . \tag{III.38}$$

Solving (III.30) to get A_0 and substituting in (III.28) and (III.29), these two inequalities become, respectively,

$$A_1 \frac{s_0 - s_1}{s^+ - s_1} \frac{s^+}{s_0 s_1} \le \frac{1}{3}\frac{p_2}{p_1} - \frac{s^+ - s_0}{s_0} A(s^+) \tag{III.39}$$

$$A_1 \frac{s_0 - s_1}{s^+ - s_1} \frac{s^+}{s_0 s_1} \ge \frac{1}{3}\frac{p_2}{p_1}\left(\frac{1 - s^+}{(1 - s_0)(1 - s_1)} \frac{s_0 s_1}{s^+} - 1\right)^{-1} - A(s^+)(s^+ - s_0)\frac{s^+}{s_0 s_1} \quad . \tag{III.40}$$

Using the same substitution in the bounding expression leads to the following form for the bound

$$\frac{s^+ - s_0}{s - s_0} A(s^+) + A_1 \frac{(s_0 - s_1)(s - s^+)}{(s - s_0)(s - s_1)(s^+ - s_1)} \quad . \tag{III.41}$$

This will be an upper (lower) bound if we maximize (minimize) it by varying A_1, s_0, s_1 subject to all the (inequality) constraints. At first, we will keep s_0, s_1 fixed and only allow A_1 to vary. In this process, one of the inequalities (III.39), (III.40) becomes an equality first, depending on the values of s, s^+, and on whether an upper or a lower bound is sought.

E.g., if $s > 1$, $s^+ < 0$, and we are looking for a lower bound, then we must increase the ℓ.h.s. of (III.39) and (III.40) by changing A_1. Therefore (III.39) becomes an equality while (III.40) remains a strict inequality. Using (III.39) as an equality to solve for A_1 and substituting the result in (III.40) and (III.41) we obtain the following forms for the remaining inequality and bounding expression, respectively

$$A(s^+)\frac{(s^+ - s_0)(s^+ - s_1)}{s_0 s_1} + \frac{1}{3}\frac{p_2}{p_1} \frac{2\,\dfrac{s^+}{1 - s^+}\dfrac{(1 - s_0)(1 - s_1)}{s_0 s_1} - 1}{\dfrac{s^+}{1 - s^+}\dfrac{(1 - s_0)(1 - s_1)}{s_0 s_1} - 1} \ge 0 \tag{III.42}$$

$$\frac{(s^+ - s_0)(s^+ - s_1)}{(s - s_0)(s - s_1)} \frac{s}{s^+} A(s^+) - \frac{1}{3}\frac{p_2}{p_1}\frac{s_0 s_1}{s^+}\frac{s - s^+}{(s - s_0)(s - s_1)} \quad . \tag{III.43}$$

We must now still let s_0, s_1 vary to as to minimize (III.43). This can be done by decreasing either s_0 or s_1. Consequently, we can decrease s_0 and s_1 until (III.42) becomes an equality. The values of s_0, s_1 that minimize (III.43) will therefore satisfy (III.42) as an equality. Therefore, if the lower bound for A(s) were a two-pole expression, then it would have to satisfy both (III.39) and (III.40) as equalities. Consequently, when translated into $A_G(t)$ language, the bounding function would become a one-pole expression $A_{Gh}(t) = A_0/(t - t_0)$ that satisfies only the analogue of (III.40) as an equality.

We now examine the possibility of a two-pole upper bound, still assuming that $s > 1, s^+ < 0$. We must now decrease the ℓ.h.s. of (III.39) and (III.40) as much as possible by changing A_1. Therefore (III.40) becomes an equality while (III.39) remains a strict inequality. Using (III.40) as an equality to determine A_1 and substituting the result in (III.39) and (III.41), we again obtain (III.42) for the remaining inequality, while the bounding expression becomes

$$\frac{(s^+ - s_0)(s^+ - s_1)}{(s - s_0)(s - s_1)} A(s^+) + \frac{1}{3}\frac{P_2}{P_1}\frac{s - s^+}{s^+}\frac{s_0 s_1}{(s - s_0)(s - s_1)}\left(\frac{1 - s^+}{s^+}\frac{s_0 s_1}{(1 - s_0)(1 - s_1)} - 1\right)^{-1} . \quad \text{(III.44)}$$

We must now still let s_0, s_1 vary so as to maximize (III.44) while satisfying (III.42). This is a more difficult task than before, because (III.44) is not a monotonic function of s_0 and s_1. Nevertheless it can be shown that if there exists a maximum value of (III.44) subject to (III.42), then it is obtained only when (III.42) becomes an equality. This again has the consequence that $A_G(t)$ is a one-pole expression that satisfies the analogue of (III.40) as an equality, while the analogue of (III.39) remains a strict inequality. It is thus the same function that we found before when looking for a two-pole lower bound.

The fact that the same two-pole function has appeared both as an upper and as a lower bound should not dismay us. It is due to the assumption, not necessarily valid, that we made at the beginning; namely, that the bound for A(s) is a two-pole function. The main result that we have shown here is that even when the bound is a two-pole function, we can still always restrict ourselves to considering only one-pole bounds either for A(s) or for $A_G(t)$.

Turning now to a discussion of these one-pole bounds, we find the situation very much simplified. Assuming the form

$$\frac{A_0}{s - s_0} \quad \text{(III.45)}$$

for the (upper or lower) bound on A(s), we first use (III.30) to eliminate A_0. We thus obtain the form

$$\frac{s^+ - s_0}{s - s_0} A(s^+) \quad \text{(III.46)}$$

for the bounding function, and the forms

$$\frac{s^+ - s_0}{s_0} A(s^+) \le \frac{1}{3} \frac{p_2}{p_1} \tag{III.47}$$

$$\frac{(2s_0 - 1)(s^+ - s_0)}{s_0(1 - s_0)} A(s^+) \le \frac{1}{3} \frac{p_2}{p_1} \tag{III.48}$$

for the inequalities (III.28) and (III.29). We must now vary s_0 to maximize (minimize) (III.46) in order to obtain an upper (lower) bound for A(s).

Assuming, as before, that $s > 1$ and $s^+ < 0$, it is clear that an upper (lower) bound will be obtained by increasing (decreasing) s_0 as far as possible, subject to (III.47) and (III.48). The process of increasing s_0 will have to stop when (III.48) becomes an equality. Similarly, the process of decreasing s_0 will have to stop when (III.47) becomes an equality. In this way we obtain the following upper bound for A(s)

$$A(s) < \frac{s^+ - s_0}{s - s_0} A(s^+) \quad , \tag{III.49}$$

where s_0 is the solution of

$$\frac{(s^+ - s_0)(2s_0 - 1)}{s_0(s - s_0)} A(s^+) = \frac{1}{3} \frac{p_2}{p_1} \tag{III.50}$$

that satisfies

$$\frac{2}{3} < s_0 < 1 \quad . \tag{III.51}$$

[Since (III.50) is a quadratic equation for s_0, it has two solutions. By using one of the Hashin-Shtrikman bounds which, for $A(s^+)$, reads

$$A(s^+) > \frac{2}{9} \frac{p_2}{p_1} \frac{1}{s^+ - \frac{2}{3}} \quad \text{for} \quad s^+ < 0 \quad , \tag{III.52}$$

we can show that one of these solutions indeed satisfies (III.51). The other solution is always negative, and hence unacceptable.] Similarly, the lower bound obtained in this way for A(s) has the form

$$A(s) > \frac{s^+ - s_0}{s - s_0} A(s^+) \quad , \tag{III.53}$$

where s_0 is now the solution of

$$\frac{s^+ - s_0}{s_0} A(s^+) = \frac{1}{3} \frac{p_2}{p_1} \quad . \tag{III.54}$$

Using (III.52), we can show that s_0 satisfies

$$0 < s_0 < \frac{2}{3} \quad . \tag{III.55}$$

The two inequalities (III.52), (III.55) are required for consistency so that the second of the two inequalities (III.47), (III.48) (i.e., the one that is not an equality) is still obeyed as a (strict) inequality.

Translated into bounds for F(s), the upper and lower bounds on A(s) become upper and lower bounds for F(s), respectively,

$$F(s) \le F_g(s) \equiv \frac{p_1(s - s_0)}{(s - s_0)(s - \frac{1}{3} p_2) - \frac{1}{3} \frac{p_2\, s_0(1 - s_0)}{2\, s_0 - 1}} , \qquad \text{(III.56)}$$

with s_0 determined by (III.50), (III.51), and

$$F(s) \ge F_e(s) \equiv \frac{p_1(s - s_0)}{s(s - s_0 - \frac{1}{3} p_2)} , \qquad \text{(III.57)}$$

with s_0 determined by (III.54).

We must also consider the one-pole bounds on $A_G(t)$. The analogue of (III.53), (III.54) leads again to the bound $F_e(s)$ for F(s). The analogue of (III.49)-(III.51) leads, however, to a two-pole bound for A(s), and finally to the following three-pole bound for F(s)

$$F(s) \le F_h(s) \equiv \frac{1}{s} \frac{p_1(s - s_0)(s - \frac{2}{3}) - \frac{1}{3} \frac{p_1 s_0(1 - s_0)}{1 - 2s_0}}{(s - s_0)(s - 1 + \frac{1}{3} p_1) - \frac{1}{3} \frac{p_1 s_0(1 - s_0)}{1 - 2s_0}} , \qquad \text{(III.58)}$$

where s_0 is the solution of

$$\frac{(s^+ - s_0)(2s_0 - 1)}{s_0(1 - s_0)} A_G(1 - s^+) = \frac{1}{3} \frac{p_1}{p_2} \qquad \text{(III.59)}$$

that also satisfies

$$0 < s_0 < \frac{1}{3} . \qquad \text{(III.60)}$$

[As we mentioned in the bracketed remarks related to (III.51) and (III.52), we can show that one of the solutions of (III.59) obeys (III.60), while the other is a negative number.] We note that the function $A_h(s)$ related to $F_h(s)$ is a two-pole upper bound on A(s), and also that it is the same function that appeared earlier in this section as a candidate for both upper and lower bounds. From the present discussion, it is clear that $F_h(s)$ is now only a candidate for the role of upper bound on F(s). The actual upper bound is in fact the greater of the two numbers $F_g(s)$ and $F_h(s)$.

It can be shown that in general (i.e., not restricting ourselves to the case $s > 1, s^+ < 0$), F_e always gives one of the bounds on F, while the other bound is given by either F_g or F_h, whichever lies further away from F_e.

Comparing these bounds with the bounds of (III.19) for ϵ_e, we find that one of the latter bounds is identical with F_e. The other bound of (III.19) is, however, different (i.e., less stringent) than either F_g or F_h. In fact, it is equivalent to

$$F(s) \le F_f(s) \equiv \frac{p_1(s - s_0)}{(s - s_0)(s - \frac{1}{3}p_2) - \frac{2}{3}p_2(1 - s_0)} , \qquad \text{(III.61)}$$

where s_0 is now obtained by treating (III.25) as an equality. Close examination shows that $F_f(s)$ violates (III.27), and that therefore it is an overly stringent bound.

As an example we now compute some of the bounds on ϵ_e in the following case

$$p_1 = p_2 = \frac{1}{2}, \quad \epsilon_1 = \epsilon_1^+ = 1, \quad \epsilon_2 = 10, \quad \epsilon_2^+ = \frac{1}{3}, \quad \epsilon_e^+ = \frac{3}{5},$$

$$s = \frac{10}{9}, \quad s^+ = -\frac{1}{2}, \quad F(s^+) = -\frac{4}{5} \quad . \qquad \text{(III.62)}$$

Listed in the table below are the Wiener bounds (W), the Hashin-Shtrikman bounds (HS), the bounds of (III.19) (B_1), and the bounds of (III.57), (III.56), and (III.58) (B_2). An upper (lower) bound is denoted by $\epsilon_u(\epsilon_\ell)$, and the last two bounds, both of them lower bounds, are distinguished by an additional suffix g or h to denote whether they arose from F_g or F_h. The difference between the actual upper and lower bounds in each category is denoted by $\Delta\epsilon$.

W			HS			B_1			B_2			
ϵ_u	ϵ_ℓ	$\Delta\epsilon$	ϵ_u	ϵ_ℓ	$\Delta\epsilon$	ϵ_u	ϵ_ℓ	$\Delta\epsilon$	ϵ_u	$\epsilon_{\ell g}$	$\epsilon_{\ell h}$	$\Delta\epsilon$
5.5000	1.8182	3.68	4.6712	2.8832	1.79	4.5357	3.2836	1.25	4.5357	3.6452	3.4367	1.20

Clearly, the bound $\epsilon_{\ell h}$ (B_2) constitutes an improvement over the bound ϵ_ℓ (B_1).

Rigorous bounds can also be constructed when one or both of ϵ_1, ϵ_2 are complex, and hence ϵ_e is in general complex too. In that case, one can proceed by separating (II.12a) into its real and imaginary parts, treating Im F(s) (for example) as given, and looking for a maximum or minimum value of Re F(s). The equation for Im F(s) is thus treated as an additional constraint. Such a program has been carried out in detail (Bergman 1980, 1981 a,b; Milton 1980, 1981), and the results are always in the form of a limited region in the complex F-plane where the allowed values of F(s) must lie. It should be noted, of course, that complex values of ϵ are usually found only by going to finite frequencies. In that case, the electrostatic theory used here becomes an approximation which is only valid as long as both the wavelength and any possible skin depth are large compared to any scale of the microscopic inhomogeneities in the composite. The case where this requirement is violated is discussed in Section VI.

IV. CALCULATION OF THE ELECTROSTATIC RESONANCES OF A TWO-COMPONENT COMPOSITE

It is often possible to describe a two-component composite as being made of a host material ϵ_2, in which are embedded separate, non-overlapping grains of ϵ_1 material. In that case, the function $\theta_1(\vec{r})$ can be written as a sum of individual grain

θ-functions

$$\theta_1(\vec{r}) = \sum_a \theta_a(\vec{r}) \qquad \text{(IV.1)}$$

where a is a grain index, and θ_a is defined by

$$\theta_a(\vec{r}) = \begin{cases} 1 & \text{for } \vec{r} \text{ inside the grain a} \\ 0 & \text{elsewhere} \end{cases} \quad . \qquad \text{(IV.2)}$$

Consequently, the operator Γ can also be written as a sum of individual grain operators

$$\Gamma = \sum_a \Gamma_a$$

$$\Gamma_a \phi \equiv \int dV' \theta_a(\vec{r}\,') \frac{\partial G(\vec{r},\vec{r}\,')}{\partial \vec{r}\,'} \cdot \frac{\partial \phi(\vec{r}\,')}{\partial \vec{r}\,'} \quad . \qquad \text{(IV.3)}$$

It is usually easier to find the eigenstates of each individual Γ_a than to find those of Γ. In that case, it is a good strategy to try to expand any solution of

$$\phi = z + u\Gamma\phi \qquad \text{(IV.4)}$$

or of the corresponding eigenvalue equation

$$s\phi = \Gamma\phi \qquad \text{(IV.5)}$$

in terms of these individual grain eigenfunctions, defined by

$$s_{a\alpha}\phi_{a\alpha} = \Gamma_a \phi_{a\alpha} \quad . \qquad \text{(IV.6)}$$

Such an expansion would have the form

$$\theta_1^+(\vec{r})\,\phi(\vec{r}) = \sum_{a\alpha} A_{a\alpha}\,\theta_a^+(\vec{r})\,\phi_{a\alpha}(\vec{r}) \quad , \qquad \text{(IV.7)}$$

where $A_{a\alpha}$ are the expansion coefficients. The θ_1-function must appear on the ℓ.h.s. because we can only hope to expand $\phi(\vec{r})$ in this way at points within ϵ_1 material. Since the scalar product of (II.4) involves integration only over the ϵ_1 volume, the completeness property of eigenstates of Γ applies only within that volume. Similarly, the functions $\phi_{a\alpha}$ for a fixed grain a, though defined everywhere, constitute a complete orthonormal set only within the volume of that grain. That is why on the r.h.s. of (IV.7) we have to restrict each of these functions to the volume of its particular grain by including θ_a. Finally, the (+) superscript attached to the θ-functions in (IV.7) means that we are to take them to equal 1 over a volume that is infinitesimally larger than the usual volume. This is done in order to prevent spurious δ-functions from appearing as a result of differentiating these θ-functions at the ϵ_1,ϵ_2 interface. The need for defining these θ^+ functions is also one of the reasons why we assumed that the different grains were non-overlapping. (It now appears that it is possible to extend this formalism to a system of overlapping grains - the price includes having to deal explicitly with these δ-functions, as well as with the fact that eigenstates of different individual grain operators are not orthogonal when the grains overlap - see Kantor and Bergman 1981 c.)

Substituting the expansion (IV.7) into (IV.4), and using the usual methods, we get a system of linear equations for the expansion coefficients

$$A_{a\alpha} = z_{a\alpha} + u \sum_{b\beta} \Gamma_{a\alpha,b\beta} A_{b\beta} \quad . \tag{IV.8}$$

Here $z_{a\alpha}$ are the expansion coefficients obtained when $\theta^{+}z$ is expanded in a series of $\theta^{+}_{a}\phi_{a\alpha}$, while the coefficients appearing in the sum are matrix elements of Γ between two individual grain eigenfunctions

$$\begin{aligned} \Gamma_{a\alpha,b\beta} &\equiv \langle \theta^{+}_{a}\phi_{a\alpha} | \Gamma | \theta^{+}_{b}\phi_{b\beta} \rangle \\ &= \int dV \int dV' \theta_a(\vec{r}) \theta_b(\vec{r}\,') \frac{\partial \phi^{*}_{a\alpha}(\vec{r})}{\partial \vec{r}} \cdot \frac{\partial^2 G(\vec{r},\vec{r}\,')}{\partial \vec{r}\, \partial \vec{r}\,'} \cdot \frac{\partial \phi_{b\beta}(\vec{r}\,')}{\partial \vec{r}\,'} \\ &= s_{b\beta} \int dV\, \theta_a(\vec{r})\, \vec{\nabla}\phi^{*}_{a\alpha} \cdot \vec{\nabla}\phi_{b\beta} \quad . \end{aligned} \tag{IV.9}$$

The last expression for the matrix element is especially convenient for practical calculations - it has the form essentially of an overlap integral of the product $\vec{\nabla}\phi^{*}_{a\alpha} \cdot \vec{\nabla}\phi_{b\beta}$ integrated over the volume of just <u>one</u> of the two grains (i.e., the grain a).

This discussion is reminiscent of the so called "tight binding approach" to the calculation of electronic states (i.e., solutions of Schrödinger's equation) in a collection of many atoms. There too, one starts from the (electronic) eigenstates of individual atoms, and then uses them as a basis in which to expand the solution of the many-atoms problem.

As in the electronic problem, here too, a significant simplification occurs if the ϵ_1 grains (analogous to individual atoms) are all identical and arranged in a periodic array. In that case, if we use a vector index $\vec{a}$ to denote the position of an individual grain, then, by a proper choice of the normalization constant, the functions $\phi_{\vec{a}\alpha}(\vec{r})$ for different grains can be made to depend only on $\vec{r} - \vec{a}$

$$\phi_{\vec{a}\alpha}(\vec{r}) = \phi_{\alpha}(\vec{r} - \vec{a}) \quad . \tag{IV.10}$$

Consequently, the $z_{\vec{a}\alpha}$ coefficients are independent of $\vec{a}$, and the $A_{\vec{a}\alpha}$ coefficients can be shown to obey a Bloch-theorem, i.e., they always have the form

$$A_{\vec{a}\alpha} = A_{\alpha}(\vec{k})\, e^{i\vec{k}\cdot\vec{a}} \quad . \tag{IV.11}$$

Equation (IV.8) is then replaced by

$$A_{\alpha}(\vec{k}) = z_{\alpha}\, \delta_{\vec{k},\vec{0}} + u \sum_{\beta} \Gamma_{\alpha\beta}(\vec{k})\, A_{\beta}(\vec{k}) \tag{IV.12a}$$

$$\Gamma_{\alpha\beta}(\vec{k}) \equiv \sum_{\vec{a}-\vec{b}} \Gamma_{\vec{a}\alpha,\vec{b}\beta}\, e^{-i\vec{k}\cdot(\vec{a}-\vec{b})} \quad . \qquad \text{(IV.12b)}$$

If we omit $z_{a\alpha}$ or z_α in (IV.8) or (IV.12a), respectively, then these become equations for the expansion coefficients of the eigenfunctions of Γ. These are just the eigenvectors of the matrix in each of these equations. With the help of these eigenvectors, we can rewrite (II.12a) in the form

$$F(s) = \frac{u}{V} \sum_{a\alpha} z_{a\alpha}^{*} A_{a\alpha} \qquad \text{(IV.13)}$$

for a general system of non-overlapping grains, or in the form

$$F(s) = \frac{uN}{V} \sum_{\alpha} z_{\alpha}^{*} A_{\alpha}(\vec{k}=0) \qquad \text{(IV.14)}$$

for a periodic array of identical grains. The factor N is the total number of unit cells in the system. Note that only the $\vec{k} = 0$ eigenvectors are needed to determine ϵ_e - the nonzero $\vec{k}$ eigenvalues have zero weight in the pole spectrum of F(s). That is why the pole spectrum remains discrete in this case, even though the eigenvalue spectrum is continuous (for systems with an infinite volume).

The procedure described in this section is especially suitable when the individual grains have very simple and highly symmetrical shapes. E.g., when the ϵ_1 grains are all spherical, the relevant individual grain eigenstates and eigenvalues can all be obtained in closed analytical form, and the same is true of the two-body overlap integrals. Thus, the matrix elements $\Gamma_{a\alpha,b\beta}$ can be set up analytically, and it is only when one wants to solve (IV.8) or (IV.12) that a numerical computation must be resorted to. This has actually been done in detail for a simple cubic array of identical spherical inclusions, with very good results (Bergman 1979 a), and also for a quasi-random system of spheres, where the results could be compared with experiments (Kantor and Bergman 1981 b).

In summary, the approach described in this section can take maximum advantage of any symmetries or simplifying features of individual grains, even when the entire composite has a much lower symmetry. We might also add that McPhedran and Milton (1981) have developed an alternative way of using the matrix $\Gamma_{a\alpha,b\beta}$ for obtaining ϵ_e. In their approach, the matrix is first used to calculate the higher moments of the pole spectrum s_n, defined in (II.15). These higher moments are then used to generate a series of successively better upper and lower bounds on ϵ_e. This has certain advantages over a direct calculation of ϵ_e, since it is unnecessary either to invert the matrix or diagonalize it. At the same time, it provides an estimate of the error in the calculation.

V. ELECTROSTATIC PROBLEMS IN MULTICOMPONENT COMPOSITES

When more than two components are involved, the bulk effective dielectric constant must be discussed in terms of an analytic function of more than one variable

$$F(s^{(1)}\dots s^{(m)}) \equiv 1 - \frac{\epsilon_e(\epsilon_1\cdots\epsilon_{m+1})}{\epsilon_{m+1}} = \sum_{i=1}^{m} \frac{u_i}{V}\int dV\ \theta_i \frac{\partial\phi}{\partial z} \quad . \tag{V.1}$$

In general, the singularities of this function lie on (m - 1)-dimensional hypersurfaces in the complex space of $s^{(1)}\dots s^{(m)}$. For a system of non-overlapping grains with an exactly known microgeometry, the method of the previous section can obviously be generalized so as to determine these hypersurfaces: Fixing the values of, say, $s^{(2)}\dots s^{(m)}$, we can find the resonance values of $s^{(1)}$ as eigenvalues of (II.1). The eigenvectors, which, like the eigenvalues, now depend on $s^{(2)}\dots s^{(m)}$, can be used to construct the weights in a pole expansion of F in the variable $s^{(1)}$.

The problem of deriving exact bounds for $F(s^{(1)}\dots s^{(m)})$ when the microgeometry is not precisely known is best handled by selecting an appropriate trajectory $s^{(1)}(s)\dots s^{(m)}(s)$. If this is done properly, the resulting function of one variable

$$F(s) \equiv F(s^{(1)}(s)\dots s^{(m)}(s)) \tag{V.2}$$

can be shown to have analytic properties similar to those that are found in the two-component problem. The methods of Section III can then be applied to this function in order to derive rigorous bounds when various types of partial information are available about the composite (see Bergman 1978 a,b).

VI. EXTENSION OF THE APPROACH TO OTHER THAN ELECTROSTATIC PROBLEMS

The basic idea - focusing on the resonances of the problem and using them in order to expand the real problem - is applicable to other bulk properties of composites, and also to scattering problems.

The problem of calculating the bulk effective elastic stiffness tensor $C^{(e)}_{ijk\ell}$ can be set up in a form that is similar to (II.1) and (II.10), namely (we assume a two-component composite),

$$\begin{aligned}
\epsilon_{ij}(\vec{r}) &= \epsilon^{o}_{ij} + \sum_{k\ell mn} \int dV'\theta_1(\vec{r}')\ G_{ijk\ell}(\vec{r},\vec{r}')\ \delta C_{k\ell mn}\ \epsilon_{mn}(\vec{r}') \\
\epsilon^{o} C^{(e)} \epsilon^{o} &\equiv \frac{1}{V}\int dV\ \epsilon^{o}\, C(\vec{r})\ \epsilon(\vec{r}) \\
C_{ijk\ell}(\vec{r}) &= C^{(1)}_{ijk\ell}\ \theta_1(\vec{r}) + C^{(2)}_{ijk\ell}\ \theta_2(\vec{r}) \\
&= C^{(2)} + \theta_1(\vec{r})\ \delta C \\
\delta C &\equiv C^{(1)} - C^{(2)}
\end{aligned} \tag{VI.1}$$

Here $\epsilon_{ij}(\vec{r})$ is the local strain tensor in the composite, while ϵ^{o}_{ij} is the (constant) strain tensor that would exist if the same boundary conditions were applied to a homogeneous solid. The C's are elastic stiffness tensors, and wherever it is convenient we have omitted the tensor indices and allowed the summations over them to be implied.

Similarly to the case of the multicomponent composite dielectric, here too, the problem depends upon more than one (complex) variable, since even in the case of an isotropic material there are two elastic constants for each component. By applying methods similar to those described in Section V for the multicomponent dielectric composites, a representation of $C^{(e)}$ in terms of elastostatic resonances can be obtained. This has been used to calculate $C^{(e)}$ for ordered two-dimensional arrays of cylindrical fibers (Kantor and Bergman 1981 a), and to derive some new exact bounds for systems with an unknown microgeometry (Kantor and Bergman 1981 c).

A different type of problem is the scattering of electromagnetic waves by a macroscopic scatterer. By that we mean that a monochromatic plane wave which is propagating through a homogeneous medium characterized by ϵ_2 (we ignore the magnetic permeability, i.e., assume that $\mu = 1$ everywhere), is scattered because a (limited) part of space is occupied by a different material - ϵ_1. When the characteristic dimensions of the scatterer are not small compared to the wavelength or the skin depth, the static approximation is inapplicable. The problem can be cast in the form of an integral equation for the electric field $\vec{E}$

$$\vec{E}(\vec{r}) = \vec{E}_0(\vec{r}) + u \int dV' \theta_1(\vec{r}\,') \overleftrightarrow{G}(\vec{r} - \vec{r}\,', k_2) \cdot \vec{E}(\vec{r}\,')$$

$$\equiv E_0 + u \, \Gamma E \tag{VI.2}$$

where $\vec{E}_0(\vec{r})$ is the incident plane wave, $k_2 \equiv \sqrt{\epsilon_2}\, \omega/c$ is the wave number in the ϵ_2 medium, $u \equiv 1/s \equiv 1 - \epsilon_1/\epsilon_2$, and where the last line of (VI.2) serves to define the linear operator Γ. The symbol $\overleftrightarrow{G}$ stands for a tensor Green's function that satisfies boundary conditions of an outgoing spherical wave at infinity

$$G_{\alpha\beta}(\vec{R}, k) \equiv (k^2 \delta_{\alpha\beta} + \nabla_\alpha \nabla_\beta) \frac{e^{ikR}}{-4\pi R} \quad . \tag{VI.3}$$

The operator Γ, while symmetric, is non-hermitian when one introduces the analogue of (II.4) as the definition of a scalar product

$$(E_1, E_2) \equiv \int dV \; \theta_1(\vec{r}) \, \vec{E}_1^*(\vec{r}) \cdot \vec{E}_2(\vec{r}) \quad . \tag{VI.4}$$

Nevertheless, we can still find the eigenfunctions (in this case they are vector fields) and eigenvalues (they are complex) of Γ, which now obey the following bi-orthogonality relations

$$\Gamma|t\rangle = t|t\rangle$$

$$\langle t^*|\Gamma = \langle t^*|t$$

$$\langle t^*|s\rangle = 0 \quad \text{for different states} \tag{VI.5}$$

If we can extend the bi-orthogonality to include also the normalization property

$$\langle t^* | t \rangle = 1 \quad , \tag{VI.6}$$

something that is not automatically ensured but must be verified in each case, then the set of eigenstates can be shown to be complete. We can then expand Γ as

$$\Gamma = \sum_t |t\rangle t \langle t^*| \quad , \tag{VI.7}$$

and use this to write the scattered part of the field E_{sc} in terms of the states $|t\rangle$

$$E = \frac{1}{1 - u\Gamma} E_0 = E_0 + \frac{\Gamma}{s - \Gamma} E_0 \equiv E_0 + E_{sc} \tag{VI.8}$$

$$E_{sc} = \sum_t |t\rangle \frac{t}{s - t} \langle t^* | E_0 \rangle \quad . \tag{VI.9}$$

It is important to realize that although the states $|t\rangle$ only form a complete set within the volume of the scatterer ϵ_1, the expansion of (VI.9) is valid everywhere. This is so because the eigenfunctions $|t\rangle$ obey the same differential equation as the total field E. Another way of perceiving this is to note that (VI.7) represents the operator Γ everywhere. Thanks to this fact, the expansion (VI.9) can be used in the asymptotic region, where it gives the asymptotic form of E_{sc} in terms of the asymptotic forms of $|t\rangle$. Note however that we need the form of $|t\rangle$ inside the scatterer in order to calculate the partial scattering amplitude into that state, i.e.,

$$\frac{t}{s - t} \langle t^* | E_0 \rangle \quad . \tag{VI.10}$$

This approach has been carried out in detail for the case of a scatterer in the form of a collection of spheres (Bergman and Stroud 1980). In that case, it is worthwhile to use the eigenfunctions of individual spheres as a basis set in Hilbert space. All the multiple scattering effects are then automatically taken into account when the matrix of Γ in that basis is diagonalized. As in the electrostatic problem, the off-diagonal matrix elements are expressible in terms of overlap intergrals between eigenstates of different spheres. Consequently, once one has determined the eigenstates and the two-body overlap integrals, the problem of scattering by three or more spheres is not significantly more difficult than the two-sphere problem.

Finally, we would like to mention the fact that the quantum scattering problem can also be cast in a similar form. Starting from the Lippmann-Schwinger integral equation for scattering by a scalar potential $V(\vec{r})$

$$\psi(\vec{r}) = e^{i\vec{k}\cdot\vec{r}} - \frac{m}{2\pi\hbar^2} \int dV' \frac{e^{ik|r - r'|}}{|r - r'|} V(\vec{r}')\psi(\vec{r}') \quad , \tag{VI.11}$$

we first multiply it by $\sqrt{V(\vec{r})}$ and rewrite it as the last of the following equations

$$\rho(\vec{r}) \equiv \sqrt{V(\vec{r})}\ \psi(\vec{r}) \tag{VI.12a}$$

$$\Gamma\rho \equiv -\frac{m}{2\pi\hbar^2}\int dV' \sqrt{V(\vec{r})}\,\frac{e^{ik|r-r'|}}{|r-r'|}\sqrt{V(\vec{r}')}\,\rho(\vec{r}')$$

$$\rho = \rho_o + \Gamma\rho \quad . \tag{VI.12b}$$

If $V(\vec{r}) \neq 0$ only in a limited region of space, which is often the case to a very good approximation, then we can characterize that region by an appropriate θ-function and define a scalar product that includes integration only over that limited region

$$(\psi_1,\psi_2) \equiv \int dV\ \theta(\vec{r})\psi_1^*(\vec{r})\psi_2(\vec{r}) \quad . \tag{VI.13}$$

We can then proceed, as in the e.m. scattering problem, to calculate the eigenstates of the operator Γ and use them as a basis set in which to expand the actual scattering problem. Because Γ is again a non-hermitian operator, bi-orthogonality relations will have to be used in manipulating the basis set. It is expected that this method will be especially useful in treating problems where multiple scattering is important, and where the symmetry of individual elements of the scatterer is high while the total scatterer has a lower symmetry.

VII. CONDUCTIVITY OR CONNECTIVITY THRESHOLDS

Composite materials that are made of a metal-insulator mixture can either be conducting or insulating, depending upon the microgeometry. Often, by changing a single microstructural parameter, e.g., the overall concentration of metal, it is possible to switch from one type of behavior to the other. One very simple (but not very practical) system where this occurs is a periodic array of conducting spheres in an insulating background. As the spheres are increased in size, such a transition will occur at the point where they just begin to touch each other. The transition in the case of a simple cubic array of spheres has been studied in some detail (Bergman 1979 a).

A somewhat different case is where the metal-insulator system has a random microstructure. It should then really be represented by an ensemble of systems with different detailed microgeometries, and the bulk effective conductivity σ_e (or any other property, for that matter) should be averaged over all members of the ensemble. The conductivity threshold is then associated with a percolation threshold of the metallic component.

In general, a conductivity threshold in a metal-insulator composite is always connected with a geometric connectivity threshold for the metallic component.

Since the function F(s) of (II.12) represents the bulk d.c. conductivity σ_e of a two-component composite as effectively as it does ϵ_e, we now focus attention upon it, recalling that the variables u,s should now be related to the pure component conductivities σ_1,σ_2

$$u \equiv \frac{1}{s} \equiv 1 - \frac{\sigma_1}{\sigma_2} \quad , \tag{VII.1}$$

while

$$F(s) \equiv 1 - \frac{\sigma_e}{\sigma_2} \quad . \tag{VII.2}$$

Clearly, if σ_2 is the metal and σ_1 is the insulator, then s = 1 and we have

$$F(1) = 1 \quad \text{if} \quad \sigma_e = 0$$

$$0 < F(1) < 1 \quad \text{if} \quad \sigma_e > 0 \quad . \tag{VII.3}$$

On the other hand, if σ_1 is the metal and σ_2 is the insulator, then we should be looking at s = 0. It is then more convenient to think of a material with the same microgeometry, but where $\sigma_1 = \infty$ while σ_2 is finite. This again leads to s = 0, and it is now clear that

$$-\infty < F(0) < 0 \quad \text{if} \quad \sigma_e < \infty$$

$$F(0) = -\infty \quad \text{if} \quad \sigma_e = \infty \quad . \tag{VII.4}$$

To be quite specific, the last line means that s = 0 is a pole of F(s) with a finite weight. Thus, the connectivity properties of $\sigma_1(\sigma_2)$ are directly related to the value of F(0) (F(1)).

In a random composite, we are really interested in the average of F(s) over the ensemble. Thus, the spectrum of poles, even if it is discrete for a specific system, gets smeared into a continuous spectrum by the averaging process. Consequently F(s) now has a branch cut on the real axis somewhere between s = 0 and s = 1.

Quite often, the system or the statistical ensemble is characterized by a parameter, such as the volume fraction p_1 or $p_2 = 1 - p_1$, that can be changed continuously until at some particular value p_c the system switches from conducting to nonconducting. We will now discuss the behavior of F(s) near such a conductivity or connectivity threshold.

From the definitions of F(s) and E(s), (II.12a) and (III.17) respectively, we can deduce the following relationship

$$(1 - F(s)(1 - E(s)) = 1 - \frac{1}{s} \tag{VII.6}$$

that holds when E and F correspond to any specified microgeometry. Since we want to use (VII.6) also for the case of a random system, where E and F correspond to ensemble averages over different microgeometries, some words of caution are in order.

Although in general, the average of E•F is not equal to the product of averages of E and of F, we can argue that equality should always hold if the ensemble is to have any physical significance. Unless the distribution of values of σ_e for systems in the ensemble is sufficiently narrow so that calculating $\langle\sigma_e\rangle$ yields the same result as calculating $\langle 1/\sigma_e\rangle^{-1}$, both values are patently useless. After all, an experiment is always performed with a single sample, and the ensemble is only a mathematical tool for representing the degree of our uncertainty about the actual microgeometry of that sample. An ensemble that yields a wide range of values for σ_e simply means that the set of samples is not well enough characterized, i.e., that its relevant properties (σ_e in this case) are irreproducible from one sample to the next.

We will now consider the conductivity or connectivity threshold of σ_2, hence we wish to take $\sigma_2 > 0$, $\sigma_1 = 0$, i.e., $s = 1$. At that point, $F(1)$ and $E(1)$ exhibit the following properties

$$E(1) = 1, \quad 0 < F(1) < 1 \quad \text{if} \quad p_2 > p_c$$

$$F(1) = 1, \quad 0 < E(1) < 1 \quad \text{if} \quad p_2 < p_c \quad . \qquad \text{(VII.7)}$$

Note that the inequalities in this equation are strict inequalities - the possibility of an equality is not allowed.

From (VII.6) it is now clear that when $p_2 < p_c$, then $1 - F(s) \sim 1 - s$ for $s \to 1$. Consequently, the derivative $F'(s)$ remains finite as $s \to 1$. As for the other derivative $E'(s)$, as $s \to 1$ it must either remain finite, or at most diverge more weakly than $1/(1-s)$ (otherwise $E(1)$ would not be finite). Taking the derivative of (VII.6) and using these results to set $s = 1$, we now get

$$F'(1)(1 - E(1)) = -1 \quad . \qquad \text{(VII.8)}$$

If we now let $p_2 \to p_c^-$, $E(1)$ must tend to 1, and consequently, $F'(1) \to \infty$.

Viewed as a function of s for $p_2 = p_c$, $F'(s)$ must diverge as $s \to 1$, but less weakly than $1/(1 - s)$ so that $F(1)$ remains finite. This means that $s = 1$ must be a singularity of $F(s)$. Consequently, in the random percolative case, the branch cut in $F(s)$ must extend all the way to $s = 1$, which is a branch point, at $p_2 = p_c$. From (VII.6) we can now deduce that, for $p_2 = p_c$, $E'(s)$ must also diverge as $s \to 1$, and that the product $E'(s)F'(s)$ diverges as $1/(1 - s)$.

I would have liked to argue that away from p_c, i.e., for $p_2 < p_c$, both $E(s)$ and $F(s)$ are analytic at $s = 1$. However, I cannot prove this assertion in general. In fact, I suspect that for particular ensembles it may not be true. However, it is clear that for a particular finite (or infinite periodic) system, with a well defined and smooth microgeometry where σ_2 is unconnected or non-percolating, the highest singularity of $F(s)$ is a pole of finite weight that appears a finite distance below $s = 1$. It is then plausible that even if an ensemble of such systems or an infinite but random system is considered, the highest singularity will still lie below $s = 1$. (A more detailed physical argument for making this plausible is given in Bergman 1978 b).

Turning to the regime above the threshold for connectivity of the σ_2 component, namely $p_2 > p_c$, we can proceed in complete analogy with the previous discussion except for switching the roles of $E(s)$ and $F(s)$. In this way we can show that $E'(s)$ must remain finite at $s = 1$, while $F'(s)$ either remains finite as $s \to 1$, or at most diverges more weakly than $1/(1-s)$. As before, physical plausibility arguments can be made to suggest that $s = 1$ is in fact a regular point of both $E(s)$ and $F(s)$ for $p_2 > p_c$ (see Bergman 1978 b).

It is worthwhile to point out that this type of predicted behavior is in complete agreement with the effective medium approximation, which is the only solvable model that exhibits a percolation-type conductivity threshold (see Bergman 1978 b). It is also in agreement with asymptotic analyses of series calculations of $F(s)$ in powers of $u = 1/s$ for a random-resistor-network (Bergman and Kantor 1981).

The fact that a conductivity threshold is invariably connected with the singularity of $F(s)$ at $s = 1$ (or at $s = 0$, depending upon whether σ_2 or σ_1 is the metallic component), as we have shown above, turns out to be a very useful tool in obtaining some important general relationships. It has thus been possible to predict that in a metal-insulator composite, the real part of the dielectric constant diverges at the conductivity threshold, as well as to relate the critical exponents for the singular behavior of the dielectric constant and the conductivity (Bergman and Imry 1977). It has also been possible to make predictions regarding the frequency dependence of the above mentioned polarization catastrophe at a metal-insulator percolative transition as well as at a normal-superconducting percolative transition (Stroud and Bergman 1981).

VIII. ACKNOWLEDGEMENTS

It is a pleasure to acknowledge useful conversations and correspondence with Graeme Milton of Cornell University and The University of Sydney. It is also a pleasure to thank Phyllis Dolan of The Ohio State University for her great devotion in getting this manuscript typed in a camera-ready form that is so pleasant to look at and read.

BIBLIOGRAPHY

Bergman, D. J. 1978a AIP Conf. Proc. No. 40, pp. 46-61.
Bergman, D. J. 1978b Physics Reports 43 377-407.
Bergman, D. J. 1979a J. Phys. C 12 4947-4960.
Bergman, D. J. 1979b Phys. Rev. B 19 2359-2368.
Bergman, D. J. 1980 Phys. Rev. Lett. 44 1285-1287.
Bergman, D. J. 1981a Phys. Rev. B 23 3058.
Bergman, D. J. 1981b, To appear in Annals of Physics.
Bergman, D. J. 1981c, Unpublished.
Bergman, D. J. and Imry, Y. 1977 Phys. Rev. Lett. 39 1222-1225.
Bergman, D. J. and Kantor, Y. 1981 J. Phys. C, to be published.
Bergman, D. J. and Stroud, D. 1980 Phys. Rev. B 22 3527.
Dantzig, G. B. 1963 Linear Programming and Extensions, Princeton Univ. Press, Princeton, New Jersey.
Hashin, Z. and Shtrikman, S. 1962 J. Appl. Phys. 33 3125-3131.
Kantor, Y. and Bergman, D. J. 1981a, Submitted to J. Mech. Phys. Sol.
Kantor, Y. and Bergman, D. J. 1981b, To appear in J. Phys. C: Solid State Physics.
Kantor, Y. and Bergman, D. J. 1981c, In preparation.
McPhedran, R. C. and Milton, G. 1981, To appear in Applied Physics.
Milton, G. 1980 Appl. Phys. Lett. 37 300-302.

Milton, G. 1981, To appear in J. Appl. Phys.
Prager, S. 1969 J. Chem. Phys. 50 4305-4312.
Schulgasser, K. 1976 J. Math. Phys. 17 378-381.
Stroud, D. and Bergman, D. J. 1981, In preparation.
Weiner, O. 1912 Abh. Math. Phys. Kl. Sächs. Akad. Wiss. Leipz. 32 509.

ELASTIC WAVES IN FLUID-SATURATED POROUS MEDIA

James G. Berryman
Bell Laboratories
Whippany, N.J. 07981/USA

1. Introduction

Porous materials arise in a variety of geophysical contexts and engineering applications. Typical examples of fluid-saturated porous materials include (1) porous, air-filled, sound-absorbing tiles used in noise control and (2) water- or oil-saturated sedimentary rocks studied by earthquake seismologists and oil-exploration geophysicists.

A general theory of elastic wave propagation in fluid-saturated porous media was developed by Biot [1-3]. Although Biot's theory was published 25 years ago, it has only been within the last few years that the slow compressional wave predicted by Biot's theory has been observed in unambiguous experiments [4-6]. Further, only recently has it been realized that *a priori* estimates of several of the phenomenological parameters in Biot's equations were available [5,7-13]. In the past, Biot's equations were used primarily as a phenomenological tool, using curve fitting methods to obtain estimates of those parameters which were difficult to measure or estimate directly [6,14-15]. Now the scientific basis for Biot's equations is more firmly established and many of the uncertainties involved in applying Biot's equations to practical problems have been resolved.

In this paper, we will review the recent progress in understanding Biot's theory, present the model equations found to be useful in applications, and finally give a brief comparison of the theory's predictions to some recent experimental data.

2. Model equations

A fluid-saturated porous material is characterized by the presence of a porous frame and a fluid which fills the interconnected pore space. For example, the porous frame might consist (1) of an open-cell rubber foam or (2) of sand grains cemented together or held together at some depth by over-burden pressure. The interconnected pore space occupies the fraction β of the total volume known as the porosity. Any part of the fluid which is trapped in pores detached from the interconnected pore space is said to occupy the "secondary porosity." Such pockets of isolated fluid are treated as fluid inclusions modifying the elastic properties of the composite porous frame. For simplicity, we generally assume the secondary porosity vanishes but this assumption is not essential to the subsequent analysis.

To establish notation: The bulk modulus and density of the fluid are K_f and ρ_f, respectively. The bulk and shear moduli of the (dry) porous frame are K^* and μ^*. If the frame is composed of a single type of solid constituent grain (as we assume here), the bulk and shear moduli and density of the grains are K_g, μ_g, and ρ_g. The average displacement vector for the solid frame is $\vec{u}$. The fluid-displacement field $\vec{u}_f$ derived from the average volume flow of the fluid through the pores. The average fluid displacement relative to the frame is $\vec{w} = \beta(\vec{u}_f - \vec{u})$. For small strains, the frame dilatation is

2.1 $$e = \vec{\nabla} \cdot \vec{u}$$

The increment of fluid content ζ [3] is defined [16] as the change in pore space per unit volume minus the change in fluid volume per unit volume, i.e.,

2.2 $$d\zeta = \beta(de - de_f) + d\beta.$$

In (2.2), e_f is the average of the microscopic fluid dilatation which must be distinguished from

2.3 $$\varepsilon = \vec{\nabla} \cdot \vec{u}_f .$$

Since $\vec{u}_f$ is a quantity derived from the average flow of the fluid, ε includes a term due to flow in addition to the average microscopic dilatation e_f. Finally, the increment of fluid content is related to the relative displacement vector by

2.4 $$\zeta = \vec{\nabla} \cdot \vec{w} = \beta(e - \varepsilon).$$

We emphasize that $e_f \neq \varepsilon$ in general.

With these definitions for a linear medium isotropic on the macroscopic scale, Biot [3] shows that, if a strain energy functional E exists, it is a quadratic function of the strain invariants [17] $I_1 = e$, I_2 and of ζ having the form

2.5 $$2E = He^2 - 2Ce\zeta + M\zeta^2 - 4\mu^* I_2 ,$$

where

2.6 $$I_2 = e_x e_y + e_y e_z + e_z e_x - \frac{1}{4}(\gamma_x^2 + \gamma_y^2 + \gamma_z^2).$$

The normal strain components are e_x, e_y, e_z while the shear strain components are γ_x, γ_y, γ_z. Using *gedanken* experiments to analyze the effects of applying an isotropic hydrostatic pressure to the saturated medium [15,18,19], the coefficients H, C, and M have been shown to be given by

2.7 $$K \equiv H - \frac{4}{3}\mu^* = K^* + \sigma C,$$

2.8 $$C = \left[\left(\frac{\sigma-\beta}{K_g} + \frac{\beta}{K_f}\right)\Big/\sigma\right]^{-1},$$

2.9 $$M = C/\sigma$$

where

2.10 $$\sigma = 1 - K^*/K_g.$$

Equation (2.7) is Gassmann's equation [20]. The derivation of (2.7) requires the additional assumption that the saturated porous medium is homogeneous on the microscopic scale as well as on the macroscopic scale. Brown and Korringa [21] have shown that Gassmann's equation can be generalized by removing the assumption of microhomogeneity. Korringa [16] has shown recently that (2.7)-(2.9) may also be generalized by removing this assumption. The resulting formulas have the same form as (2.7)-(2.9) but C and the parameter σ are no longer given by (2.8) and (2.10). For our present purposes, we will retain the assumption of micro-rohomogeneity and use (2.7)-(2.10).

To complete the model for the coefficients, either experimental or theoretical estimates of the frame moduli are required. It has been pointed out recently [5,7] that K^* and μ^* may be estimated theoretically using effective medium theory for elastic composites. Since the moduli K^* and μ^* are moduli of the (dry) porous frame, it follows that these moduli can be estimated by treating the frame as a two phase composite: one phase is a vacuum $(K_1, \mu_1, f_1) = (0, 0, \beta)$ while the other phase consists of the solid grains $(K_2, \mu_2, f_2) = (K_g, \mu_g, 1-\beta)$. The effective medium theory [7-10] then predicts that the moduli can be estimated using the coupled equations (summation convention assumed for i, j)

2.11 $$\sum_{n=1}^{2} f_n(K_n - K^*) \frac{1}{3} T^{(n)}_{iijj} = 0$$

and

2.12 $$\sum_{n=1}^{2} f_n(\mu_n - \mu^*) \frac{1}{5} (T^{(n)}_{ijij} - \frac{1}{3} T^{(n)}_{iijj}) = 0$$

where the f_n's are the volume fractions. The values of the scalars $T^{(2)}_{iijj}$ and $T^{(2)}_{ijij}$ depend on K_g, μ_g, K^*, μ^* and also on the assumed shapes of the inclusions. The scalars have been computed for ellipsoidal inclusions by Wu[22,10]. We have obtained good results by assuming the pores are approximated by needle-shaped inclusions [5]. Other authors prefer to use different ellipsoidal shapes to approximate the pore space [7].

If the strain energy functional is written in the form

2.13 $$E = \frac{1}{2} (\tau_{ij} e_{ij} + p\zeta)$$

where p is the pressure in the fluid and τ_{ij} is the average stress tensor for the saturated porous medium, it follows that the Fourier transformed equations of motion for elastic waves take the form

2.14 $$\tau_{ij,j} = (H-\mu^*)e_{,i} + \mu^* u_{i,jj} - C\zeta_{,i}$$
$$= -\omega^2(\rho u_i + \rho_f w_i)$$

2.15 $$-p_{,i} = Ce_{,i} - M\zeta_{,i}$$
$$= -\omega^2(\rho_f u_i + q w_i).$$

A subscript following a comma indicates a derivative and we have assumed a time dependence of the form $\exp(-i\omega t)$. The new coefficients in (2.14) and (2.15) are the average density

2.16 $$\rho = (1-\beta)\rho_g + \beta\rho_f$$

and

2.17 $$q(\omega) = \alpha\rho_f/\beta + iF(\xi)\nu\rho_f/\kappa\omega$$

where

2.18 $$\xi = (\omega a^2/\nu)^{1/2}.$$

The kinematic viscosity of the fluid is ν; the permeability of the porous frame is κ; the parameter a is some characteristic length. The factors α and $F(\xi)$ are dimensionless numbers of magnitude greater than or equal to unity which depend on the geometry of the porous material. To gain insight into the general form of α and F, it is instructive to consider some limiting cases where exact results are known.

In the high porosity limit, Chase [23] has derived equations equivalent to (2.14) and (2.15) with the factor q given by

2.19 $$q(\omega) = \rho_f[1 + i\Delta(\omega)]/\beta$$

where the drag-interaction coefficient Δ for spherical particles of radius a oscillating in the fluid is given by [24]

2.20 $$\Delta(\omega) = \frac{9}{4}(\beta^{-1}-1)[1+z - iz(1+2z/9)]z^{-2}$$

where $z = 2^{-1/2}\xi$. Equating (2.17) and (2.19), we find

2.21 $$\alpha = 1 + \frac{1}{2}(\beta^{-1}-1)$$

and, if we take the permeability of a suspension of spheres to be $\kappa = 2a^2\beta^2/9(1-\beta)$ [25,26], then

2.22 $$F(\xi) = 1 + 2^{-1/2}\,\xi(1-i).$$

The result (2.21) for the structure factor agrees with the general form $\alpha = 1 + r(\beta^{-1}-1)$, where $r = \frac{1}{2}$ for spheres and $0 \leq r \leq 1$ for arbitrary ellipsoids, derived in [5]. Equation (2.22) generalizes that result to include frequency dependent effects.

Equations (2.21) and (2.22) are reasonable estimates of α and $F(\xi)$ in the high porosity limit. For lower porosity, we may expect significant deviations from (2.21) and (2.22) to occur. To obtain an estimate of the structure factor for all porosities, consider a recent result of Brown [11] showing that

2.23 $$\alpha = \beta \mathcal{F}$$

where $\mathcal{F}$ is the low-frequency electrical-resistivity formation factor. Furthermore, Sen, Scala, and Cohen [13] have recently shown both theoretically and experimentally that, for spherical glass beads,

2.24 $$\mathcal{F} = \beta^{-3/2}$$

to a very good approximation over a wide range of porosities ($.01 < \beta < 1$). Combining (2.23) and (2.24) gives the estimate

2.25 $$\alpha = \beta^{-1/2}.$$

Defining $\delta = 1 - \beta$ and comparing the Taylor series expansion of (2.21) and (2.25) at high porosity ($\delta \ll 1$), we find in both cases that

2.26 $$\alpha \cong 1 + \frac{1}{2}\delta$$

for small δ. We conclude that either (2.21) or (2.25) may be used at high porosity but that (2.25) is preferred for lower porosity materials.

To obtain estimates of $F(\xi)$ valid at lower porosities, Biot [2] did two model calculations to determine the effective frequency dependence of the fluid viscosity (1) between parallel plates and (2) in cylindrical tubes. He also discussed the deviations from these forms for other pore shapes. In both model calculations, Biot found functions whose limiting forms are similar to (2.22). In particular, for cylindrical tubes of radius a and with our choice of sign for the frequency dependence Biot's result is

2.27 $$F(\xi) = \frac{1}{4}\,\frac{\xi T(\xi)}{1+2T(\xi)/i\xi}$$

where

2.28 $$T(\xi) = \frac{\mathrm{ber}'(\xi) - i\,\mathrm{bei}'(\xi)}{\mathrm{ber}(\xi) - i\,\mathrm{bei}(\xi)} .$$

The functions $\mathrm{ber}(\xi)$ and $\mathrm{bei}(\xi)$ are the real and imaginary parts of the Kelvin function. The limiting forms of (2.27) are

2.29 $$F(\xi) \to 1 \qquad \text{for } \xi \ll 1 \text{ and}$$

2.30 $$F(\xi) \to 2^{-1/2}\,\xi(1-i)/4 \qquad \text{for } \xi \gg 1.$$

The corresponding permeability for a porous medium with straight capillaries of radius a is $\kappa = \beta a^2/8$. Except for a factor of 4 at high frequencies, the limiting forms (2.29) and (2.30) agree with the limiting forms of (2.22) but the interpretation of the characteristic length a is different in the two cases: a is a typical grain radius in (2.22) while it is a typical pore radius in (2.27).

Biot [2] postulates that there exists a universal function $F(\xi)$. Since the limiting forms of (2.22) and (2.27) (and also the result of the parallel plate calculation) have the same fundamental form, Biot's postulate may be correct. However, to establish the form of this universal function requires a detailed calculation based on the microstructure. Such a calculation may be possible, e.g. using the formalism of Burridge and Keller [27], but so far such a calculation has not been done. To complete our model, we will instead assume that (2.27) gives a reasonable approximation to the universal function for some choice of a and treat the characteristic length a as a model parameter. Ideally, we should have reliable theoretical estimates for both the permeability κ and the characteristic length a. Although some progress has been made in obtaining consistent estimates of κ and a [28], at present no model known to the author is capable of making the transition from the high porosity limit to the low porosity limit in a completely satisfactory manner. The particular choices we make for the permeability and characteristic length will be discussed in the next section along with the comparison to recent experimental results.

3. Wave speeds and attenuation

One result which is easily derived from the coupled equations (2.14)-(2.15) is the existence of three distinct modes of elastic wave propagation: two compressional waves and one shear wave. The phase speed of the shear mode is given by

3.1 $$v_s^2 = \mu^*/(\rho - \rho_f^2/q).$$

The coupled equations for the two compressional wave speeds v_+ and v_- are given by the determinant

3.2 $$\begin{vmatrix} \rho - Hv^{-2} & \rho_f - Cv^{-2} \\ \rho_f - Cv^{-2} & q - Mv^{-2} \end{vmatrix} = 0.$$

It is known [1,29] that the two compressional modes correspond to an in-phase motion of the fluid and solid at speed v_+ and a second, out-of-phase, motion at speed v_-. These modes generally correspond to dispersive waves although the dispersion is weak at ultrasonic frequencies. In particular, the second compressional mode is strongly attenuated at low frequencies, having the character of a diffusion process.

In recent experiments, Plona [4] observed two distinct compressional waves in a water-saturated, porous structure composed of sintered glass beads. It has been shown [5,6] that Plona's data for the wave speeds are consistent with the predictions of Biot's theory.

The input parameters to our model [5] are $K_g = 0.407$ Mb, $\mu_g = 0.297$ Mb, $\rho_g = 2.48$ g/cc, $K_f = 0.022$ Mb, $\rho_f = 1.00$ g/cc, $\nu = 1.00$ centistoke, and $\omega = 2\pi \times 500$ kHz. The frame moduli K^* and μ^* are calculated from (2.11)-(2.12) assuming needle-shaped inclusions. For the present comparison, we use (2.25) instead of (2.21) as our estimate of the structure factor. We let the permeability obey the Kozeny-Carman relation [28]

3.3 $$\kappa(1-\beta)^2/\beta^3 = \kappa_o(1-\beta_o)^2/\beta_o^3 = \text{const}$$

which has been shown empirically to provide reasonable estimates of the dependence of permeability on porosity. We choose $\kappa_o = 9.1 \times 10^{-8}$ cm^2 and $\beta_o = 0.283$ to agree with one of Plona's measured values [4,30] of the permeability and use (3.3) to predict the value of κ for all other porosities considered. No entirely satisfactory model for the characteristic length a has been found. However, dimensional analysis suggests that a^2 must be comparable to κ, so we take

3.4 $$a^2/\kappa = a_o^2/\kappa_o = \text{const.}$$

Although this ratio is certainly a function of porosity [28], we expect the porosity dependence to be rather weak for moderate values of the porosity. At $\beta_o = 0.283$, we choose $a_o = 0.02$ mm corresponding to an average pore radius [15] $\frac{1}{5}$ to $\frac{1}{7}$ of the grain radius (the glass beads were 0.21-0.29 mm in diameter).

The speeds predicted by this model are compared to Plona's observed values in Figure 1. The theoretical results for the fast compressional wave and the shear wave agree with Plona's measurements within the experimental error (3% relative error in measured speeds and an absolute error of ±0.005 in measured porosity). Agreement is not quite so good for the slow compressional wave. The values at $\beta = 0.283$ and 0.258 are within the error bars but the value at $\beta = 0.185$ is higher than the measured value by about 10%. Nevertheless, we consider the agreement to be very good.

Biot's theory can also be used to estimate the ratio of the direct slow wave peak amplitude to the direct fast wave peak amplitude at normal incidence. To do so, we must consider the boundary conditions appropriate for an interface between a fluid and a fluid-saturated porous medium. Assuming free flow across the interface and that the exterior and interior fluids are the same, Deresiewicz and Skalak [31] have shown that the appropriate boundary conditions are

3.5 $$\tau_{nn} = -P,$$

3.6 $$\tau_{ns} = 0,$$

3.7 $$p = P, \quad \text{and}$$

3.8 $$u_{n,t} + w_{n,t} = U_{n,t},$$

where τ_{nn} (no summation convention) and τ_{ns} are the normal and shear components of the average stress and P is the pressure of the exterior fluid. The final equation (3.8) follows from the continuity of the velocity normal to the

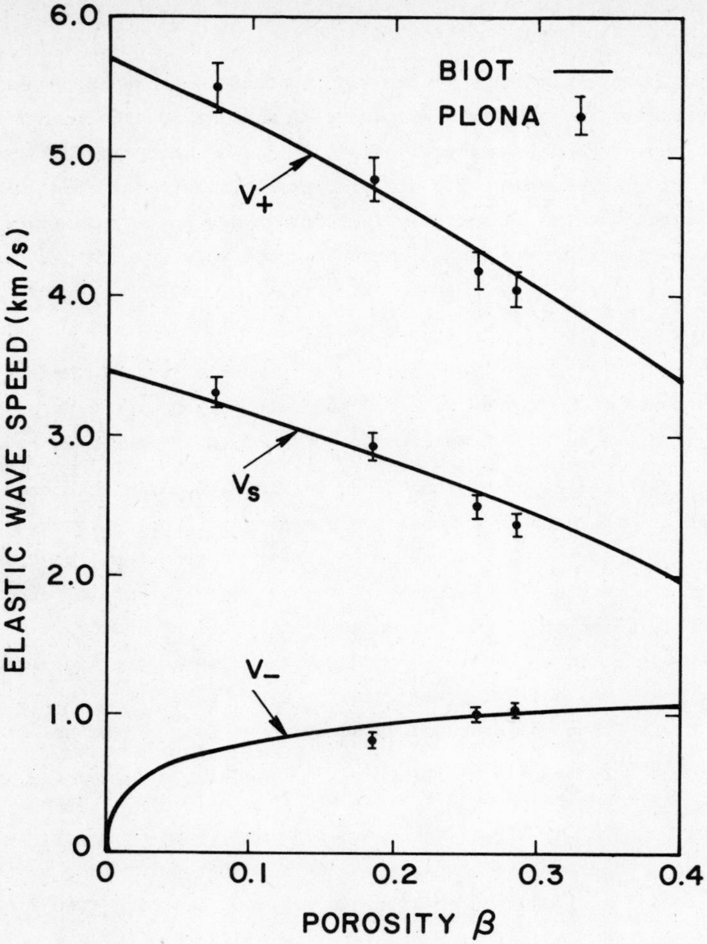

Figure 1. Comparison of theoretical predictions and experimental results for speeds of the fast compressional wave (v_+), the shear wave (v_s), and the slow compressional wave (v_-). Input parameters to the model are: $K_g = 0.407$ Mb, $\mu_g = 0.297$ Mb, $\rho_g = 2.48$ g/cc, $K_f = 0.022$ Mb, $\rho_f = 1.00$ g/cc, $\nu = 0.01$ cm^2/sec, $\omega = 2\pi \times 500$ kHz, $\kappa_o = 9.1 \times 10^{-8}$ cm^2, $a_o = 0.02$ mm, and $\beta_o = 0.283$.

interface averaged over the bulk area, $U_{n,t}$ being the normal component of the velocity of the exterior fluid.

Lack of space precludes us from presenting the details of the calculation for the amplitude ratios here. However, a brief discussion of the results of the analysis is appropriate. Assuming a plane pressure-wave of unit amplitude and frequency ω incident on a slab of saturated porous material of thickness d, we can compute the amplitudes of the fast and slow (mode converted) compressional waves transmitted into the medium. For the three largest porosities in Plona's published data, we find the amplitude of the slow wave is only about 1-3% of that of the fast wave. For $\beta = 0.075$, this ratio is less than 0.3%. When these waves reach the second interface, we find that a large fraction of the energy in the fast wave is reflected while the energy in the slow wave is much more strongly converted and transmitted into the exterior fluid. Thus, if attenuation of the ultrasonic pulses could be neglected as they traverse the porous sample, two transmitted signals of comparable amplitude would be received at a detector. If the wavenumber of a particular mode is $k = \omega/v$, then the attenuation of amplitude for that mode in a sample of thickness d is a factor of the form $\exp(-|\mathrm{Im}k|d)$. If, in addition to the effects already described, we include attenuation of both compressional waves in the calculation, we find the results displayed in Table 1. (Note that the experimental error in the measured amplitude ratios is typically 10%). The attenuation of the fast wave is nearly negligible in all four samples so that, except for a factor close to unity, Biot's theory predicts that the observed amplitude ratio is due to the attenuation of the slow wave.

It should be stressed that small differences in the assumed value of permeability lead to large differences in the computed attenuation of the slow wave. It is therefore important that accurate measurements of the permeability for these particular samples be made.

We see that the theory predicts amplitude ratios in good agreement with experiment for the porosities $\beta = 0.283$ and 0.185. For $\beta = 0.075$, the theory predicts an amplitude ratio of 0.02, which is so low that the experimental technique used would not be able to detect a slow wave of this amplitude; hence, the theory explains why no slow wave was observed in this case. The case with $\beta = 0.258$ is therefore the only one where theory and experiment disagree significantly. This discrepancy could be due to an anomalously small permeability for this sample; further experiments are required to resolve this issue.

The theory is in sufficiently good agreement with the data that we may conclude (1) that Biot's theory is a satisfactory model of wave speeds and attenuation in fluid-saturated porous media and (2) that more experiments are

Sample	1	2	3	4
Porosity	0.283	0.258	0.185	0.075
Thickness (cm)	1.880	1.793	1.646	1.461
Observed amplitude ratio (θ=0)	0.38	0.26	0.21	---
Predicted amplitude ratio (θ=0)	0.35	0.32	0.19	0.02

Table 1. Observed and predicted slow wave to fast wave amplitude ratios for first arrivals at normal incidence through water-saturated sintered glass samples of the stated porosity (β) and thickness (d). All experimental data courtesy of T. J. Plona.

needed to answer the questions raised by the few remaining anomalies. Further theoretical work is needed to find the most general form of the universal function $F(\xi)$ and to establish more satisfactory estimates of the characteristic length a. More experimental work is needed to check the dependence of wave attenuation on sample permeability and to determine whether Biot's theory correctly predicts the reflection and transmission characteristics of waves incident at non-normal angles.

References

[1] M. A. Biot, Theory of propagation of elastic waves in a fluid-saturated porous solid. I. Low-frequency range, J. Acoust. Soc. Am. 28, 168 - 178 (1956).

[2] M. A. Biot, Theory of propagation of elastic waves in a fluid-saturated porous solid. II. Higher frequency range, J. Acoust. Soc. Am. 28, 179 - 191 (1956).

[3] M. A. Biot, Mechanics of deformation and acoustic propagation in porous media, J. Appl. Phys. 33, 1482 - 1498 (1962).

[4] T. J. Plona, Observation of a second bulk compressional wave in a porous medium at ultrasonic frequencies, Appl. Phys. Lett. 36, 259 - 261 (1980).

[5] J. G. Berryman, Confirmation of Biot's theory, Appl. Phys. Lett. 37, 382 - 384 (1980).

[6] N. C. Dutta, Theoretical analysis of observed second bulk compressional wave in a fluid-saturated porous solid at ultrasonic frequencies, Appl. Phys. Lett. 37, 898 - 900 (1980).

[7] J. Korringa, R. J. S. Brown, D. D. Thompson, and R. J. Runge, Self-consistent imbedding and the ellipsoidal model for porous rocks, J. Geophys. Res. 84, 5591 - 5598 (1979).

[8] J. G. Berryman, Theory of elastic properties of composite materials, Appl. Phys. Lett. 35, 856 - 858 (1979).

[9] J. G. Berryman, Long-wavelength propagation in composite elastic media I. Spherical inclusions, J. Acoust. Soc. Am. 68, 1809 - 1819 (1980).

[10] J. G. Berryman, Long-wavelength propagation in composite elastic media II. Ellipsoidal inclusions, J. Acoust. Soc. Am. 68, 1820 - 1831 (1980).

[11] R. J. S. Brown, Connection between formation factor for electrical resistivity and fluid-solid coupling factor in Biot's equations for acoustic waves in fluid-filled porous media, Geophysics 45, 1269 - 1275 (1980).

[12] D. L. Johnson, Equivalence between fourth sound in liquid He II at low temperatures and the Biot slow wave in consolidated porous media, Appl. Phys. Lett. 37, 1065 - 1067 (1980).

[13] P. N. Sen, C. Scala, and M. H. Cohen, A self-similar model for sedimentary rocks with application to the dielectric constant of fused glass beads, Geophysics 46, 781 - 795 (1981).

[14] R. D. Stoll and G. M. Bryan, Wave attenuation in saturated sediments, J. Acoust. Soc. Am. 47, 1440 - 1447 (1970).

[15] R. D. Stoll, Acoustic waves in saturated sediments, in *Physics of Sound in Marine Sediments*, edited by L. Hampton (Plenum, New York, 1974), pp. 19 - 39.

[16] J. Korringa, On the Biot-Gassmann equations, to appear.

[17] A. E. H. Love, *A Treatise on the Mathematical Theory of Elasticity* (Dover, New York, 1944), pp. 43, 62, 102.

[18] M. A. Biot and D. G. Willis, The elastic coefficients of the theory of consolidation, J. Appl. Mech. 24, 594 - 601 (1957).

[19] J. Geertsma, The effect of fluid pressure decline on volumetric changes of porous rocks, Trans. AIME 210, 331 - 340 (1957).

[20] F. Gassmann, Über die elastizität poröser medien, Veirteljahrsschrift der Naturforschenden Gesellschaft in Zürich 96, 1 - 23 (1951).

[21] R. J. S. Brown and J. Korringa, On the dependence of the elastic properties of a porous rock on the compressibility of the pore fluid, Geophysics 40, 608 - 616 (1975).

[22] T. T. Wu, The effect of inclusion shape on the elastic moduli of a two-phase material, Int. J. Solids Structurers 2, 1 - 8 (1966).

[23] D. M. Chase, Wave propagation in liquid-saturated open-cell foam, J. Acoust. Soc. Am. 65, 1 - 8 (1979).

[24] L. D. Landau and E. M. Lifshitz, *Fluid Mechanics* (Pergamon Press, London, 1959), p. 96.

[25] T. S. Lundgren, Slow flow through stationary random beds and suspensious of spheres, J. Fluid Mech. 51, 273 - 299 (1972).

[26] J. M. Hovem, Viscous attenuation of sound in suspensions and high-porosity marine sediments, J. Acoust. Soc. Am. 67, 1559 - 1563 (1980).

[27] R. Burridge and J. B. Keller, Poroelasticity equations derived from microstructure, to appear.

[28] J. M. Hovem and G. D. Ingram, Viscous attenuation of sound in saturated sand, J. Acoust. Soc. Am. 66, 1807 - 1812 (1979).

[29] J. Geertsma and D. C. Smit, Some aspects of elastic wave propagation in fluid-saturated porous solids, Geophysics 26, 169 - 181 (1961).

[30] T. J. Plona and D. L. Johnson, Experimental study of the two bulk compressional modes in water-saturated porous structures, in *Proceedings of the November, 1980, IEEE Ultrasonics Symposium*, to appear.

[31] H. Deresiewicz and R. Skalak, On uniqueness in dynamic poroelasticity, Bull. Seismol. Soc. Am. 53, 783 - 788 (1963).

BIOT'S POROELASTICITY EQUATIONS BY HOMOGENIZATION*

Robert Burridge
Courant Institute of Mathematical Sciences, New York University
New York, NY 10012

and

Joseph B. Keller
Departments of Mathematics and Mechanical Engineering
Stanford University
Stanford, CA 94305

Abstract

Equations are derived which govern the linear macroscopic mechanical behavior of a porous elastic solid saturated with a compressible viscous fluid. The derivation is based on the equations of linear elasticity in the solid, the linearized Navier-Stokes equations in the fluid, and appropriate conditions at the solid-fluid boundary. The scale of the pores is assumed to be small compared to the macroscopic scale, so that the two-space method of homogenization can be used to deduce the macroscopic equations. When the dimensionless viscosity of the fluid is small, the resulting equations are those of Biot, who obtained them by hypothesizing the form of the macroscopic constitutive relations. The present derivation verifies those relations, and shows how the coefficients in them can be calculated, in principle, from the microstructure. When the dimensionless viscosity is of order one, a different equation is obtained, which is that of a viscoelastic solid.

1. Introduction

Poroelasticity, the mechanics of porous elastic solids with fluid-filled pores, has received attention recently for several reasons: pore fluids in the ground are believed to play a role in the triggering of earthquakes; liquid waste disposed of underground seeps into pores; the enhanced recovery of oil, gas, and geothermally heated water depends upon flow in porous strata; underwater acoustics involves propagation in the water-saturated porous bottom of the ocean; etc. The poroelastic equations derived by Biot [1-6] have long been regarded as standard and have formed the basis for solving particular problems in

*Research supported in part by the Office of Naval Research, the Air Force Office of Scientific Research, the Army Research Office, and the National Science Foundation, Divisions of Mathematics and Earth Sciences.
Presented at O.N.R. Conference on Ocean Bottom Effects; Miami, April 1979.

poroelasticity, e.g. Rice and Cleary [7], Burridge and Vargas [8]. Recently, however, the validity of these equations has been questioned, in particular by Cleary [9,10].

In view of their importance, and because of doubts about them, we have derived anew the governing equations of linear poroelasticity by starting with the detailed microstructure of the pores, the linearized equations of elasticity and the linearized equations of fluid dynamics. The equations we obtain coincide with Biot's provided that the dimensionless viscosity of the fluid is small. Our analysis also confirms the form of the constitutive relations which he assumed. In addition it shows how the coefficients in those relations can be calculated in principle from the microstructure of the medium. However when the dimensionless viscosity of the pore fluid is large, we obtain a different equation and different constitutive relations. They characterize the porous medium as a viscoelastic solid, in agreement with the suggestion of Cleary [9,10]. In this case also we show how to calculate the coefficients in the equation and in the constitutive relations in terms of the microstructure.

Our derivation is based upon the two-space method of homogenization, which was developed and studied by Sanchez-Palencia [11], Keller [12], Bensoussan, Lions and Papanicolaou [13], Larsen [14], and others. It applies to media with microstructure on a scale much smaller than the macroscopic scale of interest. The method provides a systemative procedure for deriving macroscopic dynamical equations from the basic dynamical equations which govern the behavior of the medium on the microscale. Thus it involves an appropriate form of averaging to eliminate the rapid variations associated with the small scale. It applies both to linear problems, such as we consider here, and to nonlinear problems. Sanchez-Palencia [11] and his co-workers have also derived the poroelastic equations by the two-space method, but their results were not identified with Biot's equations. Keller [15] used it to treat finite velocity flow of a compressible fluid through a rigid porous medium. Lehner [16] used a different method to derive Darcy's law for the flow of an incompressible fluid in a rigid porous medium.

In this summary we shall merely list our starting equations and results. For full details of the derivation see Burridge and Keller [17].

2. Formulation of the equations

We consider an elastic solid permeated by pores containing a compressible viscous fluid. We assume that the pore configuration has a scale length h which is small compared to a typical macroscopic scale length H. The material is fine grained, and appears locally homogeneous on the macroscopic scale. The small ratio h/H plays a key role in our analysis; we denote it by ε:

$$\varepsilon = h/H . \tag{2.1}$$

To make precise the fact that the medium varies rapidly on the small scale h and may also vary slowly on the larger scale H, we assume that every property of the medium is of the form $f(x,y)$ where

$$y = \varepsilon^{-1}x . \tag{2.2}$$

Here $x = (x_1,x_2,x_3)$ is the position vector of a point in cartesian coordinates and $y = (y_1,y_2,y_3)$ is the vector of stretched coordinates. Thus y varies by an amount of order H as x varies by the small amount h. The first argument in $f(x,y)$ accounts for the slow or gradual variation of any material property, and the second argument accounts for the rapid variation. For example $f(x,y)$ may be a periodic function of y, an almost periodic function of y, a stationary random function of y, etc. In particular the characteristic or indicator function $\chi_s(x,y)$ of the solid region D_s is of this form. It is defined to have the value $\chi_s = 1$ at a point in the solid region D_s and the value $\chi_s = 0$ at a point outside D_s. Since we assume that the pores are filled with fluid, a point of the medium outside the solid region D_s is in the fluid region D_f, so $\chi_s = 0$ at a point in D_f.

We consider motions of the fluid and solid which are small enough to be governed by the linearized Navier-Stokes equations and the linearized equations of elasticity respectively, together with linearized interface conditions. For time harmonic motions with angular frequency ω, these equations are:

$$i\omega\rho_f v = \nabla\cdot\sigma , \qquad \text{in } D_f , \tag{2.3a}$$

$$\sigma = -pI + \tilde{\mu}D\nabla v \qquad \text{in } D_f , \tag{2.3b}$$

$$i\omega p = -\kappa\nabla\cdot v \qquad \text{in } D_f , \tag{2.3c}$$

$$v = i\omega u \qquad \text{on } \partial D_f = \partial D_s , \tag{2.3d}$$

$$n\cdot\sigma = n\cdot\tau , \qquad \text{on } \partial D_f = \partial D_s , \tag{2.3e}$$

$$-\omega^2 \rho_s u = \nabla \cdot \tau \ , \qquad \text{in } D_s \ , \tag{2.3f}$$

$$\tau = C\nabla u \ , \qquad \text{in } D_s \ . \tag{2.3g}$$

Here the fluid occupies the domain D_f and has velocity v, pressure p, stress tensor σ, density ρ_f, viscosity $\tilde{\mu}$, and bulk modulus κ. The solid occupies the domain D_s and has displacement u, density ρ_s and elastic constants given by the fourth rank tensor C(x,y) which operates on ∇u to yield the stress tensor τ, and n is the unit normal to $\partial D_f = \partial D_s$ pointing into the solid. In (2.3b) D is the operator defined by

$$De = \frac{1}{2} \left(e + e^T - \frac{2}{3} I \, \mathrm{tr}\, e\right) \ . \tag{2.4}$$

Thus De is the symmetrized deviatoric part of the second rank tensor e.

We consider the unknown field quantities $v(x,y), p(x,y), u(x,y), \sigma(x,y)$ and $\tau(x,y)$ to be functions of the two sets of space variables, just as we did for the material properties. However we consider these two sets of variables to be independent, so we are extending the definitions of the field quantities from three to six-dimensional space. Then the actual physical values of the field quantities are obtained by restricting these functions to the physical diagonal defined in (2.2). Also we extend the equations (2.3) by replacing ∇ by $\nabla_x + \varepsilon^{-1}\nabla_y$. This replacement is made because on the physical diagonal, where y is related to x by (2.2), we have

$$\nabla f(x,y) = \nabla f(x, \varepsilon^{-1} x) = \nabla_x f + \varepsilon^{-1} \nabla_y f \ . \tag{2.5}$$

We assume that the dimensionless viscosity $\tilde{\mu}/\omega\rho_f h^2$, appropriate to the small scale, is of order unity. Therefore the dimensionless viscosity appropriate to the large scale, $\tilde{\mu}/\omega\rho_f H^2$ is small of order ε^2. To make this evident we replace $\tilde{\mu}$ by $\varepsilon^2\mu$ in (2.3). In this case we ultimately obtain equations similar to those introduced by Biot [1-6]. We also treat the case in which $\tilde{\mu}/\omega\rho_f H^2$ us of order unity (see [17]).

We write each field quantity f in the form

$$f(x,y,\varepsilon) = f_0(x,y) + \varepsilon f_1(x,y) + \frac{\varepsilon^2}{2} f_2(x,y) + o(\varepsilon^2) . \tag{2.6}$$

where the $f_i(x,y)$ are bounded as functions of y. This procedure groups the terms of equations (2.3) into different orders (powers) of ε. We seek equations governing the leading coefficients $v_0, \sigma_0, p_0, u_0, \tau_0$. To accomplish this it is necessary to consider also terms of first order in ε but these are eliminated by an averaging process over the fast variable y. See [17] for details.

3. Final equations and comparison with Biot's equation

Denoting a spatial average in the y variable by an overbar our final equations may be written

$$-\omega^2(\bar{\rho}u_0 + \rho_f\bar{w}) = \nabla_x\cdot(\bar{\tau}_0 - V_f p_0 I) , \tag{3.1a}$$

$$-\omega^2\rho_f u_0 + \bar{W}^{-1}\bar{w} = -\nabla_x p_0 . \tag{3.1b}$$

$$\begin{aligned}\bar{\tau}_0 - V_f p_0 I &= [\bar{C} + \overline{CLC} + (V_f I - \overline{CQ})M\ \mathrm{tr}(V_f - \overline{LC})]\nabla_x u_0 \\ &\quad + (V_f I - \overline{CQ})M\nabla_x\cdot\bar{w} .\end{aligned} \tag{3.1c}$$

$$p_0 = -M\ \mathrm{tr}(V_f - \overline{LC})\nabla_x u_0 - M\nabla_x\cdot\bar{w} . \tag{3.1d}$$

where the operators Q, L, W are defined in [17]. V_f, V_s are the fluid and solid volume fractions.

However, Biot [6] gives the poroelastic equations of motion in the form

$$\frac{\partial\tau_{ij}}{\partial x_j} = \rho\ddot{u}_i + \rho_f\ddot{w}_i , \tag{3.2a}$$

$$-\frac{\partial p_f}{x_i} - \rho_f\ddot{u}_i = \bar{Y}_{ij}(p)\dot{w}_j , \tag{3.2b}$$

$$\tau_{ij} = A_{ij}^{\mu\nu} e_{\mu\nu} + M_{ij}\zeta , \tag{3.2c}$$

$$p_f = M_{ij}e_{ij} + M\zeta . \tag{3.2d}$$

It will be seen that our equations (3.1) agree exactly with (3.2) if the following identifications are made:

Present Work	Biot [6]	
$\bar{\rho}, \rho_f, M$	ρ, ρ_f, M	(3.3a)
$\bar{C} + \overline{CLC} + (V_f I - \overline{CQ})M\ \mathrm{tr}(V_f - \overline{LC})$	$(A_{ij}^{\mu\nu})$	(3.3b)
$i\omega$	$i\omega = p = d/dt$	(3.3c)

Present Work	Biot [6]	
$\bar{\bar{W}}^{-1}$	$p\bar{Y}_{ij}(p)$	(3.3d)
$u_0, \bar{w}, p_0$	u, w, p_f	(3.3e)
$M(V_f I - \overline{CQ}) = M\,\mathrm{tr}(V_f I - \overline{LC})$	$(-M_{ij})$	(3.3f)
$\frac{1}{2}[\nabla_x u_0 + (\nabla_x u_0)^T]\ ,\ \nabla\cdot\bar{w}$	$e_{ij}\ ,\ -\zeta$	(3.3g)

Not only do our equations agree with Biot's in form, but the identification (3.3) gives a detailed prescription (see [17]) for calculating the coefficients from the micro-structure. Some of these are merely averages, such as the mass density of the bulk material $\bar{\rho} = \bar{\rho}_s + \bar{\rho}_f$, while others are far from obvious, such as (3.3b). Notice that in defining averages over quantities defined in $D_{s,f}$ an integration over $D_{s,f}$ was performed but the result was divided by the full bulk volume. However our quantities L, Q and W can be obtained only after the solution of special problems in elasticity or in fluid mechanics. They could be solved numerically for media with prescribed periodic microstructure, for instance.

When $\tilde{\mu}$ of (2.3b) is taken to be of order 1, rather than of order ε^2, we do not obtain Biot's equations but find that the bulk material behaves like a viscoelastic solid. For further details the reader is referred to [17].

References

1. M.A. Biot, General theory of three-dimensional consolidation, J. App. Phys. 12, 155-164 (1941).

2. _________, Theory of elasticity and consolidation for a porous anisotropic solid, J. App. Phys. 26, 182-185 (1955).

3. _________, General solutions of the equations of elasticity and consolidation for a porous material, J. App. Mech. 23, 91-95 (1956).

4. _________, The theory of propagation of elastic waves in a fluid-saturated porous solid, I. Low frequency range, II. Higher frequency range, J. Acoust. Soc. Am. 28, 168-178, 179-191 (1956).

5. _________, Mechanics of deformation and acoustic propagation in porous media, J. App. Phys. 33, 1482-1498 (1962).

6. _________, Generalized theory of acoustic propagation in porous dissipative media, J. Acoust. Soc. Am. 34, 1256-1264 (1962).

7. J.R. Rice and M.P. Cleary, Some basic stress-diffusion solutions for fluid saturated elastic porous media with compressible constituents, Rev. Geophys. Space Phys. 14, 227-241 (1976).

References, continued

8. R. Burridge and C.A. Vargas, The fundamental solution in dynamic poroelasticity, Geophys. J. Roy. Astro. Soc. 58, 61-90 (1978).

9. M.P. Cleary, Fundamental solutions for fluid-saturated porous media and applications to localized rupture phenomena, Ph.D. Thesis, Brown University, Division of Engineering, 1975.

10. __________, Elastic and dynamic response regimes of fluid-impregnated solids with diverse microstructures, Int. J. Solids Structures 14, 795-819 (1978).

11. E. Sanchez-Palencia, Non-homogeneous Media and Vibration Theory, Lecture Notes in Physics 127, Springer-Verlag, 1980.

12. J.B. Keller, Effective behavior of heterogeneous media, in Statistical Mechanics and Statistical Methods in Theory and Application, U. Landman, ed., Plenum, New York, 1977, pp. 631-644.

13. A. Bensoussan, J.-L. Lions, and G.C. Papanicolaou, Asymptotic analysis for periodic structures, in Studies in Mathematics and Its Applications, Vol. 5, North-Holland, 1978.

14. E.W. Larsen, Neutron transport and diffusion in inhomogeneous media, II, Nucl. Sci. Eng. 60, 357-368 (1976).

15. J.B. Keller, Darcy's law for flow in porous media and the two-space method, in Nonlinear Partial Differential Equations in Engineering and Applied Science, R.L. Sternberg, A.J. Kalinowski and J.S. Papadakis, eds., Marcel Dekker, New York, 1980, 429-443.

16. F.K. Lehner, A derivation of the field equations for slow viscous flow through a porous medium, I & EC Fundamentals 18, 41-45 (1979).

17. R. Burridge and J.B. Keller, Poroelasticity equations derived from microstructure, J. Acoust. Soc. Amer. 70, 1140-1146 (1981).

APPROXIMATIONS OF BRINKMAN TYPE

Stephen Childress*
New York University
Courant Institute of Mathematical Sciences

1. Introduction

In 1947 Brinkman [1] introduced an approximate method for computing the permeability k of a bed of identical randomly placed rigid spheres. Brinkman's method is based upon an ingenious use of an "effective medium" and a single test sphere. Since k can be derived from the expected force experienced by a single sphere within the bed, and since such a test sphere will be surrounded by a "cloud" of other spheres, it is tempting to compute an approximate value of the expected force by surrounding the test sphere by a homogeneous porous medium having the as yet unspecified permeability k. By solving the Stokes equations for flow past a sphere in such a medium, a force F(k) can be found, in which case the permeability is obtained by solving the algebraic equation

$$k^{-1} = nF(k) \tag{1.1}$$

where n is the number density of spheres in the bed.

This approach, which can be modified to deal with more complicated random geometries, is similar in spirit to certain (quasi crystalline) approximations in multiple scattering theory [2,3,4], and may generally be compared with any example of "renormalization" in deriving bulk properties of random materials. The value of k obtained from (1.1) agrees quite well with experiments [5], even outside the range of validity one might expect on physical grounds, in spite of the fact that the only statistical parameter which appears is the number density n.

One unfortunate aspect of this approximation is that the error is difficult to estimate analytically. From the viewpoint of multiple scattering theory (1.1) is obtained from an infinite partial summation of a series which is formally divergent. From a practical standpoint, however, such estimates are useful since in many cases the statistics of the random material will be known only partially or imprecisely.

---

*Research supported by the National Science Foundation under Grant MCS-79-02766.

In the present paper we study Brinkman's approximation in a one-dimensional model problem which has many of the features of a permeability problem. In this model a number of questions can be studied rather easily: (i) How does Brinkman's formula compare with upper and lower bounds on permeability? (ii) What is the effect of statistical parameters other than n? (iii) How does the formula compare with Monte Carlo simulation with uniform statistics? (iv) Is there a hierarchy of Brinkman-type models of increasing accuracy? At the end of the paper we compare this study with some known results for the permeability of a random bed of spheres.

2. A One-dimensional Model Problem

We shall consider the mechanical system shown in Figure 1.

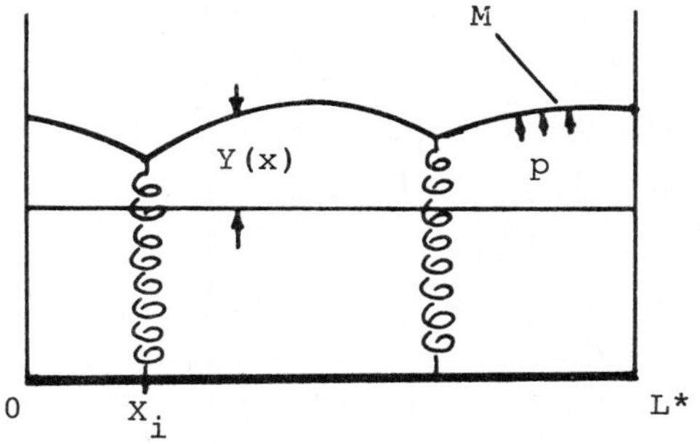

Figure 1. An elastic membrane is supported from below by a pressure p and attached at random points to linear springs.

An elastic membrane M, carrying constant tension T, is supported by a pressure difference p, and held by attachment at points X_i, i $1,2,\cdots,N$ in the interval $(0,L^*)$, to linear springs with spring constant K, the rest position of the point of attachment being on the x-axis. Outside the interval $(0,L^*)$ the membrane height Y(X) is extended as a continuous periodic function with period L^*.

Assuming that the membrane slope remains small, the function Y(X) satisfies, for $-\infty < X < \infty$,

$$(2.1)\quad Y(X)'' = -p/T + k \sum_{i=1}^{N} \delta(X-X_i)Y(X_i) \;, \qquad Y(X+L^*) = Y(X) \;.$$

Since k and p/T are inverse lengths, it is convenient to adopt k^{-1} and $pT^{-1}k^{-2}$ as units of length in the horizontal and vertical directions, respectively, so that (2.1) can be written (with $Y = ypT^{-1}k^{-2}$,

$X = xk^{-1}$, $L = L^*k^{-1}$),

$$(2.2) \qquad y''(x) = -1 + \sum_{i=1}^{N} \delta(x-x_i)y(x_i) , \qquad y(x+L) = y(x) .$$

We wish now to consider (2.2) for arbitrary N, (x_i) and L.

Since the points are to be chosen randomly, we now let $(\tilde{x}_i)$ be an unordered sequence of points in (0,L) and define $f_N(\tilde{x}_1,\cdots,\tilde{x}_N)$ to be the associated joint probability distribution and define the reduced distributions as

$$(2.3) \qquad f_m(\tilde{x}_1,\ldots,\tilde{x}_m) = \left(\frac{1}{L}\int_0^L\right)^{N-m} f_N d\tilde{x}_{N-m+1}\cdots d\tilde{x}_N ,$$

with f_1 normalized to be unity. We also assume the process to be stationary, so that all distribution functions are completely symmetric in all arguments. We then define

$$(2.4) \qquad \langle Q\rangle_{1,2\cdots m} = \left(\frac{1}{L}\int_0^L\right)^{N-m} \frac{f_N}{f_m} Q d\tilde{x}_{N-m+1}\cdots d\tilde{x}_N$$

as the conditional expectation of Q holding m points fixed. For the special case of a completely uniform distribution we have $f_m = 1$, all m. If N and L are allowed to approach infinity in such a way that $N/L = n$ has a nonzero limit, then it is known [6] that the sequence $(d_i = x_i - x_{i-1})$, where (x_i) is the ordering of $(\tilde{x}_i)$, constitutes a Poisson process, and this provides a convenient case for Monte Carlo simulation.

No doubt numerous probabilistic tools may be brought to bear to study various properties of the realizations of (2.2), but we shall confine attention to elementary estimates of the ratio

$$(2.5) \qquad \gamma = \frac{\langle y(\tilde{x}_1)\rangle_1}{\langle y\rangle}$$

in the limit $N , L \to \infty$. We note that, upon integrating (2.2) over the interval we obtain

$$(2.6) \qquad 0 = -L + \sum_{i=1}^{N} y(\tilde{x}_i) .$$

If now we compute the expectation $\langle y\rangle$ by multiplication with f_N and integration with respect to all x_i there results

(2.7) $$\langle y\rangle'' = -1 + \frac{N-1}{L} f_1(x) \quad \langle y(\tilde{x}_1)\rangle_1 = -1 + \frac{N-1}{L} \langle y(\tilde{x}_1)\rangle_1 \quad .$$

But $\langle y\rangle$ is a constant and so (2.6) and (2.7) yield, if $N >> 1$,

(2.8) $$\langle y(\tilde{x}_1)\rangle_1 = \frac{1}{N} \sum_{i=1}^{N} y(\tilde{x}_i) = \frac{1}{n} \quad .$$

Thus γ^{-1} is just the expected mean deflection of the membrane (which intuitively should be small when the springs are distributed densely) times the number of springs per unit (K^{-1}) of length.

We remark that a related one-dimensional, reduced scattering problem has been studied in considerable detail by Bazer [7]. In that problem, the term -1 on the right of (2.2) is replaced by -y, and this change allows the probability distribution for y under Poisson statistics to be studied in a scalar Markov model [8]. This approach does not appear to be possible for (2.2).

Brinkman's approximation. If we think of $y(x)$ as an expectation and replace the distribution of springs by a continuous elastic bed, then $n\gamma y$, where γ as defined by (2.5) is the average spring elongation divided by y, determines the mean force exerted by the bed. Introducing a "test spring" at the origin we consider the problem

(2.9) $$y'' = -1 + \delta(x)y(0) + n\gamma y \ , \qquad y \to \langle y\rangle \quad \text{as} \quad x \to \infty \ ,$$

with solution

(2.10) $$y = \langle y\rangle - y(0)\ (4n\gamma)^{-\frac{1}{2}} e^{-(n\gamma)^{\frac{1}{2}}|x|}$$

From (2.10) we may compute the deflection y(0) of the spring, and for this value to be self-consistent with the γ determining the continuous bed we must have

(2.11) $$\frac{y(0)}{\langle y\rangle} = \left(1 + (4n\gamma)^{-\frac{1}{2}}\right) = \gamma \quad .$$

Thus

(2.12) $$\gamma = \frac{1 + 8n - (1+16n)^{\frac{1}{2}}}{8n}$$

is the Brinkman expression for $\gamma(n)$.

This procedure can be made somewhat more formal by considering the conditional expectation of (2.2) holding one point fixed. We have, upon multiplication by f_2/f_1 and integration with respect to $\tilde{x}_2, \tilde{x}_3, \cdots$, in the limit $N, L \to \infty$, $N/L \to n$,

$$(2.13) \qquad \langle y \rangle_1 = -1 + \langle y(\tilde{x}_1) \rangle_1 \quad \delta(x - \tilde{x}_1) + n \langle y(\tilde{x}_2) \rangle_{12} \Big|_{x_2 = x} ,$$

and so (2.9) and (2.13) become equivalent with the assumption

$$(2.14) \qquad \langle y(\tilde{x}_2) \rangle_{12} \Big|_{\tilde{x}_2 = x} = \gamma \langle y(x) \rangle_1 .$$

That is, the approximation (2.12) follows from the assumption that the one-point conditioned expectation of y responds to a second spring precisely as the unconditioned expectation, a step which clearly neglects any "interaction" between the two springs.

Since $1/n$ is the average spacing between uniformly distributed points, the solution (2.10), and the notion of a continuous elastic bed, would appear to be reasonable provided that $(n\gamma)^{-\frac{1}{2}} \gg n^{-1}$ or

$$(2.15) \qquad (1+16n)^{\frac{1}{2}} \ll 1 + 4n ,$$

implying that n should be large.

<u>Upper and lower bounds on γ</u>. To compute an upper bound on γ for a given realization of points, we consider the class of functions $f(x)$ which are smooth on $(0,L)$ except for discontinuities in f' at the x_i, with periodic extension outside this interval, and with $\frac{1}{N} \sum_i f(x_i)$ equal to $1/n$. It is easily seen that the Euler-Lagrange equation for

$$(2.16) \qquad F[f] = \frac{1}{2} \int_0^L f'^2 dx + \frac{1}{2} \sum_{i=1}^{N} f^2(x_i) - \int_0^L f dx$$

is satisfied by the solution y of (2.2), and that

$$(2.17) \qquad F[f] \geq F[y] = -L/2\gamma n .$$

If we take

$$(2.18) \qquad f = 1 - \lambda(x - x_{i-1})(x - x_i) , \qquad x_{i-1} < x < x_i ,$$

as trial function, (2.17) yields an optimal bound when $\lambda = \frac{1}{2}$:

$$(2.19) \qquad \gamma \leq (1 + \beta_3/12n)^{-1} , \qquad \beta_3 = n^3 \cdot \frac{1}{N} \sum_{i=1}^{N} d_i^3 .$$

For a Poisson distribution of the d_i we have, for large N,

$$(2.20) \qquad \beta_3 = n^4 \int_0^\infty e^{-ns} s^3 ds = 3! .$$

To compute a lower bound, consider the class of functions $g(x)$ which satisfy $g' = -1$ except at the x_i, where discontinuities $[g]_i$ may occur, with again periodic extension outside $(0,L)$. It is then easily seen that

$$(2.21) \qquad G[g] = -\frac{1}{2} \int_0^L g^2 dx - \frac{1}{2} \sum_{i=1}^{N} [g]_i^2$$

satisfies

$$(2.22) \qquad G[g] \leq G(y) = -L/2\gamma n .$$

Choosing the trial function

$$(2.23) \qquad g = x - d_i/2 , \qquad x_{i-1} < x < x_i ,$$

we then obtain

$$(2.24) \qquad \gamma \geq (\beta_2 + \beta_3/12n)^{-1} , \qquad \beta_2 = \frac{n^2}{4N} \sum_{i=1}^{N} (d_{i+1} + d_i)^2 .$$

For the uniform distribution and large N, $\beta_2 = 3/2$ and therefore the two bounds yield, for the uniform distribution,

$$(2.25) \qquad (3/2 + 1/2n)^{-1} \leq \gamma \leq (1 + 1/2n)^{-1} .$$

<u>An upper bound on $\langle\gamma\rangle$</u>. Since the upper bound involves a relatively unconstrained variational problem, it is of interest to examine the possibility of a sharper result motivated by the form of Brinkman's test field (2.10). We therefore consider functions f^*, satisfying the conditions on the functions in (2.16) but now with $\langle f^* \rangle = 1/n$ replacing $\frac{1}{N} \sum_i f(x_i) = 1/n$. Let

$$(2.26) \qquad F^*[f^*] = \frac{1}{2L}\int_0^L (f^{*\prime})^2 dx + \frac{1}{2L}\sum_{i=1}^N f^*(\tilde{x}_i))^2 \quad .$$

Among such functions F^* is minimized by γy, where y solves (2.2), and we have

$$(2.27) \qquad F^*[f^*] \geq F^*[\gamma y] = \gamma/2n \quad .$$

As a trial function we now take

$$(2.28) \qquad f^* = \frac{1}{n}\left[1 + A \sum_{i=1}^N \phi(x,\tilde{x}_i)\right] ,$$

where

$$(2.29) \qquad \phi(x,\tilde{x}_i) = \cosh \mu(x-\tilde{x}_i) + \mathrm{sgn}(\tilde{x}_i - x)\tanh(\tfrac{1}{2}\mu L)\sinh \mu(x-\tilde{x}_i) + \frac{2}{\mu^2 L}\tanh(\tfrac{1}{2}\mu L) , \qquad 0 \leq x \leq L .$$

Note that this function satisfies $\phi_{xx} + \mu^2\phi = 0$ except at (x_i), and $\langle\phi\rangle = 0$. Introducing (2.29) into (2.27) and taking the <u>expectation</u> of γ we obtain:

$$\langle\gamma\rangle \leq \frac{N(N-1)A^2}{nL^3}\int_0^L\int_0^L\int_0^L f_2(\tilde{x}_1,\tilde{x}_2)\phi_x(x,\tilde{x}_1)\phi_x(x,\tilde{x}_2)d\tilde{x}_1 d\tilde{x}_2 dx$$

$$(2.30) \qquad + \frac{NA^2}{nL^2}\int_0^L\int_0^L f_1(\tilde{x}_1)\phi_x^2(x,\tilde{x}_1)d\tilde{x}, dx + \frac{1}{L}\int_0^L f_1(\tilde{x}_1)(1+A\phi(\tilde{x}_1,\tilde{x}_1))^2 dx_1$$

$$+ \frac{N(N-1)}{NL^3}\cdot\int_0^L\int_0^L f_2(\tilde{x}_1,\tilde{x}_2)(1+A\phi(\tilde{x}_1,\tilde{x}_1))A\phi(\tilde{x}_2,\tilde{x}_1)d\tilde{x}_1 d\tilde{x}_2$$

$$+ A^2\frac{N(N-1)}{nL^3}\int_0^L\int_0^L f_2(\tilde{x}_1,\tilde{x}_2)\phi^2(\tilde{x}_2,\tilde{x}_1)d\tilde{x}_1 d\tilde{x}_2$$

$$+ \frac{A^2 N(N-1)(N-2)}{nL^4}\int_0^L\int_0^L\int_0^L f_3(\tilde{x}_1,\tilde{x}_2,\tilde{x}_3)\phi(\tilde{x}_3,\tilde{x}_1)\phi(\tilde{x}_2,\tilde{x}_1)d\tilde{x}_1 d\tilde{x}_2 d\tilde{x}_3 \quad .$$

In the infinite limit $\phi(x,\tilde{x}_i)$ may be replaced by $e^{-\mu|x-\tilde{x}_i|}$ in (2.30), yielding an upper bound involving f_1, f_2, f_3, A, and μ. This expression can then be minimized with respect to μ and A to obtain an optimal bound involving f_1, f_2, and f_3. For the uniform distribution we have

$$\gamma \leq A^2\mu + (1+A)^2 + \frac{A^2 n}{\mu} \quad . \tag{2.31}$$

and the optimization yields

$$\gamma \leq \left(1 + \tfrac{1}{2}\sqrt{n}\right)^{-1} = 1 - \frac{1}{2\sqrt{n}} + \frac{1}{4n} + \ldots \quad . \tag{2.32}$$

The first two terms on the right of (2.32) are equal to the leading terms in the expansion of the Brinkman value (2.12) for large n, while the 1/n terms are not the same. Also, the bound (2.32) is worse than the previous upper bound if $n < 1$. Thus the sharper result for large n (assuming the accuracy there of the Brinkman value) is obtained at the expense of estimates at small n.

A Case of Second-order Brinkmanship. The "first-order" Brinkman approximation (2.14) closes the hierarchy of averaged equations with (2.13). We suppose now that (2.13) is retained and that the solution of (2.13), if the right-hand side is known in terms of $\langle y\rangle_1$, would be of the form

$$\langle y\rangle_1 = \langle y\rangle + \langle y(\tilde{x}_1)\rangle_1 G(x-\tilde{x}_1) \;, \qquad G(-x) = G(x) \;, \qquad G(0) = G_0 \;. \tag{2.33}$$

This ansatz will be used at the next level to obtain a second-order closure. Averaging with two points fixed yields

$$\langle y(x)\rangle_{12} = -1 + \langle y(\tilde{x}_1)\rangle_{12}\delta(x-\tilde{x}_1) + \langle y(\tilde{x}_2)\rangle_{12}\delta(x-\tilde{x}_2) + n\langle y(\tilde{x}_3)\rangle_{123}\Big|_{\tilde{x}_3 = x} \;. \tag{2.34}$$

We assume that the solution of (2.34) is well approximated in terms of G by the expression

$$\langle y(x)\rangle_{12} \simeq \langle y\rangle + \langle y(\tilde{x}_1)\rangle_{12} G(x-\tilde{x}_1) + \langle y(\tilde{x}_1)\rangle_{12} G(x-\tilde{x}_2) \;. \tag{2.35}$$

Noting (from (2.33)) that

$$1 - G_0 = 1/\gamma \;, \tag{2.36}$$

we may solve (2.35) for $\langle y(x_2)\rangle_{12}$ to obtain

$$(2.37)\qquad \langle y(\tilde{x}_2)\rangle_{12} = \frac{\gamma\langle y\rangle}{1-\gamma G(\tilde{x}_2-\tilde{x}_1)} .$$

Using the last expression in (2.13), we see that G now satisfies the nonlinear equation

$$(2.38)\qquad G''(x) = \delta(x) + \frac{Gn\gamma}{1-\gamma G} .$$

With the conditions $G(\infty) = 0, G'(0+) = 1/2$, we find upon integration of (2.38) that

$$(2.39)\qquad \frac{1}{2}(G'(0))^2 = \frac{1}{8} = -nG_0 - \frac{n}{\gamma}\,\ell n(1- G_0)$$

yielding

$$(2.40)\qquad n^{-1} = \frac{8}{\gamma}\,(1-\gamma-\ell n(2-\gamma)) ,$$

a second-order Brinkman approximation to $\gamma(n)$. It is easy to check that if the series for the logarithm is approximated by

$$(2.41)\qquad \ell n(2-\gamma) \simeq 1-\gamma-\frac{1}{2}(1-\gamma)^2 ,$$

then we recover the first-order result (2.12).

Summary of estimates in the model problem. We show in Figure 2 the various bounds and approximations to γ obtained above, all for the completely uniform distribution, and compared with Monte Carlo simulation with 200 points and 100 trials (see Appendix). The standard deviations obtained in the simulations are shown in Table 1.

n	.5	1	2	4	8	16	32	64	128
γ	.436	.563	.676	.766	.835	.885	.921	.946	.964
Standard Deviation	.047	.047	.042	.035	.029	.021	.016	.013	.010

Table 1. Monte Carlo simulation in the model Problem (2.2) for a Poisson distribution of d_i, using 200 points x_i and 100 trials.

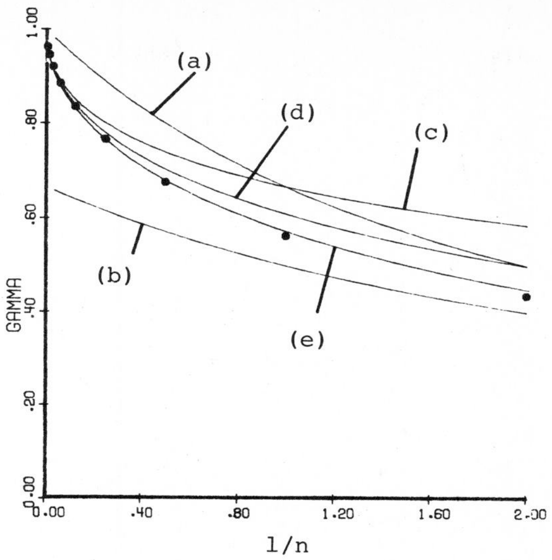

Figure 2. Estimates for γ with a uniform distribution of springs for the model problem (2.2). (a), (b) are the upper and lower bounds given by (2.25). (c) is the upper bound (2.32). (d) is the first-order Brinkman estimate (2.12). (e) is the second-order result (2.40).

The following observations can be made: (a) The bounds (2.25) do not jointly imply a good approximation to γ except for <u>small</u> n, where they converge to the asymptotic expression $\gamma \sim 2n$. (b) First-order Brinkman theory actually exceeds the upper bound (2.25) for $n < .5$. (c) The second-order value is remarkably close to the simulations down to $n = .5$.

We remark that no lower bound to go with (2.32), which might yield a proof of the asymptotic validity of the Brinkman approximation for large n, has been given here, although a suitable choice of trial function g in (2.21) may establish such a result. Also it is not immediately obvious that a useful third-order theory can be found in a hierarchy of Brinkman approximations, since the one-point function G satisfies a nonlinear equation, indicating that the effective medium is not, at that level of approximation, a linear elastic bed. Of course such "higher-order" extensions are not uniquely defined and there is perhaps a procedure different from the one used here which would clarify, at least in principle, how an "nth order" Brinkman approximation should be defined.

3. Attachments at Finite Intervals.

The condition that $y(x)$ be determined as above through a set of point attachments at x_i is unrealistic when modeling finite porosity, so we consider now the Brinkman approximation when the one-dimensional model is modified as shown in Figure 3. Each spring is now to be attached at the center of a horizontal rigid element of length d. It is assumed that the pressure difference across the membrane also acts on these rigid elements (since this mimics the Archimedean force experienced by a finite body in a permeable bed when there is a pressure gradient through the bed).

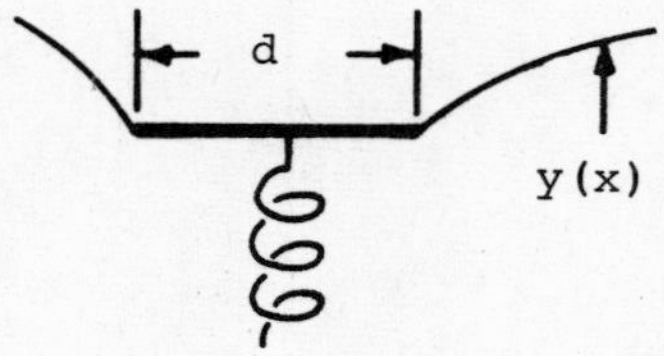

Figure 3. Spring attachment to a rigid element.

Monte Carlo simulation with $d > 0$ may in fact be used to obtain data for $d = 0$ by the following transformation: If $\gamma_0(n_0)$ corresponds to $d = 0$, then

$$(3.1)\qquad \langle y_1\rangle_1 = \frac{1}{n_0} + d = \frac{1+n_0 d}{n_0} = \frac{1}{n},$$
$$\langle y\rangle = d + \frac{dn}{n_0} + \frac{n}{n_0^2}$$

as can be seen from the definitions of the averages applied to the collection of membrane and rigid elements. Thus, using the values in Table 1 for n_0 and γ_0,

$$(3.2)\qquad \gamma = \frac{(1+n_0 d)^2\gamma_0}{1+\gamma_0 n_0 d+\gamma_0 n_0 d(1+\gamma_0 n_0 d)}$$

will yield Monte Carlo values for a Poisson distribution of gaps with density $n_0 = n/(1-nd)$.

The simplest second-order Brinkman approximation which might be applicable here assumes the two-point distribution function f_2 to have values 0 or 1. If the test spring attaches to the elastic bed after a free gap of width a, then $G(x)$ satisfies

$$(3.3)\qquad G'' = -n, \qquad 0 < x < a$$
$$= \frac{n\,G}{(1-G)}, \qquad x > a,$$

$$G(0) = \frac{\gamma-1}{\gamma} , \qquad G'(0) = \frac{1}{2}(1-nd) ,$$

G and G' continuous at $x = a$.

Thus

$$(3.4) \qquad G(a) = \frac{\gamma-1}{\gamma} + (1-nd)\frac{a}{2} - \frac{a^2 n}{2} ,$$

$$\frac{1}{8}[1-n(d+2a)]^2 + nG(a) + \frac{n}{\gamma}\ln(1-\gamma G(a)) = 0 .$$

In this case it is seen that γ becomes unity, corresponding to "blocking," at a finite value of n.

In Table 2 we give some values obtained from (3.4), compared with the computations. Note that the choice $a = d/2$ tends to overestimate

	d = .1			d = .5		
n	a = 0	.05	mc	a = 0	.25	mc
.5	.458	.470	.460	.490	.562	.547
1.0	.588	.612	.609	.638	.770	.743
2.0	.704	.746	.750	.765	.970	.892
4.0	.799	.865	.865			
8.0	.872	.965	.943			

Table 2. Comparison of Monte Carlo (mc) values with those obtained from (3.4).

γ. This suggests that the correct 2-point distribution function, corresponding to independent allocation of segments with no overlap, will improve the agreement. In one dimension this distribution function can be described explicitly [9]. With this modification we have, in place of (3.3),

$$(3.5) \qquad G'' = g(x-d/2)L(G) , \qquad L(G) = \gamma G/(1-\gamma G) ,$$

where

$$(3.6) \qquad g_d(x) = \sum_{k=0}^{\infty} \frac{n_0^{k+1}}{k!} H(x-kd)e^{-n_0(x-kd)}(x-kd)^k ,$$

and $H(x)$ is the Heaviside function. We also are led to pose the following question: Is there a choice of $L(G)$ in (3.5) which is in some sense optimal in yielding values of γ which obey or nearly obey (3.2)? We have not as yet solved (3.5), (3.6) for $\gamma(n)$, but such a computation would be useful in conjunction with any systematic study of a hierarchy of Brinkman approximations.

4. The Role of Dimension

The one-dimensional case is formally analogous to a class of higher-dimensional problems involving disordered media, although there are implications of the dimension, and of the specific operators involved, when invoking Brinkman's approximation. As Batchelor [10] has noted, a large number of elliptic problems involving Dirichlet or Neumann data at random boundaries are formally equivalent, and we shall restrict attention here to the permeability problem for slow viscous flow.

Indeed, (2.2) is equivalent to a rather artificial problem of this kind. If $y(x)$ is interpreted as a velocity of a viscous fluid in a direction orthogonal to x the x-axis, and if the planes $x = x_i$ are imagined to be porous sheets, then the velocity $y(x_i)$ will be nonzero at the sheets owing to the stress, and it is reasonable to postulate that the jump in y' will be proportional to the slip velocity. Then (2.2) is obtained and γ is equivalent to an inverse permeability. In addition the constant -1 on the right of (2.2) may be identified with the pressure gradient needed to sustain the flow.

Two dimensions. Since Monte Carlo simulation should be feasible for some two-dimensional problems, the following example is of interest: consider parallel viscous flow orthogonal to the x-y plane, in the domain exterior to randomly placed, nonoverlapping circles of diameter d. If $w(x,y)$ denotes the velocity, and g is the pressure gradient in the z direction, then when the fluid adheres to the boundaries we have

$$\mu\nabla^2 w = g\,, \qquad w = 0 \quad \text{on} \quad C_i\,, \quad i = 1,2,\cdots\,. \tag{4.1}$$

where μ is the fluid viscosity and C_i denotes the ith boundary. Then

$$F_i = \int_{C_i} \frac{\partial w}{\partial n}\, ds \tag{4.2}$$

is the force per unit length down the ith cylinder, and the problem is to determine

$$\gamma = \frac{\langle F_i \rangle}{\mu \langle w \rangle}\,. \tag{4.3}$$

This problem has been treated by Howells using Brinkman's approximation [11]. The equivalent diffusion problem has recently been analyzed in considerable detail by Talbot and Willis [12], who give upper and lower bounds on γ analogous to those of Section 2.

Three dimensions. Viscous flow past a random bed of non-overlapping spheres has been discussed by Tam [13], Childress [14], Howells [11], Saffman [15], and Hinch [16]. For small values of the volume density c of the array, and for spheres of fixed diameter d, the expected force F, divided by the force F_0 of a single sphere in the same mean flow, is found to have an expansion

$$\gamma = 1 + \frac{3}{\sqrt{2}} c^{\frac{1}{2}} + \frac{135}{64} c \, \ell n c + c h(f_2) + 0(c^{3/2} \ell n c) , \tag{4.4}$$

where we indicate the first dependence of the two-point distribution function for the centers of the spheres. For an f_2 which is zero for overlap, one otherwise, there results $h \simeq 16.5$. Unfortunately, the expansion (4.4) is less useful than the Brinkman value for c of practical interest. Childress [14] proposed the second-order extension of Brinkman's approximation used in Sections 2 and 3 above, in the context of Stokes's equations and point particles, and noted that it then contains the terms in (4.4) not specifically due to the finite size of spheres. However no application was made to finite c.

It should also be noted that an equivalent "second-order" scattering theory in three dimensions was proposed for the coherent field by Twersky [17].

Our reason for listing these problems here is to point out differences with the one-dimensional case. In the first place the relevant fundamental solutions for the case of point "scatterers" are singular in the higher dimensions. Thus, in the three-dimensional point particle problem, the terms $y(x_i)$ in (2.2) are replaced by regular parts evaluated at the centers. (The two-dimensional problem is special owing to the logarithmic behavior of the fundamental solution.)

In the second place, there is the essential difference in the expected range of validity of Brinkman's approximation. In one dimension, the expected interparticle distance is $0(n^{-1})$ while the "shielding" distance in Brinkman's theory is $0((\gamma n)^{-\frac{1}{2}})$. Thus, as noted previously, the continuous bed is reasonable when $n >> 1$, where in fact $\gamma \simeq 1$. In dimension N the first of the above estimates becomes $n^{-1/N}$. Thus for $N = 2$ there is no preferred range of n, except as provided by the value of γ. In three dimensions, by the same reasoning, Brinkman's result should apply to small n, or more exactly to small volume density c ; indeed, the first two terms on the right of (4.4) are obtained by expanding the solution of (1.1) for small c [13,14].

Nevertheless, the point seems to be that the only reason for using these approximations is to obtain useful expressions outside their expected range of validity, and to test these ideas we need more data over a range of values of volume fraction, the important asymptotics being in the limit of dense packing.

The author would like to thank George Papanicolaou, Jerry Percus, and Charles Peskin for helpful discussions.

REFERENCES

1. Brinkman, H.C., Appl. Sci. Res. A1, 27, 1947.
2. Foldy, F.L., Phys. Rev. 67, 107, 1945.
3. Twersky, V., J. Res. Nat. Bur. Standards 64D, 715, 1960.
4. Lax, M., in "Stochastic Differential Equations," SIAM-AMS Proceedings, Vol. VI, 1973.
5. Happel, J. and Brenner, H., Low Reynolds Number Hydrodynamics, Prentice-Hall, 1965.
6. Karlin, S., A First Course in Stochastic Processes, Academic Press, 1966.
7. Bazer, J., J. Soc. Indus. Appl. Math. 12, 539, 1964.
8. Frisch, H.L. and Lloyd, S.P., in Mathematical Physics in One Dimension, Academic Press, 1966.
9. Percus, J.K., in Studies in Statistical Mechanics, edited by Lebowitz and Montroll, North Holland, Amsterdam, 1981.
10. Batchelor, G.K., An. Rev. Fluid Mech. 6, 227, 1974.
11. Howells, I.D., J. Fluid Mech. 64, 449, 1974.
12. Talbot, D.R.S. and Willis, J.R., Proc. Roy. Soc. Lond. A, 370, 351, 1980.
13. Tam, C.K.W., J. Fluid Mech. 64, 449, 1974.
14. Childress, S., J. Chem. Phys. 56, 2527, 1972.
15. Saffman, P.G., Stud. Appl. Math. 52, 115, 1973.
16. Hinch, E.J., J. Fluid Mech. 83, 695, 1977.

APPENDIX

In the interval $x_{i-1} \le x \le x_i$ we have

$$\text{(A.1)} \qquad y = -\tfrac{1}{2}(x-x_{i-1})^2 + a_i(x-x_{i-1}) + y_{i-1} ,$$

so that

$$\text{(A.2)} \qquad y_1 = y_{i-1} + a_i d_i - \tfrac{1}{2}d_i^2 , \qquad d_i = x_i - x_{i-1} ,$$

$$\text{(A.3)} \qquad a_i = a_{i-1} + y_i - d_{i-1} .$$

We combine the last two equations to obtain

$$\text{(A.4)} \qquad d_i^{-1}y_i + 1 - d_i^{-1} - d_{i-1}^{-1}\ y_{i-1} + d_{i-1}^{-1}y_{i-2} + \tfrac{1}{2}(d_i+d_{i-1}) = 0 ,$$

$$d_0 = d_N , \quad y_0 = y_N .$$

The Monte Carlo simulation involves, for each trial, the inversion of a tridiagonal matrix plus two corner entries to take care of the periodicity condition. Two tridiagonal inversions are thus needed. The value of γ is then given by

$$\text{(A.5)} \qquad \gamma = \left(\sum_{i=1}^{N} g_i\right)\left(\sum_{i=1}^{N} di^3/12 + (y_i+y_{i-1})d_i/2\right)^{-1} .$$

TOPOLOGY, GEOMETRY, AND PHYSICAL PROPERTIES OF POROUS ROCKS

Morrel H. Cohen
James Franck Institute and Dept. of Physics
The University of Chicago
Chicago, IL. 60637

AND

Charlotte Lin
Schlumberger-Doll Research Center
P.O.Box 307
Ridgefield, CT. 06877

ABSTRACT

By use of the concepts of skeletization and deformation retract, we characterize the geometry and topology of a porous sedimentary rock in particularly simple ways. We briefly introduce the underlying topological concepts and present a skeletization procedure which leads to clear definitions of such concepts as grain, contact, pore chamber, channel, and throat. We apply this procedure in developing novel formulations of the problems of nuclear magnetic relaxation within the pore space, of steady flow through the pore space, and of the frame moduli. We show how the ambiguity between pore chambers and channels can be exploited for the NMR and flow problems. In particular, we find that a flow problem can be reduced to a resistance network problem, but the network is not a deformation retract of the pore space. The frame moduli problem can be mapped into the long wavelength the limit of a random "lattice" - vibration problem.

I. INTRODUCTION

In the present state of rock physics one can formulate an immense number of significant, unanswered questions about the physical properties of detrital sedimentary rocks. For example, what does the magnetic relaxation of protons in the pore water of such rocks tell about the pore space geometry? Why does the self-similar model account so well for the dc electrical conductivity associated with electrolyte in the pore space?[2,3,4] How could one obtain comparable accuracy in calculating the frame modulus in the Biot theory?[5] How could one calculate the permeability for flow through the pore space[6] to such accuracy? What would the relation of the permeability and the conductivity be in such calculations?[6] The solutions to these illustrative problems all require explicit quantitative knowledge of the geometry and topology of the porous rocks.

Consider the question of the accuracy of the selfsimilar model. It involves two sets of approximations, one set approximating the geometry and topology and one set introduced in calculating electrical conductivity for a given geometry. Without deeper study of both, one cannot understand why the model should work. The calculation procedure used is the effective medium approximation, the convergence of which is relatively easy to study. What remains is the accuracy of the description of the geometry and the topology within the selfsimilar model. Addressing that question entails learning more about the geometry and topology.

A very powerful means of doing so is through reconstructive modelling from serial sections using quantitative characterization of the topology and geometry. We describe procedures for this elsewhere.[7] Here we briefly summarize the required topology and a simple procedure for modelling the grain and pore spaces which leads to sharp definitions of grains, contacts, pores, channels, and throats. We next illustrate how these ideas can be useful in solving various problems which arise in the study of the physical properties of sedimentary rocks.

II. TOPOLOGY

Consider the simplest type of sedimentary rock, a clean sandstone. An ideally clean sandstone is a two-phase material composed of connected rock grains and the associated intergranular pore space. Each grain can be represented by a point (node vertex) within it. Each contact between two grains can be represented by an arc (branch, edge) connecting their two representative points. The graph comprised of the vertices and edges partitions any surface on which it lies into faces. The faces partition the three-dimensional space occupied by the sandstone into cells which fill the space. This construct of vertices, edges, faces, and cells is a complex which represents the grain space, G. G may also be represented by a standard topological solid, X, such as a solid torus (doughnut), an n-torus (doughnut produced with an N-holed doughnut cutter), an n-torus with internal holes which are themselves m-tori, etc. X^c is all of three-dimensional space not in X, including the m-torus holes in X. The analogous representation for the pore-space P has vertices in the pore chambers and edges passing through the channels between them. Note that X can represent either G or P and X^c either P or G, respectively and that $G^c = P$.

The graph introduced above is a deformation retract D_G of G. A subset D of a set X is a deformation retract of X if, given the function f(x) mapping X onto D, there is a function $g(x,t)$ which is continuous in x and t such that

$$g(x,t) = x, \quad \text{if } t = o$$
$$= f(x), \quad \text{if } t = 1$$
$$= \text{some element of } X \text{ for all } t \text{ in the interval } [1,0]$$

That is, g shrinks X continuously in space and time onto D, which is a skeleton of X. In the next section we discuss specific methods of generating deformation retracts D_G of P. In reconstructive modelling of G and P, it is far easier to specify D_G and D_P than G and P themselves.

We use the Betti numbers to characterize the grain and pore spaces topologically. The definition of the Betti numbers requires a nontrivial amount of topology which is summarized by Barret and Yust.[8] "The nth Betti number of a complex . . . (is) . . . the maximum number of homology-independent N-cycles".[8] The Betti numbers β_i of X are the same as those of any deformation retract D of X. It is easier to determine the β_i on D than on X.

It is enough for our purposes to know what β_o, - - -, β_3 tell us about X. The $zero^{th}$ Betti number β_o of X gives the number of separate components comprising X. The grain space G of a well-consolidated sandstone has only one component, $\beta_o(G) = 1$, unless there are N loose grains or grain clusters inside when $\beta_o(G) = N + 1$. Similarly, $\beta_o(P) > 1$ can indicate the presence of porosity which does not contribute to permeability and which would be measured by logging techniques such as hydrogen detection but missed by a porosimeter measurement. An instance would be an isolated ring formed by dissolution of a grain or grains, followed by encapsulating cementation or quartz overgrowth. The Alexander duality theorem states that the first Betti number or cyclomatic number $\beta_1(X)$ of X equals that of X^C, $\beta_1(X) = \beta_1(X^C)$. Determining $\beta_1(G)$ determines $\beta_1(P)$ as well. β_1 equals E - V + 1, where E is the number of edges and V the number of vertices in our graph D_G. It measures the number of "circuits" or closed loops in the graph, and by the Alexander duality theorem it suffices to determine either $\beta_1(D_G)$ or $\beta_1(D_P)$ if the representations are true deformation retracts. When D_G is large, β_1 can be approximated by E - V so that

$$C = \beta_1/\nabla = E/\nabla - 1$$

Defining Z as the average coordination number of the graph, we have

$$Z = 2E/\nabla = 2(1 + C)$$

For the second Betti number $\beta_2(X)$, we have the relation $\beta_2(X) = \beta_o(X^C) - 1$ so that

$\beta_2(G) = \beta_0(P) - 1$. Since $\beta_0(P)$ is the number of separate pore subsystems, $\beta_2(G)$ gives the number of internal surfaces. The third Betti number $\beta_3(X)$ is zero for the three-dimensional spaces we expect to encounter.

The Euler-Poincare formula equates the alternating sum of the first four Betti numbers to the Euler characteristic X, which is the alternating sum of the numbers of vertices, edges, faces, and cells in our complex. The latter numbers and the Betti numbers are topological invariants; any distortion of the net without the cutting of edges and vertices leaves them unchanged. The Euler characteristic is therefore a topological invariant; it is the same for complexes with the same "connectivity". Thus the connectivity of G, say, can be characterized by careful deformation to a graph D_G, by determination of the Euler characteristic of D_G which classifies it, and by subsequent mapping to an appropriate solid X with Betti numbers which give more detailed information about the components, coordination number, and internal surfaces of G.

The notion of the genus of a surface is sometimes substituted for the first Betti number of the complex containing the graph lying on the surface. While it is a technical misstatement to say that the genus and β_1 are identical, their two values are equal for graphs lying on surfaces in complexes. Careful specification of the graph at the exterior boundary of the sample will insure that the graph lies on a surface without boundary as is required for surfaces in complexes, and it is a theorem that topologically the only such surfaces are spheres, tori, and n-hole tori. Rhines[9] has used genus normalized to unit volume. One can also use genus per vertex which equals C defined above and relates directly to the coordination number of the graph.

III. SKELETIZATION AND MODELLING

The preceding section shows the convenience and utility of characterizing the topology and geometry of the grain and pore spaces through their deformation retracts or skeleton graphs D_G and D_P. In the present section we describe a particular way to construct these deformation retracts, that is, a particular skeletization,[10-15] which leads, as mentioned in the introduction, to sharp definitions of grains, contacts, pore chambers, channels, and throats.

Center on each point of the grain-pore interface an arbitrarily small sphere of radius ε. Erode the grain-space G by deleting its intersection with the collection of spheres, producing a new grain-pore interface and opening the pore space. Iterating the process ultimately causes G to disappear. The intermediate effects are to smooth small-scale roughness in the interface and, more importantly, progressively to disconnect G at progressively larger local volumetric minima.[16,17] A region of grain space which becomes isolated by these disconnections is either a grain or a cluster of grains in contact which subsequently disconnect. When an isolated grain has become convex, its centroid can be taken as the grain center.

Return to those instants during the erosion process at which manifest separation appears between two regions of G previously in contact. Run the erosion process backwards and reestablish contact between the regions. Each point of first contact represents the "center" of an intergranular contact. The minimum-area cross section which contains the contact center can be taken as the contact surface between the two regions at any stage of erosion prior to disconnection. The grain space is now partitioned into individual grains bounded by the contact surfaces and the pore-grain interface.

Return now to the original G. Embedded within G is a set of grain centers and contact surfaces. To represent G by a network take each grain center as a vertex. Connect them by edges which are arcs through each contact surface. This skeleton of G is a deformation retract D_G because it can be obtained from G by reversing an opening process which utilize spheres of arbitrarily small radius centered on each point of D_G and iterates on the resulting surface until each point of the original grain-pore interface is reached.

Continue the reversed erosion of G beyond the original pore-grain interface, thus eroding P. Having identified individual grains, we can recognize the point-by-point instances of one grain's pore-grain interface growing into contact with that of another grain. The collection of interface contact surfaces formed by these events upon P's disappearance is a deformation retract of P, D_P. The interface contact surfaces are continuations of the grain contact surfaces. The total collection of surfaces thus partitions 3-space into cells, each enclosing an individual grain. Deforming each cell surface into a topologically equivalent polygonal surface produces polyhedral cells which are topologically equivalent to Voronoi polyhedra.

The notions of pore "chambers", "channels", and "throats" can now be developed. The partitioning of space into cells by the contact surfaces defines a network of vertices and edges which is also a deformation retract of P, D_P. Edges and vertices then are the line segments and points, respectively, formed by the intersections of three or more finite contact surfaces. The point intersections occur in the pore chambers; that is, the pore chambers are the spaces contiguous to the vertices of D_P. The line intersections occur in the channels connecting pore chambers; that is, the channels are the spaces contiguous to the edges of D_P not already included in the pore chambers. Throats are the locally minimal area cross-section loci in the channels, and their locations along the edges delineating the channels are particular to the local geometry. The geometric features specified by these definitions are clearly and simply displayed by sphere packs. The construction of D_G and D_P and the definitions of grains, contacts, pore chambers, channels, and throats can be used as a basis for modelling the pore space in different ways appropriate to the study of different physical properties. Examples are given in the next section.

IV. MODELLING THE GRAIN AND PORE SPACES.

In this section the utility of the ideas in the two previous sections is illustrated by accurately modelling the pore and grain spaces as appropriate to the calculation of several different physical properties of sedimentary rocks.

A. Nuclear magnetic relaxation.

Cohen and Mendelson[1] have shown that the nuclear magnetization M in the direction of a magnetic field H which is associated with a density N_b of protons in the pore water obeys the following equations

$$\frac{\partial M}{\partial t} = D\,\nabla^2 M - \frac{M - M_\infty}{T_{1b}} \tag{1}$$

$$\frac{\partial M(s)}{\partial t} = -\frac{D}{\ell}\,\hat{n}.\vec{\nabla}\,M(s) - \frac{M(s)-M_\infty}{T_{1s}} \tag{2}$$

$$M = M_o \text{ at } t=o \tag{3}$$

In Eq. (1), $M_\infty = \chi H$, where χ is the bulk susceptibility of protons in water D, is the proton diffusion coefficient in water, and T_{1b} is the bulk relaxation time (of order sec). In Eq.(2), M(s) is the limiting value of M as a point on the rock-water interface is approached from within the water. ℓ is the ratio N_s/N_b of the surface density N_s of protons on the hydrated rock surface to the bulk proton density N_b (of order an Angstrom unit). $\hat{n}$ is the unit normal vector pointing outward from the fluid. T_{1s} is the relaxation time of protons at the rock-water interface (of order 10^{-4} sec).

We now take advantage of the ambiguity that exists in the partitioning of the pore space into pore chambers and channels described in § III. No boundaries have been placed between the chambers and the channels. In the present instance it is convenient to eliminate the channels completely, dividing all of the pore space into separate pore chambers, or simply pores, by passing a surface through the throat of each channel. Each pore i, i = 1 ---N, has volume Vi and surface S_i. S_i consists of a portion of the rock-water interface, S_i , and the boundary surfaces through the throats separating i from its neighbors j, S_{ij}

$$S_i = S_i^r + \sum_j S_{ij} \tag{4}$$

Cohen and Mendelson show that under the conditions which prevail in the rocks of interest, the magnetization is nearly uniform within each individual pore. This permits simplification of Eq.(1) and (2) to

$$\frac{dM_i}{dt} = - \frac{M_i - M_\infty}{T_i} + \frac{1}{V_i} \sum_j K_{ij} (M_j - M_i) \tag{5}$$

In (5) M_i is the mean magnetization in pore i, and T_i is the relaxation time for M_i when coupling to the neighboring pores can be ignored,

$$\frac{1}{T_i} = \frac{1}{T_{1b}} + \frac{\ell S_i^r}{V_i} \left(\frac{1}{T_{1s}} - \frac{1}{T_{1b}} \right) \tag{6}$$

The sum on j is over all contiguous pores j, and the coupling coefficients are given by

$$K_{ij} = K_{ji} = D\, S_{ij} / L_{ij}, \tag{7}$$

where the length L_{ij} is defined implicitly by the solution to (1) - (3). In the cases of interest, the throat dimensions are small compared to the channel dimensions and L_{ij} is approximately the throat diameter.

Equation (5) defines a problem in which there is a quantity M_i associated with each vertex i of the deformation retract D_P. The time dependence of the M_i is governed by a matrix with diagonal elements T_i^{-1} associated with each vertex and off-diagonal elements $V_i^{-1} K_{ij}$ associated with the directed edge ij of D_P. The values of the T_i^{-1} are random, as are the values of the $V_i^{-1} K_{ij}$. The connectivity of the matrix is also random; that is, the network D_P is topologically disordered.

Analysis of (5) is greatly facilitated by the fact that it has the form of equations extensively studied in the microscopic theory of disordered materials. Drawing upon that experience Cohen and Mendelson have shown the following:

1. In the limit that the $V_i^{-1} K_{ij}$ are negligible compared to the variation of the T_i^{-1}, the relaxation function of the magnetization is the Laplace transform of the probability distribution of the T_i^{-1}. According to (6) the probability distribution of the pore surface to volume ratios S_i^r/V_i can then be determined up to a scale factor.

2. In the limit that the $V_i^{-1} K_{ij}$ dominate the variation of the T_i^{-1}, the relaxation of the magnetization is exponential, with a T_i which contains the surface to volume ratio of the whole pore space.

3. In the intermediate case, the nuclear magnetic relaxation depends on the distributions of the S_i^r/V_i and the S_{ij}/L_{ij} and on the topology of D_p. Disentangling these dependences can be carried out with the aid of the microscopic theory of disordered materials.

B. Steady flow problems

The steady flow of fluid through a porous rock in response to a pressure head[18] or the steady flow of electricity through a conducting fluid within the pore space[4] is more conveniently addressed with the opposite resolution of the pore-chamber/pore channel ambiguity. Consider the case of the steady flow of electric current. The equations governing the flow are

$$\nabla^2 \psi = 0 \tag{8}$$

$$\hat{N}. \vec{\nabla} \psi (s) = 0 \tag{9}$$

$$\vec{E} = -\vec{\nabla} \psi \tag{10}$$

$$\vec{j} = \sigma_w \vec{E} \tag{11}$$

with appropriate boundary conditions on the electrostatic potential ψ at the outer boundaries of the rock. Here $\vec{E}$ is the internal electric field, $\vec{j}$ the current density, and σ_w the conductivity of the electrolyte. Suppose that Eqs.(8) and (9) are solved subject to the external boundary conditions so that ψ, $\vec{E}$, and $\vec{j}$ are known everywhere within the pore fluid.

Consider now a vertex i of D_p at which r edges intersect. There will be r channels entering the corresponding pore chamber. Suppose that current flows into the pore chamber through p of the r channels and out through q of them , p + q = r. The current flow is normal to the surfaces of constant electrostatic potential, which decreases in the direction of flow. Above some value ψ_{ni} of ψ, there will be p distinct pieces of each surface of constant ψ within the p channels with inward flowing current, and below some other value ψ_{1i} of ψ, there will be q distinct pieces within the q channels with outward flowing current. Let $\alpha_i = (m_i, n_i)$ specify the number of pieces of the surfaces of constant ψ near the vertex i for ψ in $[\psi_{ui}, \psi_{\ell i}]$, m_i being the number across which there is inward flow and n_i that for outward flow. α_i changes from (p,o) at ψ_{ui} to (o,q) at $\psi_{\ell i}$ in a number of steps larger than one in general. At each such change, $\psi = \psi_{\sigma i}, \sigma = 1, --- \nu_i$ $(\psi_{1i} = \psi_{ui}, \psi_{\nu_i i} = \psi_{\ell i})$ defines a critical surface separating a region of pore space with $\alpha_i = \alpha^{\sigma}_i$ from one with $\alpha_i = \alpha_i^{\sigma+1}$. We can use these critical surfaces to partition the pore chamber i. This can be repeated for all pore chambers until the entire pore space is so partitioned.

The pore space is thus divided entirely into channels by successive critical surfaces along the lines of current flow. Each critical surface defines a vertex, and each channel defines an edge connecting it to another vertex. The network F so constructed, however, is not topologically equivalent to D_P and therefore does not represent P. For each vertex i in D_P there are ν_i vertices in F. F reduces identically to D_P only in the improbable event that all ν_i are unity. However, if most of the voltage drops $\psi_{ui} - \psi_{\ell i}$ are small compared to the differences in the means $(\psi_{\sigma i} - \psi_{\sigma j})$ between adjacent vertices in D_P, one can ignore the fine structure associated with each vertex, lump the channels between ψ_{ui} and $\psi_{\ell i}$ with those leading into ψ_{ui} or out of $\psi_{\ell i}$ for convenience and arrive at a network equivalent to D_P.

The current flow within each channel as defined above is governed by the boundary conditions $\hat{n} \cdot \vec{j} = o$ at the rock fluid interfaces and $\hat{n} \times \vec{j} = o$ at the critical surface of constant voltage. We can thus associate with each vertex i of F (rigorously or of D_P approximately) a voltage ψ_i and with each edge ij a current

$$I_{ij} = \int j\, ds, \tag{12}$$

where the integration is over the area of any surface of constant voltage within the channel ij. Ohm's law holds for the channel and may be written in the form

$$I_{ij} = G_{ij} \, (\psi_i - \psi_j), \tag{13}$$

where the conductance G_{ij} is given by

$$G_{ij} = \sigma_w \int \vec{E} \cdot d\vec{s} / \int_j^i \vec{E} \cdot d\vec{r} \tag{14}$$

Consideration of the geometry of the channel can yield rough estimates of G_{ij}, for example $G \simeq \sigma_w A/L$ where A is an appropriate mean channel crossection and L an appropriate mean channel length. Integrating $\vec{\nabla} \cdot \vec{j} = o$ within a volume within the pore chamber which encloses the critical surface yields Kirchoff's law

$$\sum_j I_j = \sum_j G_{ij} \, (\psi_i - \psi_j) = o \tag{15}$$

We thus have a random resistance network problem with a voltage ψ_i at each vertex i and a resistance G_{ij}^{-1} along each ij of the network D_P.

Having made contact with random resistance networks, we have at our disposal a substantial body of work which we can carry over to the problem of the dc conductivity of porous rocks. One can proceed with the analysis of fluid flow problems on the same basis, establishing a set of Kirchoff equations for the above network F containing the permeabilities of the individual channels K_{ij} and then going on to derive the permeability of the rock.

C. The frame moduli.

Turning now to the grain space, a challenging problem is the calculation of the frame moduli, the elastic moduli of the rock in the absence of a pore fluid. When the grain contact areas are small relative to the grain size, stress concentration is such that elastic distortion occurs primarily in the vicinity of the contacts. Neglecting overlap of the displacement fields associated with different contacts on the same grain, and assuming sound wavelengths much larger than the grain size, one can derive straightforwardly a set of equations of motion for each grain:

$$M_i \frac{d^2\vec{u}_i}{dt^2} = \sum_j \overleftrightarrow{K}_{ij} \cdot (\vec{u}_j - \vec{u}_i) \qquad (16)$$

Here m_i is the mass of grain i and $\vec{u}_i$ its displacement, and K_{ij} is the force-constant tensor coupling grain i to grain j across the contact area. Evaluation of K_{ij} requires a detailed study of the elastic properties of the contact and requires specification of its geometric details. However, estimations are readily made. We have thus mapped the problem of wave propagation in and the elastic properties of a complex continuum into the same problem for the random network D_G with masses M_i associated with the vertices and force constants $\overleftrightarrow{K}_{ij}$ associated with the edges. The problem has been studied in detail in the microscopic context of phonons in disordered materials. The normal mode spectrum, sound velocities, and the frame modulii can thus be studied with the aid of an extensive literature. However, the dispersion of the sound velocity is incorrectly given in this theory, as we show in a more detailed study.

V. CONCLUSIONS AND DISCUSSION

We have provided detailed, particular answers to such questions as what is a grain, a contact, a pore chamber, a channel, or a throat. The pore chambers and channels cannot be defined unambiguously, but the ambiguity can be exploited. In problems relating to bulk properties it may be desirable to treat the pore space as composed entirely of pore chambers, whereas in those problems relating to transport through the pore space it may be desirable to treat the pore space as comprised entirely of channels.

Whichever the more convenient partitioning of pore space, one is led back ultimately to representation of the pore space by a graph which is a deformation retract of the pore space for the case of pore chambers only, but need not be for the case of channels only. We have briefly presented the underlying topology and explicit procedures for obtaining deformation retracts both of the grain and the pore spaces. We have shown that these deformation retracts are useful not only for describing the topology of the spaces, but also for detailed quantitative studies of the macroscopic properties of detrital sedimentary rocks. Questions such as those we started with in the introduction have either been answered in part, viz. the one about nuclear magnetic relaxation, or the problems have received a new and illuminating formulation, viz. those about flow and frame moduli.

VI. ACKNOWLEDGEMENT

This work was supported in part by the National Science Foundation, Grant NSF DMR 802069. A part of the work was done while one of us (MHC) visited the Schlumberger-Doll Research Center.

REFERENCES

1. M. H. Cohen and K. Mendelson, submitted to J. Appl. Phys.
2. P. N. Sen, C. Scala, and M.H. Cohen, Geophysics 46, 781 (1981).
3. K. Mendelson and M. H. Cohen, Geophysics (in press).
4. P. N. Sen, this volume.
5. cf. D. Johnson or J. Berryman, this volume.
6. W. F. Brace, J. Geophys. Res. 82, 3343 (1977).
7. C. Lin and M. H. Cohen, submitted to J. Appl. Phys.
8. L. K. Barrett and C. S. Yust, Metallography 3, 1 (1970).
9. F. N. Rhines, Stereology, Springer-Verlag, New York (1967).
10. G. Matheron, Elements pour une theorie des Milieux Poreux, Masson, Paris (1967).
11. E. R. Davis and A.P.N. Plummer, BPRA Conference on Pattern Recognition, Oxford (1980).
12. J. Freer, J. Mol. Biol. 82, 279 (1974).
13. W. K. Pratt, Digital Image Processing, Wiley, New York (1978).
14. A. Rosenfeld, J. ACM 17, 146 (1970).
15. A. C. Shaw, J. ACM 17, 453 (1970).
16. J. Serra, Leitz Sci. Tech. Inform. Supplement 1, 4, p. 125, Wetzler, Apr. (1974).
17. M. Rink and J. R. Schopper, Pageoph 114 (1976).
18. J. Koplick, Creeping Flow in Two-Dimensional Networks, preprint.

WAVE PROPAGATION IN BUBBLY LIQUIDS

Donald A. Drew
U.S. Army Research Office and
Rensselaer Polytechnic Institute

and

Lap-Yan Cheng
Rensselaer Polytechnic Institute
Troy, New York 12181

Wave propagation in a bubbly liquid is studied from the point of view of determining the validity of two-fluid two-phase flow models and the values of various parameters appearing in them. At low frequencies, the speed of sound depends on the virtual mass coefficient. At higher frequencies, the results indicate the need for a virtual heat capacity.

I. INTRODUCTION

Bubbly flows occur in many practical technological situations, including chemical reactors and nuclear reactor cooling systems. An ability to predict system behavior during flow transients is extremely desirable. The flow geometry, including the motion and vibrations of the interfaces, is complex, but for most practical purposes only certain averaged quantities (such as average heat transfer to or from the bubbles) are of interest. For this reason, averaging methods have been applied to the equations of motion for the constituent continua with the hope that some insight can be gained into the macroscopic, or averaged, motions. The success of this approach has been modest, with few results coming from the application of averaging to a realistic flow situation. Consideration of the analogous process of deriving the equations of fluid dynamics from the Boltzmann equation suggests that success at this step is not essential for obtaining useful fluid-like models for the macroscopic motions.

The approach which seems most fruitful in obtaining a working two-phase flow model is one which combines a postulational, or constitutive, approach for the averaged motions with an empirical, or rheological, approach to verify the postulated forms and to measure the coefficients which arise in the constitutive equations. This procedure is analogous to postulating that a (single-phase) fluid is Newtonian (linearly viscous), then verifying that the shear stress is a linear function of the rotation rate in a couette viscometer, and finally extracting the viscosity from the slope of the stress-rotation curve. This procedure replaces the derivation of closure relations from microscopic considerations with deduction of appropriate relations between the macroscopic variables. Intuition and empiricism are strong components of this approach; nevertheless, there is a logical framework within which certain simplifications occur (Drew & Lahey 1979).

In this approach, the features which result from the random (or more appropriately, heterogeneous) nature of the medium are contained in the constitutive equations. Indeed, the difference between bubbly flow and some other flow, for example, droplet flow, are contained in the equations expressing how the two media interact, and in an empirical statement about range of applicability.

Because two-phase flow involves the description of the motions of two separate materials, it is more complicated than single phase flow. In addition to consideration of the flux terms (stresses and heat flux) which are also present in single phase flows, it is necessary to consider terms expressing the interactions between the two phases. Moreover, a cursory examination of virtual mass effects for bubbly flows (Drew, Cheng and Lahey 1979) shows that non-steady effects in the interaction terms can be bigger than the initial terms in the bubble momentum equation.

Propagation of infinitesimal disturbances (sound) represents one way to verify and quantify some of the nonsteady aspects of the interactions between phases. There are many studies of sound propagation, almost invariably done so that the basic flow has no motion (see, e.g. Van Wijngaarden). In subsequent sections, we shall discuss a model which will alow non-zero basic flows. We will also discuss various aspects of the constitutive assumptions; we shall linearize the equations about the state of uniform bubbly flow and calculate the propagation modes for continuous waves (CW).

II. GENERAL FRAMEWORK

The time-averaged equations of motion for two-phase flow are given by Ishii (1975) as

$$\frac{\partial \alpha_k \rho_k}{\partial t} + \nabla \cdot \alpha_k \rho_k \underline{v}_k = \Gamma_k \qquad (1)$$

$$\frac{\partial \alpha_k \rho_k \underline{v}_k}{\partial t} + \nabla \cdot \alpha_k \rho_k \underline{v}_k \underline{v}_k = -\nabla \alpha_k p_k + \nabla \cdot \alpha_k \underline{\underline{\tau}}_k + \Gamma_k \underline{v}_{ki} + \underline{M}_k^d + \underline{M}_k^p \qquad (2)$$

$$\frac{\partial \alpha_k \rho_k (e_k + \frac{1}{2} v_k^2)}{\partial t} + \nabla \cdot \alpha_k \rho_k (e_k + \frac{1}{2} v_k^2) \underline{v}_k = -\nabla \cdot \alpha_k p_k \underline{v}_k + \nabla \cdot \alpha_k \underline{\underline{\tau}}_k \cdot \underline{v}_k$$

$$-\nabla \cdot \alpha_k \underline{q}_k + \Gamma_k (e_{ki} + \frac{1}{2} v_{ki}^2) + E_k \qquad (3)$$

where α_k is the volume fraction of phase k, ρ_k is the density, $\underline{v}_k$ is the velocity, Γ_k is the rate of creation of phase k, p_k is the pressure, $\underline{\underline{\tau}}_k$ is the extra (viscous plus Reynolds) stress, $\underline{v}_{ki}$ is the interfacial average velocity, $\underline{M}_k^d + \underline{M}_k^p$ interphase force, e_k is the energy density, q_k is the heat flux, e_{ki} is the interfacial average energy density, and E_k is the interfacial heat transfer rate.

The jump conditions are

$$\Gamma_1 + \Gamma_2 = 0 , \qquad (4)$$

$$\Gamma_1 (\underline{v}_{1i} - \underline{v}_{2i}) + \underline{M}_1^d + \underline{M}_2^d + \underline{M}_1^p + \underline{M}_2^p = \underline{M}_m , \qquad (5)$$

$$\Gamma_1 (e_{1i} - e_{2i} + \frac{1}{2} v_{1i}^2 - v_{2i}^2) + E_1 + E_2 = E_m , \qquad (6)$$

where $\underline{M}_m$ and E_m are the mixture sources of momentum and energy at the interface, respectively, and represent the contributions due to surface tension.

For concreteness, we shall denote the bubble phase by a subscript g, and the liquid phase by a subscript ℓ. Furthermore, we write $\alpha_g = \alpha$, and $\alpha_\ell = 1-\alpha$.

III. CONSTITUTIVE EQUATIONS

These equations (1-6) must be supplemented with constitutive equations expressing Γ_k, $\underline{v}_{ki}$, $\underline{M}_k^d$, $\underline{\underline{\tau}}_k$, q_k, E_k, and e_{ki} in terms of α, $\underline{v}_k$, p_k and T_k (Drew and Lahey 1979). First, we ignore the macroscopic viscous terms:

$$\underline{\underline{\tau}}_k = 0 \quad . \qquad (7)$$

We further assume that

$$\Gamma_k = 0 \quad . \qquad (8)$$

We assume that the pressure contribution to the interfacial momentum transfer can be represented as

$$\underline{M}_k^p = p_{ki} \nabla \alpha_k \quad , \tag{9}$$

where p_{ki} is the interfacial averaged pressure. We assume an interfacial force model of the form

$$\underline{M}_g^d = (\underline{F}_d + \underline{F}_{vm}) \quad ,$$

where

$$\underline{F}_d = \frac{3}{8} \rho_\ell \frac{C_D}{r} |\underline{v}_g - \underline{v}_\ell| (\underline{v}_g - \underline{v}_\ell) \quad , \tag{10}$$

and

$$\underline{F}_{vm} = C_{vm} \rho_\ell \Big[\Big(\frac{\partial \underline{v}_\ell}{\partial t} + \underline{v}_g \cdot \nabla \underline{v}_\ell \Big) - \Big(\frac{\partial \underline{v}_g}{\partial t} + \underline{v}_\ell \cdot \nabla \underline{v}_g \Big)$$

$$+ (1-\lambda)(\underline{v}_\ell - \underline{v}_g) \cdot \nabla (\underline{v}_\ell - \underline{v}_g) \Big] \tag{11}$$

Here C_D is the drag coefficient and r is the average radius associated with the bubbles. The drag coefficient C_D is assumed to be a function of α and the local Reynolds number Re. Also, C_{vm} is the virtual volume coefficient, and λ is a parameter. Both C_{vm} and λ depend on α.

We shall further take

$$\underline{M}_m = -\frac{2\sigma}{r} \nabla \alpha \quad , \tag{12}$$

where σ is the coefficient of surface tension. By assuming that

$$p_{gi} - p_{\ell i} = \frac{2\sigma}{r} \quad , \tag{13}$$

equation (5) becomes

$$\underline{M}_g^d = -\underline{M}_\ell^d \quad . \tag{14}$$

The task of relating the interfacial pressures p_{ki} to the average pressures p_k remains. For the bubbles, we shall assume

$$p_g = p_{gi} \quad . \tag{15}$$

It is necessary to account for scattering of pressure waves into the liquid. For this purpose, we shall assume that the scattering effects occur at a scale smaller than the interbubble distance. That is, we assume that pressure fluctuations in the neighborhood of the bubble are spherically symmetric. This leads to the Rayleigh equation

$$p_\ell - p_{\ell i} = -\left\{\rho_\ell\left[r\,\frac{D_g^2 r}{Dt^2} + \frac{3}{2}\left(\frac{D_g r}{Dt}\right)^2\right] + \frac{\mu}{r}\,\frac{D_g r}{Dt}\right\} \quad , \tag{16}$$

where $D_g/Dt = \partial/\partial t + \underline{v}_g \cdot \nabla$, and μ is the effective viscosity of the surrounding liquid. It is essential to include the Rayleigh equation in order to capture the effects of bubble resonance.

Let us now discuss the thermodynamic assumptions. For the bubbles, we shall assume equations of state of the form

$$p_g = \rho_g R T_g \quad , \tag{17}$$

where T_g is the gas temperature, and R is the gas constant for air. Also, we write

$$h_g = e_g + p_g/\rho_g \quad . \tag{18}$$

We shall assume

$$h_g = c_g T_g \tag{19}$$

where h_g is the gas enthalpy, and c_g is the specific heat of the gas at constant pressure. For the liquid, we write

$$\rho_\ell = \rho_\ell(p_\ell, h_\ell) \tag{20}$$

with

$$h_\ell = e_\ell - p_\ell/\rho_\ell \quad . \tag{21}$$

For the sound propagation studies, we need only the linearized version of equation (20). This becomes

$$\delta\rho_\ell = \frac{\partial\rho_\ell}{\partial p_\ell}\,\delta p_\ell + \frac{\partial\rho_\ell}{\partial h_\ell}\,\delta h_\ell$$

$$= \beta_k \delta p_\ell + \phi_k \delta h_\ell \tag{22}$$

where β_k is related to the compressibility of the liquid, and ϕ_k is related to the thermal expansion. Both quantities are standard properties, and are well known for water.

We shall ignore the macroscopic heat flux:

$$\underline{q}_k = 0 \quad . \tag{23}$$

The interfacial heat transfer terms E_k needed a more complicated model than first anticipated. We assume that E_m is negligible, and write

$$E_\ell = -E_g = \alpha H(T_{\ell i} - T_\ell) \tag{24}$$

where H depends on the thermodynamic properties of the two phases and the interface geometry. For low frequencies, equation (24) worked well with a standard model for the coefficient H. For high frequencies (near bubble resonance), however, a model was derived from microscopic considerations after linearization of the fields involved. Since the analysis relates directly to the kinds of constitutive assumptions needed for an adequate description of two-phase media, it will be summarized here.

The harmonically oscillating linearized (radial) continuity, momentum and energy equation in the gas bubble were solved subject to the temperature and heat flux boundary conditions,

$$\hat{T}_\ell(r_o,t) = \hat{T}_g(r_o,t)$$

$$-\kappa_\ell \frac{\partial \hat{T}_\ell(r_o,t)}{\partial r} = -\kappa_g \frac{\partial \hat{T}_g(r_o,t)}{\partial r}$$

and the kinematic condition,

$$\hat{u}(r_o,t) = \frac{dr}{dt} ,$$

where r_o is the undisturbed bubble radius, and κ_g and κ_ℓ are the thermal conductivities of the gas and liquid, respectively. Note that $\hat{T}_k$ and $\hat{u}$ represent the exact temperatures and the interface velocity, respectively. The analysis extends the work of Plesset and Hsieh (1960) and of Prosperetti (1977). The resulting heat transfer coefficient H was complex valued, and frequency dependent.

IV. WAVE PROPAGATION

In order to study the propagation of infinitesimal disturbances, we assume that the variables depend only on z and t, and that all velocities have only non-zero components in the z-direction. Thus,

$$\underline{v}_k = u_k(z,t)\underline{e}_z \quad . \tag{25}$$

The previously derived conservation laws, which describe a two-phase mixture as a two-fluid medium with different pressures in each phase (gas/liquid) can be written as a system of nonlinear first order partial differential equations. After linearization, these conservation laws take the form

$$\underline{\underline{A}}(\underline{\phi}) \frac{\partial \delta \underline{\phi}}{\partial t} + \underline{\underline{B}}(\underline{\phi}) \frac{\partial \delta \underline{\phi}}{\partial z} = \underline{\underline{C}}(\underline{\phi}) \delta \underline{\phi} \tag{26}$$

where $\underline{\phi}$ is the vector of all the field variables: α, p_ℓ, u_g, u_ℓ, h_ℓ, $\dot{r}$, r. The perturbations in gas pressure and gas enthalpy, δp_g and δh_g, are related to δr by expressions derived in an analysis of the heat transfer between a pulsating bubble and the surrounding liquid.

If we assume a traveling wave perturbation

$$\delta \underline{\phi} = \underline{\xi}\, e^{i(kz-\omega t)} \quad , \tag{27}$$

then Equation (26) becomes

$$\left\{ \underline{\underline{A}}(\underline{\phi})[-i\omega] + \underline{\underline{B}}(\underline{\phi})[ik] - \underline{\underline{C}}(\underline{\phi}) \right\} \delta \underline{\phi} = 0 \ , \tag{28}$$

In order for the $\delta\underline{\phi}$'s to be finite, we must have,

$$\det \left\{ (\omega \underline{\underline{A}} - i \underline{\underline{C}}) - k \underline{\underline{B}} \right\} = 0 \quad . \tag{29}$$

Equation (24) is the dispersion relation from which, for the given ω's, we can solve for the k's. This gives the dispersion relationship, that is the wave number's (k) dependence on the frequency (ω),

$$k = k(\omega) \tag{30}$$

One physical problem of interest is the boundary value problem modeling the situation where a transducer is located at the bottom of the flow and sound is propagated upward through the bubbly liquid. If the transducer is driven sinusoidally, the resulting pressure disturbances should reach an equilibrium situation where the only frequency observed is the driving frequency. This continuous wave (CW) flow situation is expressed as the boundary value problem

$$\delta p_\ell(0,t) = a \cos \omega t \ , \tag{31}$$

$$\delta p_\ell(L,t) = 0 \quad . \tag{32}$$

All other perturbations at z=0 are 0, except $\delta\alpha$. Equation (31) expresses the action of the transducer, while equation (32) states that the top (z=L) is open to the atmosphere. The significance of the remaining boundary conditions is that the gas and liquid are being injected at specified velocity, but that the "flow split" (the ratio of gas to liquid mass flow rates) is not specified.

Since the channel is long, we approximate the solutions with

$L \to \infty$. This removes the possibility of resonance with the channel, but allows us to ignore modes which have $\text{Im } k < 0$.

The solutions for $\delta\underline{\phi}$ have the form,

$$\delta\phi = \sum_n C_n \underline{\xi}_n e^{ik_n z} e^{-i\omega t} + \begin{bmatrix}\text{Complex}\\ \text{Conjugate}\end{bmatrix}, \qquad (33)$$

where k_n is the (complex) wave number of the nth mode and $\underline{\xi}_n$ the corresponding eigenvector for Equation (28). Because of the condition discussed above $C_n = 0$ for all k_n's having negative imaginary parts. The relative importance of each harmonic (corresponding to each eigenvalue k) in the solution of Equation (33) is based upon the relative magnitude of C_n; the greater the value, the more the importance. Using Equations (31-33) we can solve for the C_n's. The mode which dominates will be the one which is actually measured in standing-wave experiments.

V. RESULTS AND CONCLUSION

The high frequency data which we used for comparison are those due to Silberman (1957). As can be seen in Figures 1 and 2, the dispersion model appears to predict the (wave) phase velocity and attenuation quite well for frequencies both above and below bubble resonance. In Figure 2, virtual mass is observed to have a noticeable effect on the phase velocity at frequencies around the resonance. The resonant spike seen in the attenuation plot is due to insufficient damping. Although there isn't enough data to be conclusive, it can be noted in Figure 1 that the predicted attenuation at the resonance point appears to be closer to the data when virtual mass is absent. However, it is possible that there are other dissipation mechanisms at resonance which we have not modelled (e.g., finite amplitude effects, relaxation effects, etc.). Thus, it is premature to draw strong conclusions from these plots. The low frequency data are taken from Hall's (1971) thesis. As can be noted in Figure 3, the effect of virtual mass is more significant in Hall's data than in Silberman's data, apparently because the void fractions in Hall's experiments are about 200 times higher than in Silberman's data. Note in Figure 3 that if virtual mass is not included the agreement is very poor. As can be seen in Figure 3-5, $C_{VM} = 0.5$ appears to agree with the data.

Unfortunately, there is too much scatter in the data to be able to give a good functional form for $C_{VM}(\alpha)$. Moreover, other experimentation showed no sensitivity at all to different values of λ.

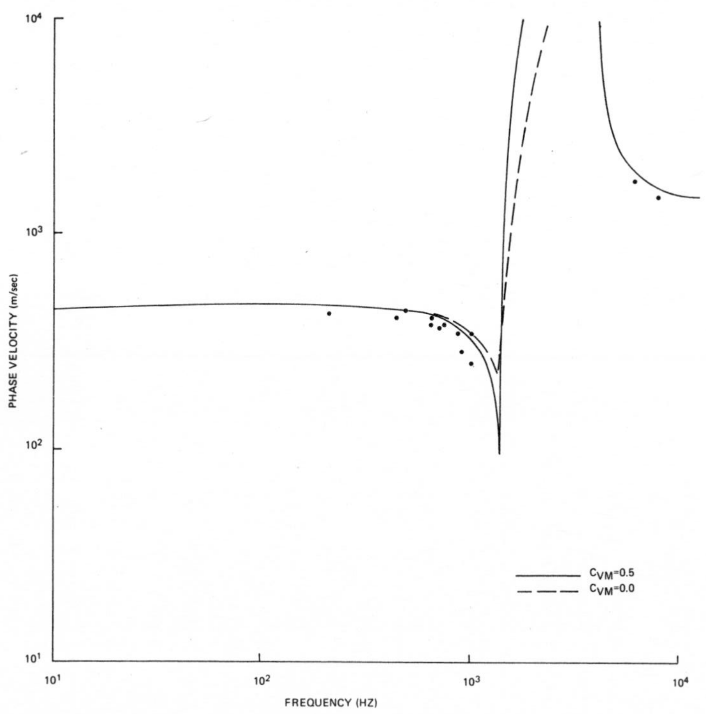

Fig. 1: Phase velocity vs. frequency, $r_o = 2.5 \times 10^{-3}$ m., $\alpha_o = 5.84 \times 10^{-4}$

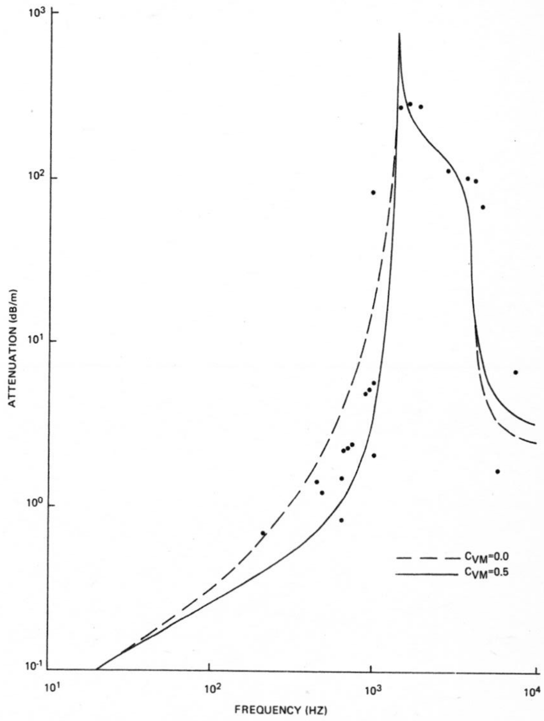

Fig. 2: Attenuation vs. frequency, same conditions as Fig. 1.

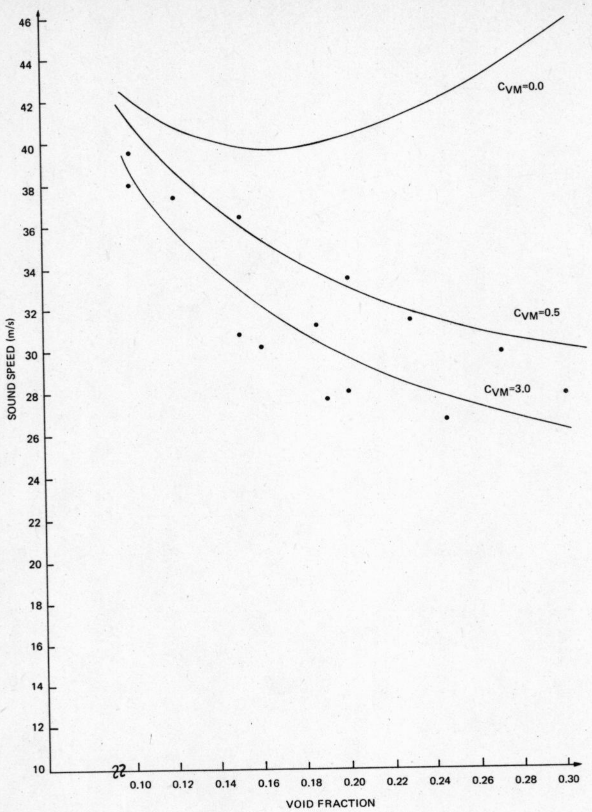

Fig. 3: Sound speed vs. void fraction. Frequency 30 Hz.

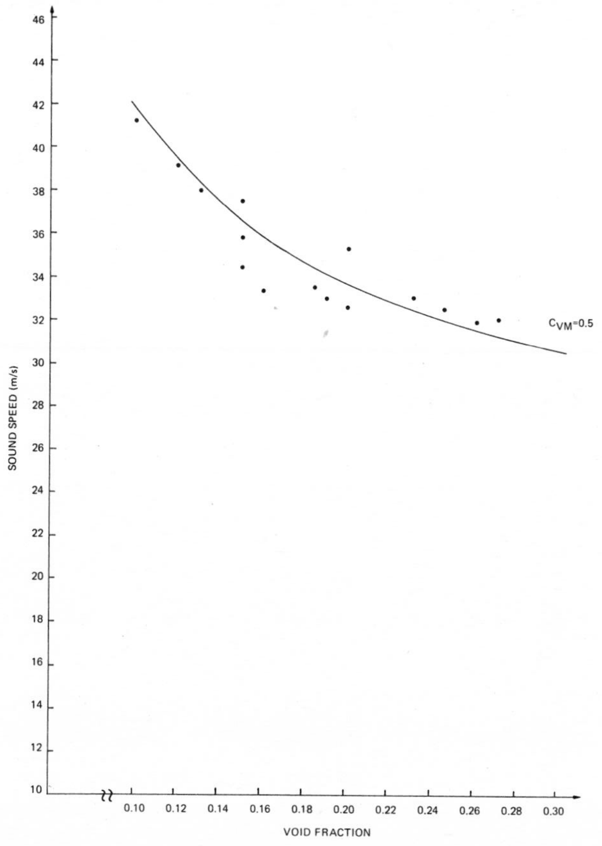

Fig. 4: Sound speed vs. void fraction.Frequency 50 Hz.

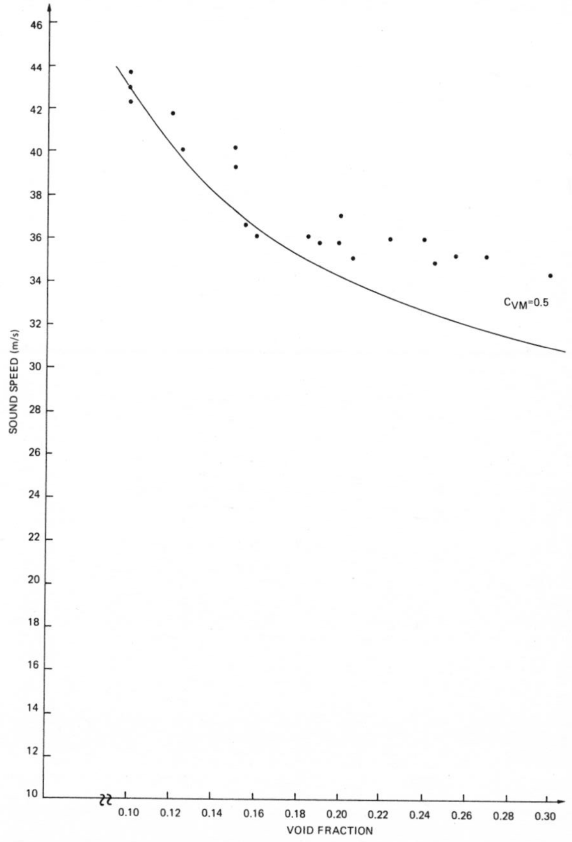

Fig. 5: Sound speed vs. void fraction. Frequency 100 Hz.

The implications of the lack of success with a steady-state model for the interfacial heat transfer parameter H have not been fully evaluated. One tentative conclusion is that a "virtual heat capacity" is needed in the interfacial heat transfer model.

REFERENCES

Drew, D.A., L.Y. Cheng, and R.T. Lahey, Jr. (1979). The analysis of virtual mass effects in two-phase flow, Int. J. Multiphase Flow 5, 233-242.

Drew, D.A. and R.T. Lahey, Jr. (1979). Application of general constitutive principles to the derivation of multidimensional two-phase flow equations, Int. J. Multiphase Flow 5, 243-264.

Hall, P. (1971). The propagation of pressure waves and critical flow in two-phase mixtures. Ph.D. Thesis, Heriot-Watt University, Edinburgh, G.B.

Ishii, M. (1975). Thermo-Fluid Dynamic Theory of Multiphase Flow, Eyrolles, France.

Plesset, M.S. and D.Y. Hsieh (1960). Theory of gas bubble dynamics in oscillating pressure fields, Phys. of Fluids 3, 6, 882-892.

Prosperetti, A. (1977). Thermal effects and damping mechanisms in the forced radial oscillations of gas bubbles in liquids, J. Acoustic Soc. Am. 61, 1, 17-27.

Silberman, E. (1957). Sound velocity and attenuation in bubbly mixtures measured in standing wave tubes, J. Acoustic Soc. Am. 29, 925-933.

Wijngaarden, L. van (1972). One dimensional flow of liquids containing small gas bubbles, Ann. Rev. Fluid Mech., vol. 4.

ELASTODYNAMICS OF POROUS MEDIA

David Linton Johnson
Schlumberger-Doll Research
P. O. Box 307
Ridgefield, CT 06877

ABSTRACT

The viscous and elastic properties of a two component medium are derived from a general 2 component Lagrangian formulation. For the case when one of the components is a fluid, there are 2 compressional waves, one of which (the slow wave) is diffusive at low frequencies and propagatory at high. Comparison is made against experiments on various porous fluid saturated systems. It is seen that the diffusive mode in polymer gels, 4th sound in HeII, the diffusion of a fluid pressure pulse through a porous medium, and the recent observation of a slow compressional propagatory wave in water saturated fused glass beads are special cases of this additional mode prediced by the theory.

It is generally recognized that an important aspect of acoustic attenuation and dispersion in porous fluid saturated media is due to the relative motion that can occur between the solid and the fluid parts. Therefore, it is important to develop a theory in which the displacements of the two components are followed separately and on an equal footing, at least in some macroscopic, average sense. A theory of this type was developed by Biot[1-4] and we will explore its consequences (below) but first let us consider the most general continuous Lagrangian formulation of wave mechanics in a two component system where it is envisioned that there exist volume elements dV small compared to the relevant wave length but large compared to the individual grains of the two components; the Biot theory will seem to be a special case of this formulation. Let $\vec{U}^a(\vec{r},t)$ be the average displacement of component a within dV at position $\vec{r}$ and time t. Define the symmetric and anti-symmetric strain tensors for each component:

$$\epsilon^a_{KL} = \frac{1}{2}\left[\frac{\partial U^a_K}{\partial x_L} + \frac{\partial U^a_L}{\partial x_K}\right] \qquad a = 1,2$$

$$\omega^a_{KL} = \frac{1}{2}\left[\frac{\partial U^a_K}{\partial x_L} - \frac{\partial U^a_L}{\partial x_K}\right]$$

If one wishes a wave-like equation of motion i.e. with no derivatives higher than 2nd order, then the Lagrangian contains terms up to first order in derivatives only. We insist that it costs no energy for a uniform rotation or translation of both components. The most general Lagrangian density for a homogeneous, isotropic system involving two components is, therefore[5]

$$\mathcal{L} = \frac{1}{2}\{\rho_{11}|\dot{\vec{U}}_1|^2+2\rho_{12}\dot{\vec{U}}_1\cdot\dot{\vec{U}}_2+\rho_{22}|\dot{\vec{U}}_2|^2$$

$$-[\gamma_1\epsilon^1_{KK}\epsilon^1_{LL}+\gamma_2\epsilon^1_{KK}\epsilon^2_{LL}+\gamma_3\epsilon^2_{KK}\epsilon^2_{LL}$$

$$+\gamma_4\epsilon^1_{KL}\epsilon^1_{KL}+\gamma_5\epsilon^1_{KL}\epsilon^2_{KL}+\gamma_6\epsilon^2_{KL}\epsilon^2_{KL}$$

$$+\gamma_7(\omega^1_{KL}-\omega^2_{KL})(\omega^1_{KL}-\omega^2_{KL})+\gamma_8(\vec{U}_1-\vec{U}_2)^2]\}$$

$$= \frac{1}{2}\{\rho_{11}|\dot{\vec{U}}_1|^2+2\rho_{12}\,\dot{\vec{U}}_1\cdot\dot{\vec{U}}_2+\rho_{22}\dot{\vec{U}}_2\cdot\dot{\vec{U}}_2$$

$$-[\alpha_1(\nabla\cdot\vec{U}_1)^2+2\alpha_2(\nabla\cdot\vec{U}_1)(\nabla\cdot\vec{U}_2)+\alpha_3(\nabla\cdot\vec{U}_2)^2$$

$$+\alpha_4|\nabla\times\vec{U}_1|^2+2\alpha_5(\nabla\times\vec{U}_1)\cdot(\nabla\times\vec{U}_2)+\alpha_6|\nabla\times\vec{U}_2|^2$$

$$+\nabla\cdot\vec{A}_T + \alpha_8(\vec{U}^1-\vec{U}^2)^2]\} \ .$$

where the set $\{\rho_{ij}\}$, $\{\gamma_i\}$, $\{\alpha_i\}$ are parameters of the system. (Each α_i is linearly related to the γ_i's.) In general, one does not know how to calculate these parameters. The relevance of the 2nd form is that the term $\nabla\cdot\vec{A}^T$ drops out identically of the equations of motion, which are:

$$\rho_{11}\frac{\partial^2\vec{U}^1}{\partial t^2}+\rho_{12}\frac{\partial^2\vec{U}^2}{\partial t^2}=\alpha_1\nabla(\nabla\cdot\vec{U}^1)+\alpha_2\nabla(\nabla\cdot\vec{U}^2)$$

$$-\alpha_4\nabla\times(\nabla\times\vec{U}^1)-\alpha_5\nabla\times(\nabla\times\vec{U}^2)$$

$$-\alpha_8(\vec{U}^1-\vec{U}^2)$$

$$\rho_{22}\frac{\partial^2\vec{U}^2}{\partial t^2}+\rho_{12}\frac{\partial^2\vec{U}^1}{\partial t^2}=\alpha_3\nabla(\nabla\cdot\vec{U}^2)+\alpha_2\nabla(\nabla\cdot\vec{U}^1)$$

$$-\alpha_6\nabla\times(\nabla\times\vec{U}^2)-\alpha_5\nabla\times(\nabla\times\vec{U}^1)$$

$$-\alpha_8(\vec{U}^2-\vec{U}^1)$$

The plane wave normal modes can almost be determined by inspection. There are two longitudinal modes and two transverse modes (of each polarization) and the dispersion relations are

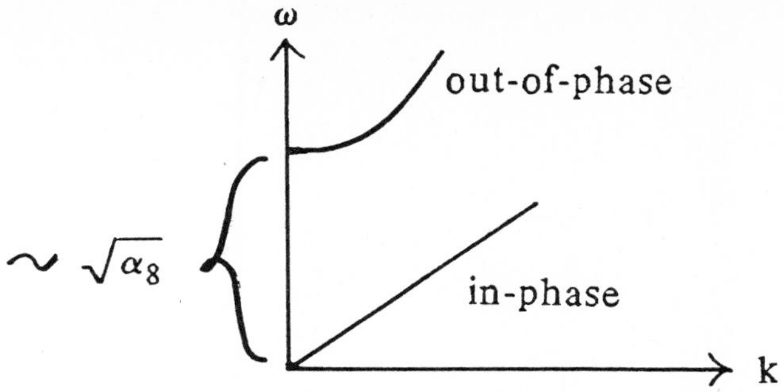

Fig. 1
Acoustic dispersion relation for either transverse or longitudinal modes of a 2 component medium.

The upper branch is the formal equivalent of an optic phonon branch whereas the lower is the acoustic branch. The frequency "cutoff" of the upper branch is a direct manifestation of the restoring force due to uniform displacement of the two components with respect to each other.

If one specializes to the case where one of the components ($\vec{U}^2$ say) is a fluid, then α_8 is identically zero because if the 2 components are uniformly displaced relative to each other, they don't try to "snap back". Moreover, we shall postulate that the fluid does not experience any shear restoring force, nor does it contribute to one on the solid ($\alpha_5 = \alpha_6 = 0$). In that case *both* longitudinal modes are non-dispersive and there is only one shear mode (the other shear mode is an identically zero frequency mode for all wave vectors).

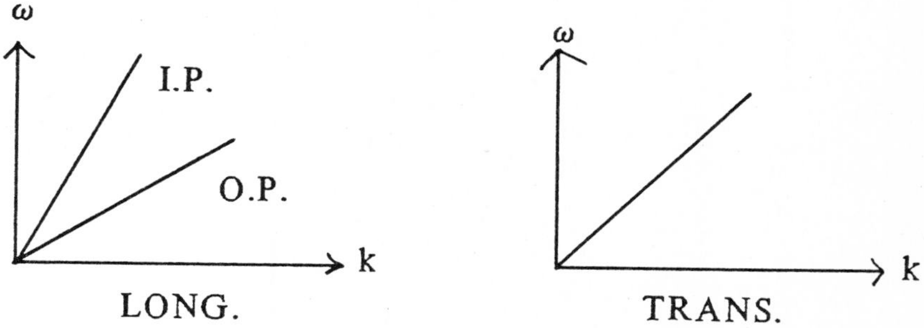

Fig. 2
Longitudinal (left) and transverse (right) dispersion relation of a fluid-solid medium. Fluid is assumed to be non-viscous.

Note that the "slow" compressional, (longitudinal) mode always corresponds to $\vec{U}^1$ and $\vec{U}^2$ moving out of phase (O.P.) with respect to each other. Up to this point an important piece of physics has been neglected, namely the viscous attenuation created by the relative fluid-solid motion. This can be introduced into the Lagrangian formation via a dissipation function but we shall, for the purposes of the rest of this paper, use the formulation due to Biot. We shall also use his notation.

In a series of papers,[1-4] Biot proposed a simple phenomenological theory of acousic propagation in porous, fluid filled, macroscopically homogeneous and isotropic media. It is assumed that there exist volumes large compared to pore/grain sizes but small compared to a wavelength and that each volume element is describable by the average displacement of the fluid $\vec{U}(\vec{r},t)$ and of the solid parts $\vec{u}(\vec{r},t)$. The equations of motion, including viscous damping, are

$$\rho_{11}\frac{\partial^2\vec{u}}{\partial t^2}+\rho_{12}\frac{\partial^2\vec{U}}{\partial t^2} = P\nabla(\nabla\cdot\vec{u})+Q\nabla(\nabla\cdot\vec{U})-N\nabla x\nabla x\vec{u}$$

$$+bF(\omega)\left(\frac{\partial\vec{U}}{\partial t}-\frac{\partial\vec{u}}{\partial t}\right) \tag{1a}$$

$$\rho_{22}\frac{\partial^2\vec{U}}{\partial t^2}+\rho_{12}\frac{\partial^2\vec{u}}{\partial t^2} = R\nabla(\nabla\cdot\vec{U})+Q\nabla(\nabla\cdot\vec{u})$$

$$-bF(\omega)\left(\frac{\partial\vec{U}}{\partial t}-\frac{\partial\vec{u}}{\partial t}\right) \tag{1b}$$

Here we have used the notation of ref. 1 as it is simpler than (although equivalent to) that of the later articles. P, Q, R are generalized elastic coefficients which can be related[1,4] to the bulk modulus of fluid K_f, the bulk modulus of solid K_s, the bulk modulus K_b of the skeletal frame ("jacketed and drained") and to N which is the shear modulus of both the skeletal frame and of the composite:

$$P = \frac{(1-\phi)\left[1-\phi-\frac{K_b}{K_s}\right]K_s+\phi\frac{K_s}{K_f}K_b}{1-\phi-\frac{K_b}{K_s}+\phi\frac{K_s}{K_f}}+\frac{4}{3}N\ , \tag{2a}$$

$$Q = \frac{\left(1-\phi-\frac{K_b}{K_s}\right)\phi K_s}{1-\phi-\frac{K_b}{K_s}+\phi\frac{K_s}{K_f}}\ , \tag{2b}$$

$$R = \frac{\phi^2K_s}{1-\phi-\frac{K_b}{K_s}+\phi\frac{K_s}{K_f}}\ . \tag{2c}$$

ϕ is the porosity (fluid volume fraction). In the so-called "jacketed and drained" gedanken test[4] the sample is stressed but the fluid is allowed to escape as needed in order to remain at ambient pressure. Therefore, K_b and N are the elastic constants of the bare skeletal frame and, in the absence of an electrochemical interfacial effect between fluid and solid, they are independent of what fluid is in the pores, including vacuum; we will make this assumption throughout the paper.[6] Equations (2) are equivalent to those given in e.g., Stoll,[7] Geertsma and Smit.[8] The density terms ρ_{ij}, are related to the density of solid, ρ_s, and fluid, ρ_f, by

$$\rho_{11} + \rho_{12} = (1-\phi)\rho_s \quad , \tag{3a}$$

$$\rho_{22} + \rho_{12} = \phi \, \rho_f \quad . \tag{3b}$$

The term ρ_{12} describes the inertial (as opposed to viscous) drag that the fluid exerts on the solid as the latter is accelerated relative to the former and vice-versa. The equation of motion of the solid part, for example, equation (1a), may be rewritten using (3a):

$$(1-\phi)\rho_s \frac{\partial^2 \vec{u}}{\partial t^2} = -\rho_{12}\left(\frac{\partial^2 \vec{U}}{\partial t^2} - \frac{\partial^2 \vec{u}}{\partial t^2}\right)$$

$$+ bF(\omega)\left[\frac{\partial \vec{U}}{\partial t} - \frac{\partial \vec{u}}{\partial t}\right] + \text{(spatial derivative terms)} \tag{1a'}$$

That is, even for a non-viscous pore fluid ($bF(\omega)\equiv 0$), there is a reactive force per unit volume on the solid [whose mass is $(1-\phi)\rho_s$] whenever one component is accelerated relative to the other. The proportionality constant, ρ_{12}, represents the induced mass tensor[9] per unit volume, assumed to be diagonal in the coordinate indices for a homogeneous isotropic system; it is always negative and is always proportional to the fluid density:

$$\rho_{12} = -(\alpha-1)\,\phi\,\rho_f \tag{3c}$$

where $\alpha > 1$ is a purely geometrical quantity independent of solid or fluid densities. Berryman[10] has considered the case of isolated spherical solid particles in the fluid to derive $\alpha = (1/2)[\phi^{-1}+1]$, for example. The remaining parameters govern attenuation; $b = \eta\phi^2/k$ where η is the fluid viscosity, k is the permeability, and $F(\omega)$ allows for the fact that the effective damping changes when the viscous skin depth $(\sqrt{2\eta/\rho_f\omega})$ becomes smaller than the pore size as the frequency ω increases.[11] It is worth noting that ρ_{12} and $bF(\omega)$ *always* appear together in the combination $\rho_{12} - \frac{ibF(\omega)}{\omega}$. i.e., equation (1a') can be rewritten

$$(1-\phi)\rho_s \frac{\partial^2 \vec{u}}{\partial t^2} = -\tilde{\rho}_{12}(\omega)\left[\frac{\partial^2 \vec{U}}{\partial t} - \frac{\partial^2 \vec{u}}{\partial t^2}\right] + \text{(spatial derivative terms)} \tag{1a''}$$

where

$$\tilde{\rho}_{12}(\omega) = \rho_{12} - \frac{ibF(\omega)}{\omega}$$

and similarly for equation (1b).

Although the linear term in the Taylors series expansion of $F(\omega)$ mimics the effects of ρ_{12} (i.e., both terms describe an ω^2 dependence of the force on the relative displacement), they are of different physical origin and it is not valid to put $\alpha=1$ (i.e., $\rho_{12}=0$) as has been done.[12] Equivalently, one could define $\tilde{\alpha}(\omega)$ by analogy with equation (3c), viz: $\tilde{\rho}_{12}(\omega) = -[\tilde{\alpha}(\omega)-1]\phi\,\rho_f$. Therefore,

$$\tilde{\alpha}(\omega) = \alpha + \frac{ibF(\omega)}{\omega\phi\rho_f} \tag{3c'}$$

The crossover between the high-frequency and low-frequency behavior of $F(\omega)$ occurs when the viscous skin depth is approximately equal to the pore size, a:

$$\omega_c = \frac{2\eta}{\rho_f a^2} \quad .$$

If $\omega >> \omega_c$, the attenuation mechanism has little effect on the velocities of the normal modes which are derived from Eqs. (1); one has[1] $\lim_{\omega\to\infty} F(\omega)\alpha\omega^{1/2}$ and therefore $\lim_{\omega\to\infty} \tilde{\rho}_{12}(\omega) = \rho_{12}$ (a constant *not* equal to zero), and $\lim_{\omega\to\infty} \tilde{\alpha}(\omega) = \alpha$ (a constant *not* equal to 1). (As was shown previously,[13] the constant α is the crucial parameter for the slow wave in systems having a very stiff frame; see Equations (8a,b) below also.)

In this high frequency limit, the non-dispersive velocities of the normal modes are

$$V^2(\text{SHEAR}) = \frac{N}{(1-\phi)\rho_s + \phi\rho_f - \frac{1}{\alpha}\phi\rho_f} \quad , \tag{4a}$$

$$V^2(\text{FAST, SLOW}) = \frac{\Delta \pm \sqrt{\Delta^2 - 4(\rho_{11}\rho_{22} - \rho_{12}^2)(PR - Q^2)}}{2(\rho_{11}\rho_{22} - \rho_{12}^2)} \quad , \tag{4b}$$

where

$$\Delta = P\rho_{22} + R\rho_{11} - 2\rho_{12} Q \quad .$$

That is, there is one shear wave, a fast compressional wave corresponding to solid and fluid moving in phase, and a slow compressional wave corresponding to solid and fluid moving out of phase. Note that these equations depend on α either explicitly or through Eq. (3c) but not on $bF(\omega)$. Indeed, using various literature approximations to $F(\omega)$ and measured values of $k(\sim 10^{-8}\text{cm}^2)$, we have verified[14] that Eqs. (4a,b) are accurate to better than 1%. (In the ultrasonics experiments considered in this article,[15,16] the viscous skin depth in water is less than 1 μm at 500 kHz whereas the pore sizes are larger than 20 μm. Thus, these experiments were made in the high frequency limit of the theory.)

First, though, we consider the opposite limit, $\omega << \omega_c$; the fast and shear waves are still propagatory but the slow wave is described by a diffusion equation,

$$C_D \nabla^2 \vec{\xi} = \frac{\partial \vec{\xi}}{\partial t} \quad , \tag{5a}$$

where $\vec{\xi}$ is a normal mode coordinate and C_D is given by

$$C_D \equiv \frac{kK_f}{\eta\phi}\left[1+\frac{K_f}{\phi\left[K_b+\frac{4}{3}N\right]}\right.$$

$$\left.\cdot\left\{1+\frac{1}{K_s}\left[\frac{4}{3}N\left[1-\frac{K_b}{K_s}\right]-K_b-\phi\left[K_b+\frac{4}{3}N\right]\right]\right\}\right]^{-1} \quad . \tag{5b}$$

(Equations 5 are also derivable from equation 4b by the previously mentioned substitution $\alpha \to \tilde{\alpha}(\omega)$ and then taking the limit $\omega \to 0$.) In ref. 17 it was shown that Eqs.(5a,b) provide the most general description of quasi-static flow in permeable media and that they subsume other theories as special cases (well-to-well pressure pulse testing in the petroleum industry, diffusion in dilute polymer gels). Two cases of special interest (in the low frequency limit) are:

1) Very stiff frame, $K_b, N >> K_f$. In this case equation 5b simplifies greatly:

$$\lim_{K_b, N >> K_f} C_D = \frac{kK_f}{\eta\phi} \tag{5c}$$

Chandler[17] has measured the diffusivity of porous water saturated media (fused glass bead samples and a Berea sandstone). He found that the permeability deduced from the approximate equation (5c) differs from the statically measured permeability by factors of 2-3 whereas if one accounts for the finite compressibility of the frame (i.e. equation 5b) agreement is much improved.

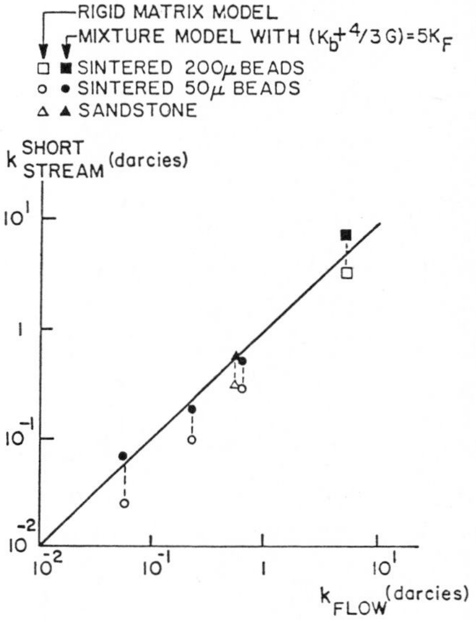

Fig. 3
Permeability deduced from diffusivity vs. static permeability.

In Fig. 3, $G = N$, the open symbols refer to permeability deduced from equation (5c) and the closed symbols refer to permeability deduced from the full theory, equation (5b). (The abscissa is the statically measured permeability whereas the ordinate is that deduced from either equation 5b or 5c.) Although one is nearly in the stiff frame limit here, it is not valid to neglect the finite compressibility of the solid matrix.

2) A cross-linked polymer gel in an aqueous solution corresponds to a fluid- solid system in the *weak* frame limit, $K_b, N, << K_f, K_s$. Since the pores are considerably less than a micron in size, one is essentially always in the low frequency limit where the slow wave is diffusive. Indeed, light scattering experiments clearly indicate a diffusive mode in addition to the usual propagatory mode.[18] Upon gelation, the individual molecules cross link, thus taking the system from the "sol" phase to the "gel" phase, and so the skeletal frame moduli grow from zero in the former to some small value in the latter. In this weak frame limit the diffusivity Equation 5b simplifies greatly[17,19]:

$$\lim_{K_b, N << K_f, K_s} C_D = \frac{k}{\eta}(K_b + \frac{4}{3}N) \tag{5d}$$

This is to be compared with an expression derived independently by Tanaka et al[18] and DeGennes[20] which is, in the present notation:

$$C_D = \frac{k}{\eta \phi^2}(K_b + \frac{4}{3}N) \tag{5e}$$

Since (5e) was derived under the assumption of a dilute concentration of molecules ($\phi \approx 1$), it is essentially the same as (5d). We may confidently conclude that the slow wave has been observed in gels. The speed and attenuation of the fast wave (ordinary sound) also change upon gelation. It is straightforward to show from Equation (4b) that the speed in the gel phase is[19]

$$V(FAST) = V_o[1 + \zeta_1 K_b + \zeta_2 N] \tag{6a}$$

where V_o is the speed in the sol phase and $\{\zeta_i\}$ are complicated functions of ϕ, ρ_f, ρ_s, K_f, K_s. This is to be compared with an expression derived by Bacri, et al.[21] which involved several simplifying assumptions:

$$V(FAST) = V_o[1 + (K_b + \frac{4}{3}N)/(2K_f)]. \tag{6b}$$

Similarly, the attenuation of the ordinary sound (fast wave) in the weak frame limit is

$$\alpha = \frac{1}{2}\frac{(\rho_T - \rho_f)^2}{V_o \rho_T}\frac{k}{\eta}\omega^2 \tag{7a}$$

where $\rho_T = (1-\phi)\rho_s + \phi\rho_f$ is the total density of the gel. This is to be compared witth an expression derived by Bacri, et al[21]

$$\alpha = \frac{1}{2}\frac{(1-\phi)^2 \rho_s^2}{V_o \rho_T}\frac{k}{\eta}\omega^2 \tag{7b}$$

which is the same as (7a) *if* one neglects the volume occupied by the polymer molecules, $\rho_T = \rho_f + (1-\phi)\rho_s$. Note that the general Biot theory, equations (1a,b) simultaneously provide a description of the 2 longitudinal modes, which these other theories do not.

Note also that the low frequency limit of the diffusive slow wave has been observed in both the stiff frame (fused glass beads, rocks) and weak frame (polymer gels) limit. It is obviously a fairly common phenomenon.

What happens when the frequency is increased to, say, the point where the viscous skin depth is much less than the pore size? According to the theory $\tilde{\alpha}(\omega) \approx \alpha$, a real constant (Equation 3c′) and all 3 modes fast, slow, and shear are propagatory with speeds given by Equations (4a,b). The slow wave had never been observed as a propagatory mode until T.J. Plona[15] of Schlumberger observed it in water saturated fused glass beads and in other porous solids (typically water filter materials). Some of his data are reproduced below[16]

Table I

	Sintered Beads #1	Sintered Beads #2	3M	Mott	Coors
pore size (μm)	-	-	55	20	55
porosity	28.3	18.5	34.5	48.0	41.5
permeability (μm^2)	9.1	1.5	8.8	5.9	8.9
velocity:fast compressional (km/sec)	4.05	4.84	2.76	2.74	3.95
velocity:shear (km/sec)	2.37	2.93	1.41	1.54	2.16
velocity:slow compressional (km/sec)	1.04	0.82	0.91	0.92	0.96
sample thickness (cm)	1.91	1.75	.93	.64	.60
sample diameter (cm)	10.4	10.3	10.8	12.7	10.2

Note that the slow wave speed is always less than that of water (which is 1.5 km/sec).

The theoretical velocities of the modes in the high frequency limit depend on three microgeometric parameters α, K_b,N (in adddition to bulk properties ρ_s,$K_s$etc.). It is particularly informative to consider the speeds as a continuous function of the frame moduli K_b,N keeping α (and all the other parameters) fixed. For example, an unconsolidated (loose) sample of glass beads corresponds to vanishing frame moduli (K_b = N = 0) whereas a light sintering of the beads obviously "stiffen up" the skeletal frame without changing the porosity ϕ or the inertial drag parameter α measurably. We compare the calculated speeds as a function of the longitudinal modulus of the skeletal frame $K_b+\frac{4}{3}N$ in Fig. 4, keeping K_b/N a constant.[22]

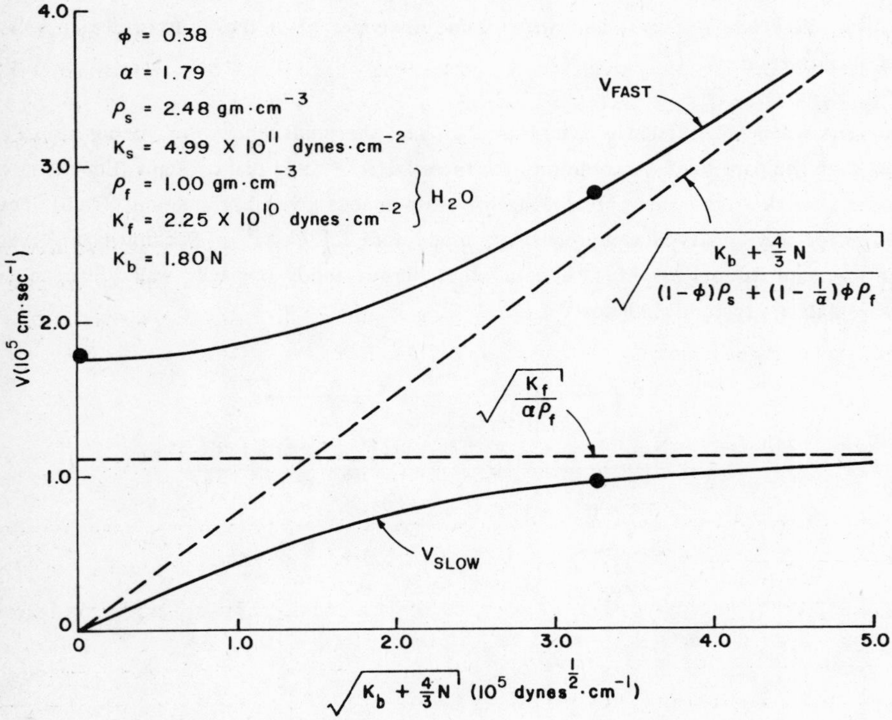

Fig. 4
Fast and slow speeds as a function of frame moduli.

We have also plotted the experimental points corresponding to a lightly fused frame ($K_b,N \neq 0$) and to a non-fused, loose collection of water saturated glass beads ($K_b = N = 0$). We see that in this latter limit both theoretically and experimentally there is only a fast compressional wave whose speeds are in good agreement with each other. We hope eventually to fill in a substantial portion of these curves by controlling the frame moduli; one might, for example, apply an overburden pressure to compress the beads together and thus drive the slow wave speed up from zero.

Notice from Fig. 4 that the complicated expressions for the compressional velocities in the high frequency limit, Equation 4a, simplify greatly if the skeletal frame is very stiff; $K_b,N >> K_f$. In this limit the speeds are[13]:

$$V(FAST) = \sqrt{\frac{K_b + (4/3)N}{(1-\phi)\rho_s + (1-\alpha^{-1})\phi\rho_f}} \quad (8a)$$

$$V(SLOW) = \frac{V_f}{\sqrt{\alpha}} \quad (8b)$$

The fast wave corresponds now to oscillation of the skeletal frame in which some, but not all, of the fluid is dragged along. The slow wave, on the other hand, corresponds to oscillation of the fluid only; the speed is renormalized down from V_f (speed of sound in fluid) because of the tortuous nature of the pore space. We see from Fig. 4 that the light fusing used to make the consolidated frame already puts us in the stiff frame limit.

Suppose one puts superfluid ^{4}He in the pore space. Since the viscosity of the superfluid fraction is identically zero, we are always in the "high frequency" limit of the theory. Let us compare the prediction of the full theory, Equations 4a,b, with the approximate formulae, Equations (8a,b). The results, for three of Plona's samples, are presented in Table II.[13]

Table II
Theoretical wave speeds for the samples of Table I under the condition of saturation with superfluid ^{4}He.

		Sample 1	Sample 2	Sample 3
	ϕ	0.283	0.258	0.185
	$\alpha = n^2$	1.81	1.94	3.00
Fast-compressional speed(km s^{-1})	full theory	4.06	4.18	4.89
	$\left(\frac{K_b + \frac{4}{3}N}{(1-\phi)\rho_s + (1-1/\alpha)\phi\rho_f}\right)^{1/2}$	4.06	4.18	4.89
Shear speed (km s^{-1})	full theory	2.44	2.57	3.00
Slow wave/ fourth-sound speed (m s^{-1})	full theory	173.514	170.797	137.357
	$\left(\frac{K_f}{\alpha\rho_f}\right)^{1/2} = \frac{C_4^0(T=0)}{n}$	173.554	170.839	137.381

Since the speeds of the slow wave are very accurately given by $V_f/\sqrt{\alpha}$, we conclude that in this limit the slow wave is identical to the phenomenon known as 4th sound.[23,24] In 4th sound, the normal fluid component (if any) is assumed to be locked relative to the porous superleak by its viscosity and only the superfluid fraction can oscillate; from the 2 fluid equations of motion one can calculate the theoretical speed of 4th sound as a function of temperature, $C_4^0(T)$. The tortuous pore geometry renormalizes the speed so that the experimentally observed speed, $C_4^E(T)$, is considerably less. The ratio of the two is the index of refraction, n, of the 4th sound:

$$C_4^E(T) = \frac{C_4^0(T)}{n} \quad (9)$$

This equation is identical to Equation (8b) which we have derived from the Biot theory, with $n^2 = \alpha$. Therefore 1) 4th sound, which has been observed since 1965,[24] is an ideal example of the propagatory Biot slow wave. 2) Any porous and permeable solid will exhibit a slow wave if it is saturated with superfluid ^4He. 3) Observations of the slow wave under condition of water saturation have enabled us to make predictions of 4th sound velocities. Conversely, 4th sound speeds very directly give us the key parameter (α) for the high frequency speed of the slow wave when other fluids are in the pore space.

It is clear that the parameter α, or more generally $\tilde{\alpha}(\omega)$, Equation (3c'), is the key parameter for the slow wave, at least in the theory presented so far. It turns out that α is also simply related to the electrical conductivity of the pore space because both are determined by the solution of Poisson's equation with the same boundary condition.[25] If the (non conducting) matrix is saturated with a conducting fluid of conductivity σ_f, the sample conductivity, σ, is simply $\sigma = \frac{1}{F}\sigma_f$, where the quantity F is a geometrical factor independent of pore fluid condutivity and is related to the acoustic parameter α by $\alpha = F\phi$. A theory,[26] described by Sen in this conference, predicts $F = \phi^{-m}$ where m depends on the aspect ratio of the grains. For spherical grains ($m = 3/2$) and this result ($F = \phi^{-3/2}$) is perfect agreement with conductivity measurements on fused glass bead samples over a wide range of porosity. Preliminary measurements on the samples used in the acoustics experiments seem to confirm this and will be reported elsewhere.[27] Meanwhile, we have applied[28] the self-similar model to the 4th sound index of refraction data in Fig. 5.

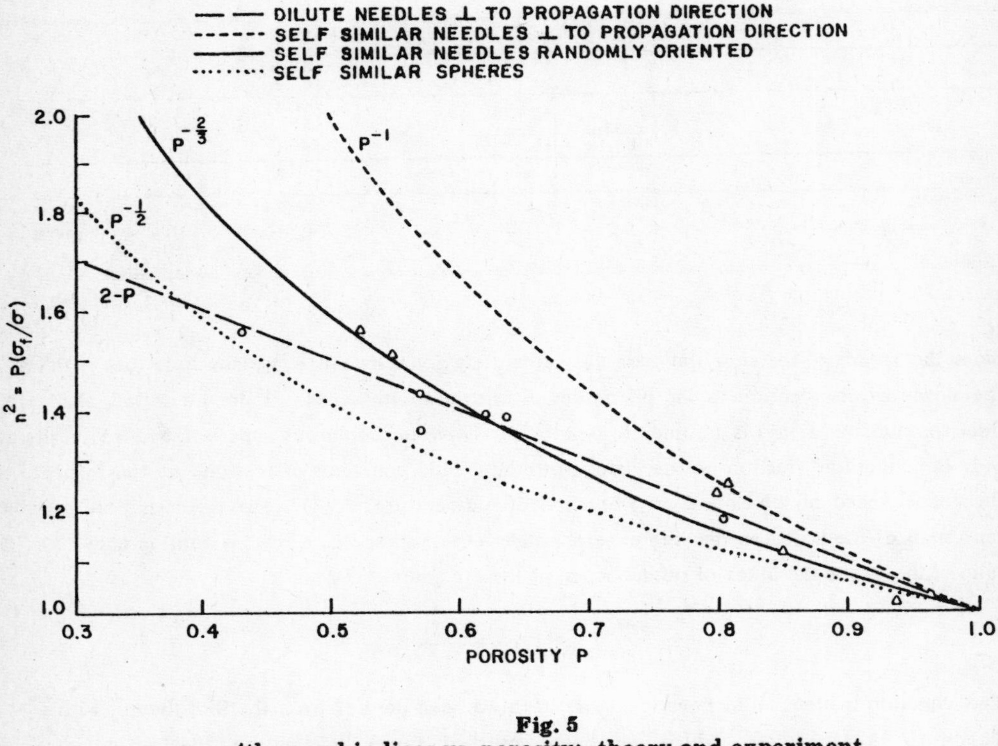

Fig. 5
4th sound indices vs. porosity: theory and experiment

For purposes of the figure, P = ϕ (porosity), $n^2 = \alpha$, and F = σ_f/σ. We see that m = 5/3 ($n^2 = P^{-2/3}$) corresponding to a random array of needles, gives a good account of the data. This is reasonable considering that the packed powder superleaks used are so fine grained that they stick together to form bridges through the very open, high porosity pore space.

To summarize, we have seen that the slow compressional wave predicted by the theory is a much more commonly observed phenomenon than previously thought. Table III summarizes the systems in which the slow wave has been observed in one form or another.

Table III
Systems in which the slow wave has been observed.

	Weak Frame	Stiff Frame
Low Freq. (diffusive) (dominance of η_f)	Polymer gels	Rocks, fused glass beads. (e.g. Pressure Pulse Testing)
High Freq. (propagatory) (dominance of ρ_f)	Unconsolidated beads under confining pressure (?)	4th sound in HeII Fused glass beads & other artificial media

As yet, a propagatory slow wave in an unconsolidate saturated system (under confining pressure) has not been observed but we expect that it will be. It should be emphasized that it makes no sense whatever to analyze data on the fast compressional and shear waves, in terms of the Biot theory, unless quantitative comparison with the slow wave is also made. Unfortunately, this has not always been the case. Since the theory always predicts a slow wave it remains to explain why it is not always observed.

References

1. M. A. Biot, J. Acoust. Soc. Am. *28,* 168 (1956); *28,* 179 (1956).
2. M. A. Biot, J. Appl. Phys. *33,* 1482 (1962).
3. M. A. Biot, J. Acoust. Soc. Am. *34,* 1254 (1962).
4. M. A. Biot and D. G. Willis, J. Appl. Mech. *24,* 594 (1957).
5. D. L. Johnson (unpublished).
6. This assumption may very well break down when the pore fluid is water and the matrix is a rock [M.R.J. Wyllie, G.H.F. Gardner, and A. R. Gregory, Geophysics *27,* 569 (1962)].
7. R. D. Stoll in *Physics of Sound in Marine Sediments,* edited by L. Hampton (Plenum, NY, 1974).
8. J. Geertsma and D. C. Smit, Geophysics *26,* 169 (1961).
9. L. D. Landau and E. M. Lifshitz, *Fluid Mechanics,* (Pergamon, NY, 1959) p. 31ff.
10. J. G. Berryman, Appl. Phys. Lett. *37,* 382 (1980).
11. Ref. 9, p. 88 ff.
12. C. H. Yew and P. N. Jogi, Exp. Mech. *18,* 167 (1978); H. D. McNiven and Y. Mengi, J. Acoust. Soc. Am. *61,* 972 (1977).
13. D. L. Johnson, Appl. Phys. Lett. *37,* 1065 (1980).
14. R. Johnson and D. Johnson (unpublished).
15. T. J. Plona, App. Phys. Lett. *36,* 259 (1980).
16. T. J. Plona and D. L. Johnson, in *1980 Ultrasonic Symposium Proceedings,* edited by J. deKlerk and B. R. McAvoy (IEEE, NY, NY 1980), pp. 868-872.
17. R. C. Chandler and D. L. Johnson, J. Appl. Phys. (to be published); R. C. Chandler, J. Acoust. Soc. Am. (to be published).
18. T. Tanaka, L. O. Hocker, and G. B. Benedek, J. Chem. Phys. *59,* 5151 (1973).
19. D. L. Johnson (unpublished).
20. P. G. DeGennes, Macromolecule *9,* 587 (1976).
21. J. C. Bacri, J. M. Courdille, J. Dumas and R. Rajaonarison, J. Physique *41,* L-369 (1980); J. C. Bacri and R. Rajaonarison, J. Physique *40,* L-5 (1979).
22. D. L. Johnson and T. J. Plona (unpublished).
23. I. Rudnick, *New Directions in Physical Acoustics,* Proceedings of the Enrico Fermi Summer School, Course LXIII (Academic, NY, 1976), p. 112.
24. K. A. Shapiro and I. Rudnick, Phys. Rev. *137,* A1383 (1965).
25. R. J. S. Brown, Geophysics *45,* 1269 (1980).
26. P. N. Sen, C. Scala and M. H. Cohen, Geophysics *46,* 781 (1981).
27. T. J. Plona, D. L. Johnson, and C. Scala (unpublished).
28. D. L. Johnson and P. N. Sen, Phys. Rev. B (to be published).

BOUNDS FOR THE EFFECTIVE CONDUCTIVITY OF RANDOM MEDIA

W. Kohler and G. C. Papanicolaou
Virginia Polytechnic Institute and State University and
Courant Institute, New York University

Abstract

We formulate the problem of calculating the effective conductivity of a random medium in a suitable manner. Then we obtain upper and lower bounds using variational principles. Some of the bounds depend on the random geometry of the medium while others (like the Hashin-Shtrikman bounds) do not.

1. Introduction

We shall describe briefly our work on the determination of bounds for the effective conductivity of a random medium. Some additional information is contained in [1].

Bounds for effective parameters received a lot of attention in the past twenty or so years primarily because of the work of Hashin and Shtrikman [2]. In [3]-[7] extensions, improvements and new results are presented, always however within the framework of variational principles. Here we shall also stay within this framework. Our aim is (i) to remove the difficulties associated with the definition of effective parameters over finite samples of materials; (ii) derive geometry dependent bounds from multiple scattering expansions; and (iii) derive geometry independent bounds also from multiple scattering expansions.

Recently Bergmann [8] has introduced another method for deriving bounds that does not rely on variational principles. It works for two component media, relies on analytic continuation and can handle, for example, frequency dependent (hence complex-valued) effective dielectric constants (or conductivity). Milton [9] has used Bergman's method and has obtained further results. In [10] we give a mathematical formulation of Bergman's method similar to the one we employ here and we obtain a number of properties of the effective conductivity (dielectric constant) as a complex-valued function of a complex variable (the ratio of the conductivities of the two components).

2. Definition of effective conductivity

Let (Ω,Σ,P) be a probability space with $\omega \in \Omega$ labeling the realizations of the microscopic conductivity $a(x,\omega)$ which is a given stationary random function such that $0 < \alpha \leq a(x,\omega) \leq \beta < \infty$ for all $x \in \mathbb{R}^3$ and $\omega \in \Omega$. Stationary means that the joint distribution of $a(x_1,\omega), a(x_2,\omega), \ldots, a(x_N,\omega)$ for any set of points $x_1, x_2, \ldots, x_N$ is the same as that of $a(x_1+h,\omega), \ldots, a(x_N+h,\omega)$ for any $h \in \mathbb{R}^3$.

We denote by $G(x,\omega)$ and $F(x,\omega)$ the negative temperature gradient and the heat flux respectively at a point $x \in \mathbb{R}^3$ for the realization ω of the medium. The vector fields F and G are defined as follows. They are stationary random fields such that

$$\int_\Omega |G(x,\omega)|^2 P(d\omega) < \infty \ , \qquad \int_\Omega |F(x,\omega)|^2 P(d\omega) < \infty$$

and

$$F(x,\omega) = a(x,\omega)G(x,\omega) \tag{2.1}$$

$$\nabla \times G(x,\omega) = 0 \tag{2.2}$$

$$\nabla \cdot F(x,\omega) = 0 \tag{2.3}$$

$$\int_\Omega G(x,\omega)P(d\omega) = e \ . \tag{2.4}$$

Here e is a fixed unit vector in $\mathbb{R}^3$. The effective conductivity $a^* = a^*(a,e,g)$ is defined by

$$a^*(a,e,g) = \int_\Omega F(x,\omega) \cdot g P(d\omega) \tag{2.5}$$

where g is another unit vector in $\mathbb{R}^3$. One can verify easily [1] that (2.1)-(2.4) has a unique solution and that a^* is well defined.

The definition of a^* is of course what one intuitively calls effective conductivity. The tensor $a^*(a,f,g)$ is the average heat flux in the direction g when unit average negative temperature gradient is prescribed in the direction e. The conducting material occupies an infinite region and its properties are statistically homogeneous. One can show [1] that the conductivity of a finite sample of material tends to a^* for almost all realizations ω as the volume of the sample tends to infinity. This is a spatial ergodic theorem that justifies the usual operational definition of conductivity.

Working with the canonical problem (2.1)-(2.4) is advantageous because properties of a^* can be obtained without the need to take limits, such as infinite volume limits.

To simplify exposition we shall assume in most of the sequel that a^* is isotropic, i.e. it is independent of e and g.

Our objective is to get bounds for a^* under various hypotheses about the values of the microscopic conductivity a and their statistical distribution. For this reason consider the following well-known variational principles.

The Dirichlet or energy principle gives

$$(2.6) \qquad a^* = \min_{\substack{\nabla\times G = 0 \\ \langle G\rangle = e}} \langle aG\cdot G\rangle$$

where G runs over all stationary square integrable random fields that are curl-free and $\langle G\rangle$ denotes average

$$\langle G\rangle = \int_\Omega G(\mathbf{x},\omega)P(d\omega) .$$

The dual variational principle or Thomson's principle gives

$$(2.7) \qquad \frac{1}{a^*} = \min_{\substack{\nabla\cdot F = 0 \\ \langle F\rangle = e}} \langle \frac{1}{a} F\cdot F\rangle .$$

The solution pair G and $(a^*)^{-1}F$ of (2.1)-(2.4) realizes the minimum in (2.6) and (2.7) respectively. Note that (2.5), (2.6) and (2.7) are all compatible since (in the isotropic case for simplicity)

$$a^* = \langle F\cdot e\rangle = \langle aG\cdot e\rangle = \langle aG\cdot G\rangle$$

because

$$\langle aG\cdot\tilde{G}\rangle = 0$$

for any test field $\tilde{G}$ with $\langle G\rangle = 0$ and $\nabla\times\tilde{G} = 0$. This follows from (2.3). We also have that

$$\begin{aligned}\langle\frac{1}{a}(a^*)^{-1}F\cdot(a^*)^{-1}F\rangle &= (a^*)^{-2}\langle\frac{1}{a}F\cdot F\rangle \\ &= (a^*)^{-2}\langle aG\cdot G\rangle \\ &= (a^*)^{-2}\langle F\cdot e\rangle = (a^*)^{-1} .\end{aligned}$$

The problem is then to find suitable trial fields G and F to insert in (2.6) and (2.7). For example, choosing $G = e$ and $F = e$ yields the very well known bounds

$$(2.8) \qquad (\langle \tfrac{1}{a} \rangle)^{-1} \leq a^* \leq \langle a \rangle .$$

These bounds are achieved by parallel plate 2-component materials.

Two questions arise in connection with bounds now:

(i) What are the best bounds one can obtain for a^* when the material is known to have specific structure such as spherical inclusions in a homogeneous matrix? It is assumed that the distribution functions of the spheres are known here.

(ii) What are the best bounds one can obtain for a^* when the geometrical structure of the composite material is very complex, for example a two-component material with neither of the components being a host or inclusion material?

Clearly in the second question we are asking for bounds that are somehow geometry-independent. The difference between the two types of bounds is quite analogous to the difference between the various kinds of approximate formulas for a^* that are commonly obtained. For example the Claussius-Mosotti formulas [11] are results responding to questions of the first kind while the effective medium theories [12] (cf. also [13]) give results responding to questions of the second kind.

We close this section by deriving in a very simple manner two geometry independent results that are quite useful. The first is due to Keller [14] for a two-component periodic structure. It was generalized by Mendelshon [15]. It is proved here in full generality. The second is an inequality which generalizes Keller's result and was obtained by Schulgasser [16]. We prove it here by a different method in full generality.

Theorem 1

Let e_1 and e_2 be the orthogonal unit vectors in $\mathbb{R}^2$ such that e_2 is obtained from e_1 by a 90° rotation. Let $a^*(a,e_j,e_j) = a^*(a,e_j)$. Then

$$(2.9) \qquad \frac{1}{a^*(\frac{1}{a},e_2)} = a^*(a,e_1)$$

Remark. In the isotropic case (2.9) reads

$$(2.10) \qquad \frac{1}{a^*(\frac{1}{a})} = a^*(a) .$$

Since for any positive constant γ, $a^*(\gamma a) = \gamma a^*(a)$, when a takes only the values $a = a_1$ or a_2 then the relation (2.10) can be written

$$1 = a^*(\frac{1}{a_1},\frac{1}{a_2})a^*(a_1,a_2)$$

or

(2.11) $$a_1a_2 = a^*(a_2,a_1)a^*(a_1,a_2) \ .$$

The form (2.10) is more appropriate since it holds in general, not just a two-component medium.

Theorem 2

Let e_1, e_2 and e_3 be an orthonormal triplet of vectors in $\mathbb{R}^3$ so that $e_1 \times e_2 = e_3$, $e_2 \times e_3 = e_1$, $e_3 \times e_1 = e_2$. Then

(2.12) $$\frac{1}{a^*(\frac{1}{a},e_i \times e_j)} \leq a^*(a,e_j) \ , \qquad i,j = 1,2,3, \quad i \neq j \ .$$

Remark: In the isotropic, two-component case (2.12) reduces to

(2.13) $$a_1a_2 \leq a^*(a_2,a_1)a^*(a_1,a_2)$$

and as Schulgasser [16] observed, strict inequality holds for the coated sphere medium (used by Bruggeman) that realizes the upper and lower Hashin-Shtrikman bounds.

Proof of Theorem 1

From (2.7)

(2.14) $$\frac{1}{a^*(\frac{1}{a},e_2)} = \min_{\substack{\nabla \cdot F = 0 \\ \langle F \rangle = 0}} \langle a(e_2+F)\cdot(e_2+F)\rangle \ .$$

Let S be the 90° rotation map that takes e_1 into e_2, i.e. $Se_1 = e_2$. Let G be a two-dimensional, zero mean stationary random field with $\nabla \times G = 0$, i.e. $\partial G_1/\partial x_2 = \partial G_2/\partial x_1$. Let

(2.15) $$\tilde{F} = SG \ .$$

Then clearly $\tilde{F}$ has mean zero is stationary and $\nabla \cdot \tilde{F} = 0$. Moreover

(2.16) $$\begin{aligned} \langle a(e_2+\tilde{F})\cdot(e_2+\tilde{F})\rangle &= \langle aS(e_1+G)\cdot S(e_1+G)\rangle \\ &= \langle a(e_1+G)\cdot(e_1+G)\rangle \ . \end{aligned}$$

Since the mapping (2.15) is one-one and onto we conclude that taking minima in (2.16) yields (2.9) in view of (2.14) and the analogous expression from (2.6).

Proof of Theorem 2

The proof is similar to the one above. The field $e_i \times e_j + e_i \times G$, where G is stationary, zero mean and $\nabla \times G = 0$, is an admissible field in the variational principle (2.7), i.e.

$$\frac{1}{a^*(\frac{1}{a}, e_i \times e_j)} = \min_{\substack{\nabla \cdot F = 0 \\ \langle F \rangle = 0}} \langle a(e_i \times e_j + F) \cdot (e_i \times e_j + F) \rangle .$$

But the class of divergence-free fields is not spanned by fields of the form $e_i \times G$, $\nabla \times G = 0$ as was the case in dimension 2. Thus

$$\frac{1}{a^*(\frac{1}{a}, e_i \times e_j)} \leq \min_{\substack{\nabla \times G = 0 \\ \langle G \rangle = 0}} \langle a(e_i \times e_j + e_i \times G) \cdot (e_i \times e_j + e_i \times G)$$

$$= \min_{\substack{\nabla \times G = 0 \\ \langle G \rangle = 0}} \langle a[e_i \times (e_j + G)] \cdot [e_i \times (e_j + G)] \rangle$$

$$= \min_{\substack{\nabla \times G = 0 \\ \langle G \rangle = 0}} \{ \langle a(e_j + G) \cdot (e_j + G) \rangle - \langle a[e_i \cdot (e_j + G)]^2 \rangle \}$$

$$\leq \min_{\substack{\nabla \times G = 0 \\ \langle G \rangle = 0}} \langle a(e_j + G) \cdot (e_j + G) \rangle$$

$$= a^*(a, e_j) ,$$

as was to be shown.

3. Geometry-independent bounds

The difficulty in applying the variational principles (2.6) and (2.7) comes of course from the fact that the trial fields must be curl-free and divergence-free, respectively. To overcome this difficulty, i.e. in order to be able to construct many interesting admissible fields, we look again at the Euler equations for (2.6) which are (2.1)-(2.4).

We introduce, as did Hashin and Shtrikman [2], a reference medium with constant conductivity a_o and rewrite (2.3) as

(3.1) $\nabla\cdot(a_o G) + \nabla\cdot(\delta a G) = 0$, $\nabla\times G = 0$,

where

(3.2) $$\delta a = a(x,\omega) - a_o \quad .$$

Let us define formally an operator Γ by

(3.3) $$\Gamma G = -\nabla(-\Delta)^{-1}(\nabla\cdot G) \quad .$$

Lemma 1

The operator Γ given formally by (3.3) is a well-defined projection operator on the Hilbert space

(3.4) $$H = \left\{ G|G \text{ is stationary vector field, } \langle G\rangle=0, \ \langle G\cdot G\rangle < \infty \right\}$$

into the subspace of curl free fields in H. In particular, Γ satisfies

(3.5) $$\Gamma^2 = \Gamma \quad .$$

Lemma 2

If G in H is in addition isotropic,

$$\langle G_i(x+y)G_j(y)\rangle = R_{ij}(|x|) \qquad i,j = 1,2,3,$$

i.e., the correlation function depends on $|x|$ only, then

(3.6) $$\langle \Gamma G\cdot G\rangle = \frac{1}{3}\langle G\cdot G\rangle \quad .$$

The proof of both lemmas is quite simple and we shall not pause to give it (cf. [1]).

With the help of the projection operator Γ we can write (3.1) in the form

(3.7) $$G + \Gamma(a_o^{-1}\delta a G - \langle a_o^{-1}\delta a G\rangle) = e \quad .$$

We observe that in fact (3.7) summarizes at once all of (2.1)-(2.4). However, (3.7) is not the Euler equation of a variational problem as it stands because the operator $\Gamma(a_o^{-1}\delta a\cdot)$ is not selfadjoint in H even though the projection operator Γ is. One way to transform (3.7) to the Euler equation of a variational problem is to let the unknown field be $a_o^{-1}\delta a G$, rather than G. Let

(3.8) $$P = a_o^{-1}\delta a G$$

so that (3.7) becomes

$$(3.9)\qquad a_o(\delta a)^{-1}P + \Gamma(P-\langle P\rangle) = e$$

Clearly (3.9) is equivalent to (3.7) and by analogy with electrostatics we call P the polarization field.

Equation (3.9) is the Euler equation of the quadratic functional

$$(3.10)\quad U(\mathcal{P}) = \frac{1}{2}\langle a_o(\delta a)^{-1}\mathcal{P}\cdot\mathcal{P}\rangle + \frac{1}{2}\langle\Gamma(\mathcal{P}-\langle\mathcal{P}\rangle)\cdot(\mathcal{P}-\langle\mathcal{P}\rangle)\rangle - \langle e\cdot\mathcal{P}\rangle$$

on $\tilde{H} = H$ + constants. Note in particular that the functional U is defined over all of $\tilde{H}$ without constraints. One verifies easily that when $a_o < a(x,\omega)$ for all x and ω, so that $\delta a > 0$, then $U(\mathcal{P})$ takes its minimum at $\mathcal{P} = P$ of (3.8) and

$$(3.11)\qquad U(P) = \frac{1}{2}\left(1 - \frac{a^*}{a_o}\right) .$$

Thus

$$(3.12)\qquad \frac{1}{2}\left(1 - \frac{a^*}{a_o}\right) \leq U(\mathcal{P}) \quad \text{when} \quad \delta a > 0$$

where $\mathcal{P}$ is an arbitrary field in $\tilde{H}$. The bound (3.12) can be rewritten as

$$(3.13)\qquad a_o(1 - 2U(\mathcal{P})) \leq a^* , \quad \text{when} \quad \delta a > 0 .$$

Note that a_o is also arbitrary provided $0 < a_o < a(x,\omega)$.

When $\delta a < 0$, i.e. $a(x,\omega) < a_o < \infty$, then the quadratic form (3.10) is negative definite in $\tilde{H}$. This is clear because $a_o(\delta a)^{-1} < -1$ and since Γ is a projection $\|\Gamma\| \leq 1$. Thus

$$(3.14)\qquad a^* \leq a_o(1 - 2U(\mathcal{P})) , \quad \text{when} \quad \delta a < 0 .$$

The next step is the selection of a trial field $\mathcal{P}$ for (3.13) and (3.14). Obviously we want a trial field as close to P as possible. Let us look at (3.9) which we rewrite as

$$(3.15)\qquad P = \frac{\delta a}{a_o}[e - \Gamma(P-\langle P\rangle)] .$$

There are several ways in which we can construct trial fields close to P depending on what we know about the medium. We single out 3 cases.

<u>Case 1</u>. $|\delta a|$ is small for some a_o, i.e. the random medium is nearly uniform.

In this case we can expand P in (3.15) in power series in $|\delta a|$ and obtain trial fields by truncating the series. For example,

$$P = \frac{\delta a}{a_o} e - \frac{\delta a}{a_o} \Gamma\left(\frac{\delta a}{a_o} e - \langle\frac{\delta a}{a_o} e\rangle\right)$$

is an admissible field, The calculations are not particularly interesting here although with the help of the analytic continuation method (for two-component media, cf. [8]-[10]), they can become significant.

<u>Case 2</u>. δa is not small for any a_o but the medium has a host-inclusion geometry and the multiple scattering formalism can be used. This is discussed further in the next section.

<u>Case 3</u>. δa is not small for any a_o and the medium is not of the simple host-inclusion type. We suppose however that we know that the conductivity $a(x,\omega)$ is statistically isotropic.

We shall pursue here this last case because it is geometry-independent and leads to interesting results.

We assume from now on that $a(x,\omega)$ is statistically isotropic.

We note first that the operator Γ on H defined by (3.3) can be written in the form

(3.16) $$\Gamma = \Gamma_0 + \Gamma_1$$

where

(3.17) $$\Gamma_0 = \frac{1}{3} I \qquad \text{and} \qquad \Gamma_1 = \Gamma - \frac{1}{3} I .$$

Because of lemma 2 and (3.6) we see that for any <u>isotropic</u> fields G_1 and G_2 in H

(3.18) $$\langle\Gamma_1 G_1 \cdot G_2\rangle = 0 .$$

To verify (3.18) one has to note that (3.6) implies that

(3.19) $$\langle\Gamma G_1 \cdot G_2\rangle = \frac{1}{3} \langle G_1 \cdot G_2\rangle$$

for any G_1, G_2 in H which are isotropic.

We now return to (3.9) and rewrite it in the form

$$(3.20)\qquad P - \langle P\rangle + \frac{\delta a}{a_0}\Gamma_0(P - \langle P\rangle) + \frac{\delta a}{a_0}\Gamma_1(P - \langle P\rangle) = \frac{\delta a}{a_0} e - \langle P\rangle$$

or

$$\left[I + \left(I + \frac{\delta a}{a_0}\Gamma_0\right)^{-1}\frac{\delta a}{a_0}\Gamma_1\right](P - \langle P\rangle) = \left(I + \frac{\delta a}{a_0}\Gamma_0\right)^{-1}\left[\frac{\delta a}{a_0} e - \langle P\rangle\right] .$$

Let

$$(3.21)\qquad L = -\left(I + \frac{\delta a}{a_0}\Gamma_0\right)^{-1}\frac{\delta a}{a_0}\Gamma_1 .$$

Then

$$(3.22)\qquad P - \langle P\rangle = \sum_{n=0}^{\infty} L^n\left[\left(I + \frac{\delta a}{a_0}\Gamma_0\right)^{-1}\left(\frac{\delta a}{a_0} e - \langle P\rangle\right)\right] .$$

We now choose as a trial field for (3.13) or (3.14) the simplest $\mathcal{P}$ compatible with (3.22), namely

$$(3.23)\qquad \mathcal{P} - \langle\mathcal{P}\rangle = \left(I + \frac{\delta a}{a_0}\Gamma_0\right)^{-1}\left(\frac{\delta a}{a_0} e - \langle\mathcal{P}\rangle\right) .$$

We thus neglect all terms in (3.22) that involve Γ_1. Taking averages in (3.23), noting that $\Gamma_0 = \frac{1}{3} I$ and rearranging we find that

$$(3.24)\qquad \langle\mathcal{P}\rangle = \left\{\left\langle\left(I + \frac{\delta a}{3a_0}\right)^{-1}\right\rangle\right\}^{-1}\left\langle\left(I + \frac{\delta a}{3a_0}\right)^{-1}\frac{\delta a}{a_0}\right\rangle e .$$

We evaluate next the functional U of (3.10) on the trial field $\mathcal{P}$ of (3.23). This gives

$$(3.25)\qquad U(\mathcal{P}) = -\frac{1}{2}\langle\mathcal{P}\rangle\cdot e .$$

To verify (3.25) we note that since a is isotropic the Γ_1 term $(\Gamma = \Gamma_0 + \Gamma_1)$ in (3.10) drops out and that (3.23) can be rewritten in the form

$$3.26)\qquad \frac{a_0}{\delta a}\mathcal{P} + \Gamma_0(\mathcal{P} - \langle\mathcal{P}\rangle) = e ,$$

which simplifies computations.

Inserting (3.24) into (3.25) and the result into (3.13) and (3.14) we obtain the bounds

$$(3.27)\qquad a_0\left\{1+\left[<\left(1+\frac{\delta a}{3a_0}\right)^{-1}>\right]^{-1}<\left(1+\frac{\delta a}{3a_0}\right)^{-1}\frac{\delta a}{a_0}>\right\}\ \le\ a^*$$

when $\delta a > 0$

and

$$(3.28)\qquad a^*\ \le\ a_0\left\{1+\left[<\left(<1+\frac{\delta a}{3a_0}\right)^{-1}>\right]^{-1}<\left(1+\frac{\delta a}{3a_0}\right)^{-1}\frac{\delta a}{a_0}>\right\}$$

when $\delta a < 0$.

Now suppose that $a(x,\omega)$ takes only two values (or any finite number) a_1 and a_2 with $0 < a_1 < a_2 < \infty$. Then the inequality (3.27) is optimized by taking $a_0 = a_1$ and (3.28) by taking $a_0 = a_2$. A simple calculation yields now the bounds

$$(3.29)\qquad a_1 + \frac{\rho_2}{\frac{1}{a_2-a_1}+\frac{\rho_1}{3a_1}}\ \le\ a^*\ \le\ a_2 + \frac{\rho_1}{\frac{1}{a_1-a_2}+\frac{\rho_2}{3a_2}}$$

which are the bounds of Hashin and Shtrikman [2]. In (3.29) we have set

$$(3.30)\qquad \rho_j = P\{a(x) = a_j\}\ ,\qquad j=1,2\quad .$$

To improve on the bounds (3.27)-(3.28), or (3.29) for the two component medium, one has to use in (3.13)-(3.14) a better trial field P. From (3.22) we see that this should be

$$(3.31)\qquad P - <P> = \left(I+\frac{\delta a}{a_0}\Gamma_0\right)^{-1}\left(\frac{\delta a}{a_0}e - <P>\right)$$

$$-\left(I+\frac{\delta a}{a_0}\Gamma_0\right)^{-1}\frac{\delta a}{a_0}\Gamma_1\left[\left(I+\frac{\delta a}{a_0}\Gamma_0\right)^{-1}\left(\frac{\delta a}{a_0}e - <P>\right)\right]\quad .$$

However this is not quite appropriate because it does not satisfy

$$(3.32) \qquad <\frac{a_0}{\delta a} \mathcal{P}> \; = \; e$$

as the exact field P does from (3.8) and (2.4). To get the appropriate higher order trial field $\mathcal{P}$ we look again at (3.15). From (3.9) we can write formally

$$(3.33) \qquad P - <P> = \left(I + \frac{\delta a}{a_0} \Gamma\right)^{-1} \left(\frac{\delta a}{a_0} e - <P>\right) .$$

Using this in (3.15) we obtain the iterated form

$$(3.34) \quad P - <P> = \frac{\delta a}{a_0} e - <P> - \frac{\delta a}{a_0} \Gamma \left[\left(I + \frac{\delta a}{a_0} \Gamma\right)^{-1} \left(\frac{\delta a}{a_0} e - <P>\right) \right] .$$

Now we take as trial field

$$(3.35) \qquad \mathcal{P} - <\mathcal{P}> = \frac{\delta a}{a_0} e - <\mathcal{P}>$$

$$- \frac{\delta a}{a_0} \Gamma \left[\left(I + \frac{\delta a}{a_0} \Gamma_0\right)^{-1} \left(\frac{\delta a}{a_0} e - <\mathcal{P}>\right) \right] .$$

This can be rewritten in the form

$$(3.36) \qquad \mathcal{P} - <\mathcal{P}> = \left(I + \frac{\delta a}{a_0} \Gamma_0\right)^{-1} \left(\frac{\delta a}{a_0} e - <\mathcal{P}>\right)$$

$$- \frac{\delta a}{a_0} \Gamma_1 \left[\left(I + \frac{\delta a}{a_0} \Gamma_0\right)^{-1} \left(\frac{\delta a}{a_0} e - <\mathcal{P}>\right) \right]$$

which shows the difference with (3.31).

The trial field (3.36) is now used in (3.13)-(3.14). Of course the terms involving the operator Γ_1 do not drop out now and it is necessary to have information about correlation functions in order to calculate explicitly the resulting bounds. The calculations are given in [1].

One can also work directly with the variational principles (2.6)-(2.7) by using trial fields based on multiple scattering expansion, for example from (3.7) (cf. [7]). If is more convenient however to work with (3.13)-(3.14) and the above method for constructing trial fields.

4. Geometry dependent bounds

Let $\{y_j\}$ be a stationary distribution of points in $\mathbb{R}^3$. A realization of the points will be denoted by $\omega = \{y_j\}$. We now define the conductivity $a(x,\omega)$ by

$$(4.1) \qquad a(x,\omega) = \begin{cases} a_2 \, , & |x-y_j| \le \delta \quad \text{for some } j \\ a_1 \, , & |x-y_j| > \delta \quad \text{for all } j \end{cases} .$$

We assume that the spheres of radius δ centered at $\{y_j\}$ do not overlap. We denote by ρ the volume fraction occupied by the spheres

$$(4.2) \qquad \rho = P\{a(x) = a_2\} = \frac{4}{3}\pi\delta^3\alpha$$

where α is the average number of sphere centers per unit volume. The ρ here corresponds to ρ_2 of the previous section.

We wish to find upper and lower bounds for the effective conductivity a^* which when ρ is small are much closer than the Hashin-Shtrikman bounds (3.29). The improvement must come of course from the knowledge we have now of the random geometry: conducting spheres distributed in a uniform medium. Unfortunately the two and three point functions of the point process $\{y_j\}$ that are necessary for our calculations are not known for realistic cases that allow dense packing with volume exclusion. To render our formulas explicit we shall use the heuristic "well-stirred" approximation which we now describe.

Let $f(x)$, $g(x)$ and $h(x)$ be smooth functions of compact support in $\mathbb{R}^3$. We calculate

$$(4.3) \quad \langle \sum_j f(y_j) \sum_k g(y_k) \sum_\ell h(y_\ell) \rangle = \alpha \int f(x)g(x)h(x)p_1(x)\,dx$$
$$+ \alpha^2 \iint f(x)g(x)h(y)p_2(x,y)\,dxdy + \alpha^2 \iint f(x)h(x)g(y)p_2(x,y)\,dxdy$$
$$+ \alpha^2 \iint g(x)h(x)f(y)p_2(x,y)\,dxdy$$
$$+ \alpha^3 \iiint f(x)g(y)h(z)p_3(x,y,z)\,dxdydz \; .$$

Here α is the average number of points $\{y_j\}$ per unit volume, $p_1(x)$ is identically equal to one (by stationarity), p_2 and P_3 are the two and three point functions respectively. By stationarity, $p_2(x,y) = p_2(0,x-y)$ and $p_3(x,y,z) = p_3(0,y-x,z-x)$. The well-stirred approximation consists in taking

$$(4.4) \qquad p_2(z) = \begin{cases} 0\,, & |z| \leq \delta \\ 1\,, & |z| > \delta \end{cases}, \qquad p_3(y,z) = \begin{cases} 0, & |y| \leq 1 \\ 0, & |z| \leq 1 \\ 0, & |y-z| \leq 1 \\ 1, & \text{all other cases.} \end{cases}$$

There are less naive models for the function p_2 (cf. [19]) which are still not entirely rigorous but have been used with success. For the function p_3 the situation is quite complicated and little seems to be known about it for realistic situations (i.e. that allow close packing).

Now our approach is to use the classical variational principles (2.6) and (2.7) directly. Trial fields are constructed by multiple scattering expansions [7,13] as follows.

Let

$$(4.5) \qquad v_j(x) = \begin{cases} 1\,, & |x-y_j| \leq \delta \\ 0\,, & |x-y_j| > \delta \end{cases}.$$

We suppose for the moment that j runs from 1 to $N < \infty$ and write (3.1) in the form

$$(4.6) \quad \nabla\cdot(a_0 G) + \sum_{j=1}^{N} \nabla\cdot\left[\left((a_2-a_1)v_j(x) + \frac{a_1-a_0}{N}\right)G\right] = 0 .$$

Now let

$$L_0 = \nabla\cdot(a_0\,\cdot\,)\,, \qquad M_0 = \sum_{j=1}^{N} V_j^{(0)}$$

$$(4.7) \qquad V_j^{(0)} = \nabla\cdot\left[\left((a_2-a_1)v_j + \frac{a_1-a_0}{N}\right)\cdot\right]$$

so that (4.6) becomes

$$(4.8) \qquad (L_0 + M_0)G = 0 \quad .$$

Define further the operator T_0 by

$$(4.9) \qquad T_0 = -(L_0 + M_0)^{-1}M_0 \quad .$$

If we set

$$(4.10) \qquad G = (I + T_0)(I + \langle T_0\rangle)^{-1}e \ ,$$

we can verify that this field G satisfies (4.6); in fact (2.1)-(2.4), if in (4.8) and the definition of T_0 we include $\nabla \times G = 0$.

From (4.7) and (4.9)

$$(4.11) \qquad T_0 = -\left[L_0 + \sum_k V_k^{(0)}\right]^{-1} \sum_j V_j^{(0)}$$

$$= -\sum_j \left[L_0 + V_j^{(0)} + \sum_{k\neq j} V_k^{(0)}\right]^{-1} V_j^{(0)}$$

$$= -\sum_j \left[I + (L_0+V_j^{(0)})^{-1} \sum_{k\neq j} V_k^{(0)}\right]^{-1} (L_0+V_j^{(0)})V_j^{(0)}$$

$$= \sum_j \left[I - \sum_{k\neq j} T_{jk}^{(0)}\right]^{-1} T_j^{(0)} \quad .$$

Here

$$(4.12) \qquad T_j^{(0)} = -(L_0 + V_j^{(0)})V_j^{(0)} \ , \qquad T_{jk}^{(0)} = (L_0 + V_j^{(0)})^{-1}V_k^{(0)} \quad .$$

The form (4.11) of the "scattering" operator T_0 is the multiple scattering form when the inverse operator in the last line of (4.11) is expanded further. If we approximate T_0 by

$$(4.13) \qquad T_0 \sim \sum_j T_j^{(0)} \ ,$$

then a trial field can be constructed by taking

$$(4.14) \qquad G = \left[I + \sum_j T_j^{(0)}\right]\left[I + \langle\sum_j T_j^{(0)}\rangle\right]^{-1} e \quad .$$

Thus we must find the explicit form of $T_j^{(0)}$ on uniform fields. This is an elementary computation that yields

$$(4.15) \qquad (T_j^{(0)}\tilde{e})(x) = -\gamma\left[v_j\tilde{e} + \delta^3(1-v_j)\nabla_x\left(\frac{(x-y_j)\cdot\tilde{e}}{|x-y_j|^3}\right)\right]$$

where $\tilde{e}$ is some uniform field and

$$(4.16) \qquad \gamma = \frac{a_2-a_1}{2a_0+a_2-a_1} \quad .$$

Here we have already passed to the limit $N \to \infty$ (cf. (4.7) and (4.12)) and, in order that $<\sum_j T_j>$ exist we assume that the field $T_j^{(0)}\tilde{e}$ is cut off smoothly when $|x-y_j| > 10\delta$ say. Because of this $<\Sigma T_j> = 0$ and hence the trial field takes the form

$$G = e + \sum_j T_j^{(0)} e$$

with $T_j^{(0)} e$ defined by (4.15).

We insert this G into (2.6) and perform the averages as indicated in (4.3) with the use of the well-stirred approximation (4.4). This is a very lengthy computation which we shall not present. At the end of the computation the constant a_0 is chosen to optimize the bound. We simply state the result of our computations

$$(4.17) \qquad \frac{a^*}{a_1} \leq 1 + \rho(K-1) - \frac{(K-1)^2\rho(1-\ )^2}{3-(2-\kappa)\rho+(K-1)(1-2)(1-\xi)\rho+\eta\rho^2)} \quad .$$

Here

$$(4.18) \qquad K = \frac{\dot{a}_2}{a_1}$$

and κ, ξ and η are constants that stand for integrals of dipole-like functions over multidimensional domains with excluded regions arising from (4.4). To three decimal point accuracy

$$(4.19) \qquad \kappa = -.832 \ , \qquad \xi = .176$$

while a crude bound for η is

$$(4.20) \qquad |\eta| \leq 101.43 \quad .$$

For $\rho << 1$, the bound (4.17) expanded to $O(\rho^2)$ is

$$(4.21)\qquad \frac{a^*}{a_1} \leq 1 + 3\left(\frac{K-1}{K+2}\right)\rho + [3.168 + .352(K-1)]\left(\frac{K-1}{K+2}\right)^2 \rho^2 + \dots .$$

An entirely analogous computation can be carried out for a lower bound based on the trial field

$$(4.22)\qquad F = (1+3\rho\gamma)^{-1}\left\{e - \gamma\sum_j \left[(1-v_j)\delta^3\nabla_x\left(\frac{(x-y_j)\cdot e}{|x-y_j|^3}\right) - 2v_j e\right]\right\}$$

which is admissible for (2.7). Here the smooth cutoff assumption explained below (4.16) is in effect and γ is given by (4.16).

Again the calculation of $\langle a^{-1}F\cdot F\rangle$ for the F of (4.22) is very lengthy. After it is completed one must choose a_0 to optimize the bound that results from (2.7). We do not give the result in general since it is a very complicated expression. We give only the expanded to $O(\rho^2)$ form analogous to (4.21) which is

$$(4.23)\qquad \frac{a^*}{a_1} \geq 1 + 3\left(\frac{K-1}{K-2}\right)\rho + \left[2.832 + .352\,\frac{K-1}{K}\right]\left(\frac{K-1}{K+2}\right)^2 \rho^2 + \dots .$$

We must now compare the bounds (4.21) and (4.23) with the expanded form of the Hashin-Shtrikman bounds (3.29). We rewrite (3.29) with the notation of (4.21)-(4.23):

$$1 + \frac{\rho}{\frac{1}{K-1} + \frac{1-\rho}{3}} \leq \frac{a^*}{a_1} \leq K + \frac{1-\rho}{\frac{1}{1-K} + \frac{\rho}{3K}} .$$

Expanding the lower bound for small ρ we get

$$(4.24)\qquad 1 + 3\left(\frac{K-1}{K+2}\right)\rho + 3\rho^2\left(\frac{K-1}{K+2}\right)^2 \rho + \dots \leq \frac{a^*}{a_1} .$$

We see therefore that the upper bound (4.21) is above the lower Hashin-Shtrikman bound but close to it while the lower bound (4.23) is actually below (4.24) but close again. Of course it is impossible to realize media with given ρ and a^* below the one given by (3.29). Our result

(4.23) reflects the particular nature of the well stirred approximation (4.4) and its limitations. We could have expressed our results (4.21), (4.23) in terms of constants that involve integrals of dipole fields, p_2 and p_3. Then the lower bound would surely have been above the Hashin-Shtrikman lower bound.

We feel that the calculations with the well-stirred approximation (4.4) are useful in showing that geometry-dependent bounds can be derived that are sharp to order ρ as $\rho \downarrow 0$ (i.e. the order ρ term in (4.21) and (4.23) are the same). This cannot be the case with geometry-independent bounds.

It would be interesting to find the form of the bounds (4.21) and (4.23) when the Percus-Yevick two-point function [19],[21] and its analogous three-point function are used.

Let us also compare the above bounds with results from selfconsistent calculations. We compare with formula (8.23) of [13]

$$(4.25) \qquad a^* = \frac{\langle a\rangle - \rho a_2\left(\dfrac{a_2-a_1}{3a^*-a_2-a_1}\right)}{1-\rho\,\dfrac{a_2-a_1}{3a^*+a_2-a_1}}$$

which for small ρ is

$$(4.26) \qquad \frac{a^*}{a_1} = 1 + 3\left(\frac{K-1}{K+2}\right)\rho + 9\left(\frac{K-1}{K+2}\right)^2\rho^2 + \ldots \; .$$

We see that the upper bound (4.21) is below (4.26) except when K is large ($K \geq 1$ here). Since (4.25) is expected to be a good approximation for the spherical inclusion composite conductor under consideration, we see again that the well-stirred approximation is quite crude when used for bounds.

The effective medium theory [11] gives for a^* the solution of

$$(4.27) \qquad \frac{a^*-a_1}{2a^*+a_1}(1-\rho) + \frac{a^*-a_2}{2a^*+a_2}\rho = 0$$

which for ρ small becomes

(4.28) $$\frac{a^*}{a_1} = 1 + 3\,\frac{K-1}{K+2}\,\rho + 9\left(\frac{K-1}{K+2}\right)^2 \frac{1}{K+2}\,\rho^2 + \ldots .$$

The Claussius-Mosotti formula [11] gives

$$\frac{a^*}{a_1} = 1 + 3\,\frac{K-1}{K+2}\,\rho + 3\left(\frac{K-1}{K+2}\right)^2 \rho^2 + \ldots$$

and Böttcher's result [17] is

$$\frac{a^*}{a_1} = 1 + 3\,\frac{K-1}{K+2}\,\rho + \rho^2\left[\frac{3(K-1)^2}{(K+2)^2} + 6\left(\frac{K-1}{K+2}\right)^3\right] + \ldots$$

All these formulas have unknown regions of validity but for small ρ, say $\rho = .1$ or $\rho = .15$, they are not too bad when compared with experimental data [20].

We conclude by noting again that computation of bounds by multiple scattering when sophisticated two and three point functions are used (and the resulting integrals are computed numerically) would produce inequalities analogous to (4.21)-(4.23), or their unexpanded form, which could be very useful.

This work was supported by the Army Research Office under Grant #DAAG29-78-G-0177.and #DAAG29-79-C-0085.

REFERENCES

[1] W. Kohler and G. Papanicolaou, Upper and lower bounds for effective conductivities, to appear.

[2] Z. Hashin and S. Shtrikman, A variational approach to the theory of effective magnetic permeability of multiphase materials, J. Applied Phys. 33 (1962), pp. 3125-3131.

[3] L.J. Walpole, On bounds for the overall elastic moduli of inhomogeneous systems I, II, J. Mech. Phys. Solids, 14 (1966), pp. 151-162 and 289-301

[4] R. Hill, Elastic properties of reinforced solids: some theoretical principles, J. Mech. Phys. Solids, 11 (1963), pp. 357-372.

[5] M. Beran, Statistical Continuum Theories, Wiley, New York, 1968.

REFERENCES, continued

[6] J.R. Willis, Bounds and self-consistent estimates for the overall properties of anisotropic composites, J. Mech. Phys. Solids, 25 (1977), pp. 185-202.

[7] P.H. Dederichs and R. Zeller, Variational treatment of elastic constants of disordered materials, Z. Physik 259 (1973), pp. 103-116.

[8] D. Bergman, The dielectric constant of a composite material -- a problem in classical physics, Phys. Rep. C, 43 (1978), pp. 377-407.

[9] G. Milton, Bounds on the transport and optical properties of a two-component composite material, J. Appl. Phys., to appear.

[10] K. Golden and G. Papanicolaou, Bounds for effective parameters of heterogeneous media by analytic continuation, to appear.

[11] R. Landauer, Electrical conductivity in inhomogeneous media. AIP Conf. Proc. #40, J. Garland and D. Tanner, editors, Am. Inst. of Phys., New York, 1978.

[12] D.A.G. Bruggeman, Berechnung verschiedener physicalischer Konstanten von heterogenen Substanzen, Annalen der Physik 22 (1935), pp. 636-679.

[13] W. Kohler and G. Papanicolaou, Some applications of the coherent potential approximation, in Multiple Scattering and Waves in Random Media, edited by Chow, Kohler and Papanicolaou, North Holland, 1981.

[14] J.B. Keller, A. Theorem on the conductivity of a composite medium, J. Math. Phys. 5 (1964), pp. 548-549.

[15] K.S. Mendelshon, Effective conductivity of a two-phase material with cylindrical phase boundaries, J. Appl. Phys. 46 (1975), pp. 917-918.

[16] K. Schulgasser, On a phase interchange relationship for composite materials, J. Math. Phys. 17 (1976), p. 378.

[17] W.F. Brown, Jr., in Handbuch der Physik 17, Springer Verlag, Berlin, 1956.

[18] C.J.F. Bötcher, Theory of Electric Polarization, Elsevier, Amsterdam, 1973.

[19] G. Stell, G.N. Patey, J.S. Hoye, Dielectric constants of fluid models: statistical mechanical theory and its quantitative implementation, Adv. in Chem. Phys. 48, J. Wiley and Sons, New York, 1981.

[20] Z. Hashin, Theory of composite materials, in Mechanics of Composite Materials, edited by F.W.Wendt, H. Lebowitz and N. Perrone, Pergamon Press, 1970, pp. 201-242.

[21] J.K. Percus and G.J. Yevick, Hard-core insertion in the many-body problem, Phys. Rev. 136 (1964), pp. B290-B296.

STRUCTURAL DESIGN OPTIMIZATION, HOMOGENIZATION AND RELAXATION OF VARIATIONAL PROBLEMS

Robert V. Kohn
Courant Institute of Mathematical Sciences

Gilbert Strang
Massachusetts Institute of Technology

We propose to survey some relationships between three distinct areas of investigation:

I) The design of engineering structures to optimize a given performance criterion.

II) The determination of "optimal" bounds for the effective properties of a composite material, in terms of the volume fractions of the components.

III) The determination of the lower-semicontinuous hull of a non-convex variational problem.

A separate section is devoted to each area. In section 1 we survey briefly - and rather idiosyncratically - some problems that have been the focus of attention in the structural optimization literature, and what is known about them. Section 2 explains why finding sharp bounds for effective equations is essentially equivalent to solving a broad class of geometry optimization questions. This idea is not new to experts; it is implicit in recent work by Murat, Tartar, Olhoff, Cheng, Lurie, and others. We believe, however, that it has not been made sufficiently explicit elsewhere, so we have tried to explain it here. Section 3 describes a program, initiated recently by the authors, for using the relaxation of non-convex variational problems as a tool in the theory of optimal design. Both sections 2 and 3 include a number of explicit open problems.

We acknowledge here with gratitude a great debt to many colleagues who have shared their expertise and insight over a period of several years. Particular thanks go to John Taylor and Niels Olhoff regarding structural optimization, to Luc Tartar and George Papanicolaou regarding homogenization, and to Roger Temam regarding convex duality.

This work has been supported in part by NSF grants MCS-7919146 A01 (to R.V.K.) and MCS81-02371 (to G.S.), and also by Army research contract DAAG29-80-K0033 (G.S.)

§1: Structural Optimization

We discuss a number of different design optimization questions, always with this basic form: given a design problem in which the shape or composition of the structure is to some extent variable, but the loads it must withstand are fixed, choose a design that "performs optimally" subject to constraints on its cost.

A simple version of this problem - in the context of buckling of columns - was considered as early as 1770 by Lagrange [1]; a fundamental paper was written by Michell in 1904 on the optimal layout of trusses [2]. Most of the now vast literature, however, developed in the past 25 years - stimulated, it seems, largely by the aerospace industry, where weight is critical, and by the availability of computers for carrying out optimization algorithms.

Reviewing this extensive literature has become an almost hopeless task; for reviews of various parts the reader may look to [3,4,5,6], especially the article by Haug in [3]. Our more limited goal here is to summarize some of the themes running though this work. Following common practice, we restrict our attention to the linearized models of behavior: linear elasticity and linearly elastoplastic behavior. Most of the questions discussed herein have analogues for more realistic, nonlinear models, but it is difficult at this point to say much about the corresponding optimization problems.

There are many different criteria that one might use to gauge the efficiency of a structure in withstanding a given load; certain of these have dominated the recent literature. For problems of linear elastostatics, one typically seeks a design that minimizes

1.1a) The work done by the load upon the structure (its "compliance" under the load);

or

1.1b) The maximum stress produced by the load (taken pointwise, using a suitable norm upon symmetric tensors);

or

1.1c) The maximum displacement produced by the load.

For elasto-plastic materials (the model of choice for ductile metals) a corresponding performance criterion is to maximize

1.1d) The limit multiplier of the load (by definition, the largest multiple of the load which the structure can withstand without developing global plastic flow).

In addition to these static performance criteria, there are many important problems relating to eigenvalues of linearly elastic structures, interpreted as buckling modes or as modes of vibration. One might attempt to maximize the lowest eigenvalue; or, in a vibration problem, to create a large gap in the spectrum near a particular frequency. For structures made from elastoplastic materials, dynamic optimization problems arise also in avoiding fatigue under cyclic loading conditions.

In many problems - particularly those of aerospace design - the "cost" of the structure may be interpreted as the total weight. One might seek either designs that perform optimally for fixed cost, or ones that minimize the cost for fixed strength.

What kind of "structures" do we have in mind? Perhaps the simplest problem conceptually is that of fully three dimensional <u>material distribution optimization</u>: the data are

1.2a) a domain $\Omega \subset \mathbb{R}^3$

1.2b) a finite collection of linearly elastic materials, each homogeneous, characterized by the stress-strain laws $\sigma = A^{(r)}(\varepsilon)$ $(r=1,\ldots,N)$ and by costs per unit volume $c_r \geq 0\,(r=1,\ldots,N)$.

Here $\varepsilon = \varepsilon(u) = \frac{1}{2}(\nabla u + \nabla u^t)$ is the linear strain associated to a deformation $u:\Omega \to \mathbb{R}^3$, and each $A^{(r)}$ is a fourth-rank tensor, which must be positive and symmetric when viewed as a linear map upon second order symmetric matrices. The class of "structures" is then the set of measurable partitions of Ω into N sets, $\Omega = \Omega_1 \cup \ldots \cup \Omega_N$. To each such partition one associates an operator with piecewise constant coefficients:

$$Lu = \mathrm{div}(\sigma(u)) = \left\{ \sum_{j=1}^{3} \frac{\partial \sigma_{ij}}{\partial x_j} \right\}_{i=1}^{3}$$

$$(1.3) \qquad \sigma(u) = A(x) \circ \varepsilon(u)$$

$$A(x) = A^{(r)} \quad \text{if } x \in \Omega_r$$

representing the elliptic system that characterizes the elastostatics of the corresponding structure. A cost constraint has the form

$$(1.4) \qquad \sum_{r=1}^{N} c_r \,\mathrm{Vol}(\Omega_r) \leq \text{constant} \quad .$$

Given a body load $f:\Omega \to \mathbb{R}^3$ and a boundary load $F:\partial\Omega \to \mathbb{R}^3$ such that $\int_\Omega f + \int_{\partial\Omega} F = 0$, the corresponding displacement is the solution u to

$$L\bar{u} + f = 0 \text{ on } \Omega, \qquad \sigma(\bar{u})\cdot\nu = F \qquad \text{on } \partial\Omega,$$

where ν is the unit normal on $\partial\Omega$; the <u>compliance</u> under the load is

$$\int_{\partial\Omega} \langle \bar{u}, F \rangle + \int_{\Omega} \langle \bar{u}, f \rangle = \int_{\Omega} \langle \bar{\sigma}, \bar{\varepsilon} \rangle$$

where $\bar{\sigma} = \sigma(\bar{u})$, $\bar{\varepsilon} = \varepsilon(u)$ are the stress and strain of the solution.

If one were to take $A^{(0)} = 0$ in (1.2b), the problem would become one of shape optimization: the "zero'th material" is actually the absence of material, and one seeks not only the distribution of material but also the distribution of holes in the fixed domain Ω. The hole boundaries carry the natural boundary condition $\sigma \cdot \nu = 0$: This is the most natural choice for structural design problems. (One could, of course, impose other boundary conditions; doing so would alter the problem radically.)

The reader will have no difficulty formulating analogous classes of "structures" in the context of two-dimensional models and for other types of material behavior. In case the available materials vary continuously (instead of discretely as in 1.2b) the piecewise-constant stress strain law (1.3) must be replaced by a variable law

$$(1.3)' \qquad \sigma = A(x) \circ \varepsilon$$

where $x \mapsto A(x)$ is a measurable, tensor-valued function on Ω with values in the set of available stress-strain relations: now it is the L^∞ function $A(x)$, rather than a partition of Ω, which is the control parameter. The cost constraint (1.4) would be replaced by a condition

$$(1.4)' \qquad \int_{\Omega} G(A(x))dx \leq \text{constant}$$

where $A \mapsto G(A)$ assigns to each stress-strain law its cost per unit volume.

Although the fully 3-dimensional problems are in a sense the most fundamental ones, they are by no means the most studied. Significant work has been done, both on three-dimensional problems and on two-dimensional analogues relating to plane problems and rods in torsion, e.g. [7-12]; but the bulk of the literature addresses problems that are in various senses "reduced". For example, much effort has been devoted to optimizing the thickness distribution of beams, plates, or shells, by working directly with the corresponding fourth-order thin-structure equations. This approach has been particularly successful in one-dimensional problems - e.g. [13] - and in certain two-dimensional contexts where the limit multiplier optimization problem carries a convex structure - see [14]. In some cases, however, such analyses yield "optimal" thickness distributions with vanishing cross-sections or rapidly varying thickness. In these cases the true three-dimensional behavior is not correctly represented by the thin-structure equations.

We must add to this list of types of structures the "truss-like continua" of Michell and Prager [5] and the "grillage-like continua" of Prager and Rozvany [6]. These structures represent an idealized limit of trusses in space or stiffeners in the plane - limits, it must be said, in a sense that has never been made fully precise. The optimal layout problem in this context is among the oldest and most studied problems in the field; even here, however, much remains to be done regarding the description and effective computation of optimal structures.

The state of the art in the solution of geometry optimization problems may be summarized as follows. There are essentially three tools available: convex analysis, gradient-flow optimization, and the direct solution of first-order necessary conditions for optimality. For the most part, all these methods are restricted in their utility to integral objectives such as (1.1a,d); the sup-norm objectives (1.1b,c) have been treated successfully only in special cases (e.g.[15]).

Convex analysis is relevant to certain problems - particularly regarding limit multiplier optimization - in which the admissible materials vary continuously, and their "cost" is in some sense a convex function of their strength. A typical example is this: suppose that $\|\cdot\|$ is a norm on stresses, and one has available for each $\lambda > 0$ a material with plastic yield criterion "$\|\sigma\| = \lambda$", at cost $G(\lambda)$. Given a domain Ω and a boundary load f, the minimum cost of a structure occupying domain Ω with limit multiplier greater than or equal to one is

$$\inf_{\substack{\operatorname{div}\sigma=0 \\ \sigma\cdot\nu=f \text{ on } \partial\Omega}} \int G(\|\sigma\|) .$$

If $\lambda \mapsto G(\lambda)$ is increasing and convex, this is a convex variational problem.

Most structural design optimization problems do not, however, have such a formulation - for example, a problem that deals with a discrete set of materials can not be cast in this form. (We shall see in section 3 that certain such problems may be equivalent, via relaxation, to convex ones.) In the absence of a convex structure, one may resort to the gradient - flow method, sometimes also called "optimal redesign" ([12],[16]). Given any design, a sensitivity analysis determines whether it is a stationary point in design-space for the relevant optimization problem; if not, then by following the gradient (in design space) of the objective one can iteratively improve it. Unfortunately, such methods often require a great deal of computation time, and they sometimes place implicit constraints on the geometry of the structure; on the other hand they are of great utility, since in practice it may be more important to improve a known design slightly than to find the true optimum of an abstract problem.

The third tool is the direct solution of first-order conditions. It is usually not difficult to formulate necessary conditions for a design to be optimal [3]. In the context of (1.2) - (1.4), for example, the condition that a partition be optimal for compliance under the loads f,F amounts to an extra "free boundary condition" for the displacement $\bar{u}$ at the partition boundaries $\{\partial\Omega_r\}_{r=1}^N$. There is, of course, no guarantee that this free boundary problem should have a solution. Nonetheless, the direct solution of these stationarity conditions - often called "optimality criteria" in the structures literature - has often proven fruitful. Important developments in mathematical programming have been achieved in this direction - see e.g. [17] and the "fixed-point method" described in [12].

A critical gap in all this work, however, has been the <u>absence of appropriate existence and convergence theorems</u>. While existence may be trivial for the discrete version of a given optimization problem, no convergence analysis can be expected unless one understands the behavior of the underlying continuous case; such analyses have been notably lacking. For the "generalized structures" represented by truss-like or grillage-like continua, optima apparently do exist - though even this has yet to be studied with a satisfactory level of mathematical rigor. For many other problems, however, such as the shape or material distribution problem discussed earlier, <u>there need not exist an optimal geometry</u>, at least not in the sense so far described. This realization, only gradually reached by the structural optimization community ([8,11,18 19,41,42],see also the "nonexistence" proofs in [20]), is fundamental. In order to design computationally efficient algorithms for geometry optimization, one must first analyze the continuous problem, to determine what "generalized structures" or "effective materials" arise naturally there; then the result of <u>this</u> analysis may be approximated by discrete models, for instance using finite elements. Only by following such a procedure can rapidly oscillatory solutions - and corresponding numerical difficulties-be avoided even in the discrete problems.

§2. Homogenization and "optimal" bounds.

From the perspective of the theory of homogenization, it is clear why optimal geometries, as understood in section 1, need not exist: the problem is simply that the systems defined by (1.3), ranging over all measurable partitions of Ω, are <u>not closed</u> in the appropriate sense. This fact is precisely the starting point of the subject of homogenization. Given a sequence of such systems, corresponding perhaps to increasingly complicated geometries, the theory guarantees the existence of a subsequence for which the displacements converge weakly in $H^1(\Omega)$; the limit solves a new elliptic system, whose symmetries identify it as the system corresponding to an anisotropic, inhomogeneous, linearly elastic material. The subsequence can be chosen so that this convergence is uniform for loads f, F ranging in compact subsets of $H^{-1}(\Omega)$ and $H^{-1/2}(\partial\Omega)$, respectively, and so that the compliances of the solutions converge. The limiting system is called the "effective equation" or the limit of the subsequence in the sense of "homogenization" or "G-convergence". A recent review of the theory related to this sort of limiting procedure is [22]; the existence of G-limits is proven in [21]. The main point here, however, is simply this: although the original sequence arose from partitions of Ω as in (1.2,1.3), the G-limit need not correspond to any measurable partition of Ω.

Because compliances converge under passage to the G-limit, a solution to the problem of material distribution for minimal compliance will certainly exist if one extends the class of admissible structures to include the G-closure of the original class. The same is true, in fact, for any design that is preserved under G-convergence, for example the sup-norm objectives (1.1b,c). This enlargement of the class of structures is perfectly reasonable from a physical point of view: it represents the recognition that all composites which can be made from the materials on hand should be included among the admissible structures.

For problems involving shape optimization, thickness optimization of plates and shells, or elasto-plastic materials we can not be quite so precise: in these contexts the general form of the effective equations is not known. (One can, however, analyze the case of periodic local variation for some of these cases, see [23], [24], [25], [39]). The main point, however, is unaltered: in all such problems, one must allow for the possible introduction of new "effective materials" representing composites or generalized structures that can be approximated by the original class.

We have not yet demonstrated that homogenization will occur in the sort of structural optimization problem discussed in section 1; so far we have only explained why it might occur. It is not hard to construct <u>ad hoc</u> problems, for example minimizing $\int_\Omega |u-u_0|^2$ for suitable u_0, that display homogenization (see [20]); but proving that it occurs in a "natural" problem requires a finer analysis. This has so far been carried out primarily in two-dimensional, scalar equation model problems. The earliest, to our current knowledge, is in [8]: there one finds an analysis of the first-order optimality conditions for the distribution of material in a plane domain, so as to maximize the torsional rigidity of the corresponding rod, under various cost contraints. For certain such problems this procedure leads to a "contradiction": the reason, explained only somewhat vaguely in [8], is that for those formulations the optimal configuration is a homogenized one. The recent paper [11] reexamines the problem, extending the class of admissible structures to laminar composites (still not the full G-closure), and shows that the new first-order optimality conditions so obtained can indeed be satisfied.

Recent work by the authors ([26],[29]; see also section 3 below) deals with some model problems involving the distribution of holes or of several materials in a plane domain, to optimize the torsional rigidity, compliance, or limit multiplier of the corresponding rod under given loading. For these problems one finds that homogenization almost always occurs; the optimal structures are a sort of fibered composite.

Homogenization has also been detected in problems involving the thickness distribution of an elastic plate. In numerical work [18], and in the associated analyses [19] and [11], Cheng, Olhoff, and Lurie, et al. have shown that for certain compliance optimization problems under transverse loads, with constraints on the maximum and minimum thickness and on the total weight, an "optimal" structure must be a generalized one.

There are, it should be noted, apparently similar problems where homogenization does not occur: for example, the holes constructed in [28] represent, according to the results of [15], configurations that minimize the stress concentrations in certain problems of plane stress. One does not yet know how to predict, for a particular problem, whether or not homogenization will occur. There is, however, one characteristic shared by many of the problems now known to exhibit it: in them, the "unknown boundaries" carry Neumann or natural transmission boundary conditions. Another class of geometry problems is often studied in the context of heat conduction, electrostatic capacity, and cavitation: these problems place a Dirichlet condition on the "unknown boundary". Existence theorems (without homogenization) for some problems of this type may be found in [29], [30], and the references therein.

We can now make explicit the relationship between "optimal bounds" for effective materials and the optimization of structural geometry.

To find optimal configurations in a problem where homogenization can occur, one needs to know

i) The G-closure of the original class of equations (without cost constraints)

ii) The minimum cost per unit volume of each element of this G-closure.

It certainly sufficies for this to know sharp bounds (traditionally called "optimal bounds") for the relevant effective equations, as a function of relative volume fractions. Such bounds would specify, for given $\{\rho_j\}_{j=1}^N$ $(0\leq\rho_j\leq 1, \sum\rho_j=1)$, which anisotropic, homogeneous, elastic effective materials can be obtained using volume fraction ρ_j of the jth component.

In special cases one might be able to use an invariance in the problem to reduce the information required. For example, according to the analysis in [31], one could maximize the torsional rigidity of a rod by knowing the minimum cost per unit volume, as a function of the <u>largest eigenvalue</u> of the "stress-strain law" (which in this context is a symmetric 2×2 matrix).

The idea that one can create new effective materials by mixing given ones on a microscopic scale is certainly not new. The problem of determining optimal bounds, too, is an old one: major contributions have been made in recent years by Hashin, Shtrikman, Bergman, Beran, Hill, and Willis, to name but a few. For the study of geometry optimization problems, however, it is very important that no geometric constraints - such as periodicity or randomness - be placed upon the admissible configurations; and it is crucial that the bounds be attainable.

The best tool for proving fully rigorous bounds without geometric hypotheses seems to be the machinery of compensated compactness introduced by Murat and Tartar [33,34,35]. Recent work by Tartar (presented at this conference but not yet published) includes fully rigorous proofs of such bounds, and proofs of their attainability, in the case of the scalar Laplace equation, for mixtures of two isotropic dielectrics. Further results on more complex problems are sure to come.

Much remains to be done, however, to bridge the gap between homogenization theory and structural optimization practice. The following list of open problems represents a summary of part of the job that remains

1) Find sharp bounds for effective equations in the context of plane strain and three dimensional elastostatics.

2) Explore the effect on the optimization problem of replacing the correct bounds on effective equations by approximate ones (which might, for example, be produced

by an algorithm for solving question 1).

3) Determine the effective equations for composites that include holes or elasto-plastic inclusions; also, for the variational inequality problems that arise from the optimal design of bearings, see e.g. [36].

4) What regularity can one expect for the coefficients in the effective equation that represents an optimal configuration?

5) Design an efficient finite element scheme for using sharp effective-equation bounds to compute locally optimal configurations.

6) Develop a mathematically rigorous theory of Michell's "truss-like continua" and Prager-Rozvany's "grillage-like continua," including a proof of the existence and regularity of optimal geometries and a description of their behavior.

§3. Structural Optimization and Relaxation.

We have recently introduced another method for answering certain questions of geometry optimization. Our approach, which involves the relaxation of an appropriate nonconvex variational problem, is less general than the one sketched in section 2; when it applies, however, it gives more explicit information about the optimal configurations.

In principle the approach applies to a variety of elasticity problems, involving either shape optimization or optimal material distribution, in the presence of compliance constraints under each of a finite number of loads. It applies equally well to a plasticity problem in which the constraints are on the limit multipliers.

In practice, however, we have been able to carry it out only for a much more restricted class of problems: those modeled by a scalar equation, in the presence of a single load. Even these simple cases lead to interesting results; a full discussion is currently in preparation, and examples may be found in [26,27]. Here we present a brief and nontechnical exposition of the method in a special case, the <u>shape optimization of a rod in antiplane shear</u>. With that discussion as motivation, we then describe some open problems whose resolution would allow the extension of the method to a more practical context.

Let Ω be a bounded domain in $\mathbb{R}^2$, and $f: \partial\Omega \to \mathbb{R}$ with $\int_{\partial\Omega} f = 0$. We imagine that Ω is the cross-section of an infinite elastic rod, with load f directed lengthwise along its boundary, uniformly along the length of the rod. (See figure 1.) After some reductions, and ignoring

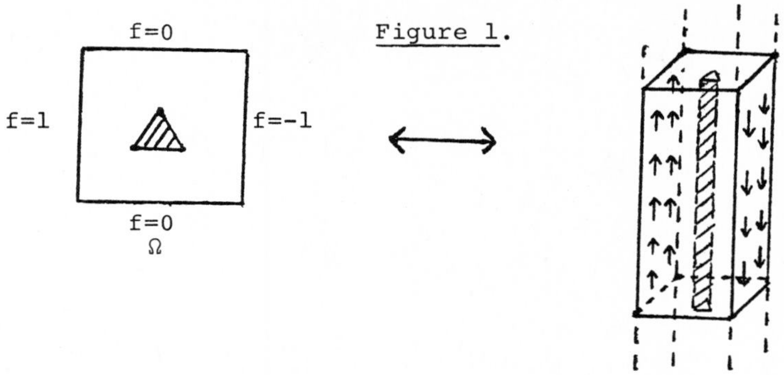

Figure 1.

elastic constants, the compliance of the rod per unit length can be expressed as

$$c(\Omega) = \inf_{\sigma \in X(f)} \int_\Omega |\sigma|^2 ,$$

where

$$X(f) = \{\sigma \in L^2(\Omega; \mathbb{R}^2): \quad \operatorname{div} \sigma = 0 \text{ on } \Omega,\ \sigma\cdot\nu = f \text{ on } \partial\Omega.\}.$$

We assume that Ω, f are regular enough that X(f) be nonempty.

The shape-optimization problem is this: how should one remove material from the interior of Ω so as to increase the compliance (i.e. weaken the structure) as little as possible? Whenever a hole (a closed set) $H \subset \Omega$ is removed, the newly formed boundary ∂H is to be unloaded, so the resulting structure has compliance

$$c(\Omega\sim H) = \inf_{\substack{\sigma \in X(f) \\ \sigma = 0 \text{ a.e. on } H}} \int_\Omega |\sigma|^2 .$$

The constraint "$\sigma = 0$ a.e. on H" serves to impose the condition $\sigma\cdot\nu = 0$ on ∂H. Introducing a parameter $\alpha > 0$, we consider the modified problem

$$(3.1) \qquad \inf_{H \subset \Omega} \quad \text{Area}(H) + \alpha\, c(\Omega\sim H) ,$$

where H ranges over closed subsets of Ω. If there is an extremal set H_0 for (3.1), then it certainly must minimize $c(\Omega\sim H)$ for fixed Area(H);

This would be an optimal geometry in the conventional sense. If there is no extremal, a minimizing sequence for (3.1) must in any case consist of sets that nearly minimize $c(\Omega\sim H)$ for their areas, and we can expect to find an optimal "generalized structure." (One can also treat α formally as a Lagrange multiplier, in order to find minimum - area geometries with fixed compliance; this more technical aspect will be presented elsewhere.)

Ignoring the fact that $\{x:\sigma(x)=0\}$ need not be closed for arbitrary $\sigma \in X(f)$ (this technical point, too, will be addressed elsewhere), one finds that (3.1) is essentially the same as

$$(3.2) \qquad \inf_{\sigma\in X(f)} \int_\Omega (1_{\{\sigma\neq 0\}} + \alpha|\sigma|^2) \quad ,$$

where 1_Y denotes the characteristic function of a set $Y \subset \Omega$. This, then is the starting point of our method: the shape optimization problem (3.1 can be reformulated as a variational problem over $X(f)$ with the discontinuous integrand

$$G_\alpha(\sigma) = \begin{cases} 1 + \alpha|\sigma|^2 & \text{if } \sigma \neq 0 \\ 0 & \text{if } \sigma = 0 \end{cases}$$

If in (3.2) σ ranged over all L^2 vector fields then the infimum would be unchanged by replacing G_α with its convex hull

$$F_\alpha(\sigma) = \begin{cases} 1 + \alpha|\sigma|^2 & |\sigma| \geq \alpha^{-1/2} \\ 2\alpha^{1/2}|\sigma| & |\sigma| \leq \alpha^{-1/2} \end{cases}$$

(see figure 2). Being convex, $\int F_\alpha$ is weakly lower semicontinuous, so it has a minimizer in $X(f)$; given any minimizer for $\int F_\alpha$, one can construct a minimizing sequence for $\int G_\alpha$ by introducing appropriate small-scale oscillations in σ: this passage from $\int G_\alpha$ to its lower semicontinuous hull $\int F_\alpha$ is called relaxation (see e.g. [38]). In the present context more care is required: we must respect the constraints $\sigma\cdot\nu=f$ and $\operatorname{div} \sigma = 0$. However, using methods related to those of [38, chapter X] one may show that the constraints do not affect the relaxation. Hence (3.2) is equivalent to the relaxed problem

$$(3.3) \qquad \inf_{\sigma\in X(f)} \int_\Omega F_\alpha(\sigma) \quad .$$

A more careful examination of the relaxation procedure shows that to any extremal σ_0 of (3.3) one can associate a minimizing sequence

$\{\sigma_i\}_{i=1}^{\infty}$ for (3.2) by choosing - roughly speaking - σ_i parallel to σ_0, with $\sigma_i = \sigma_0$ where $|\sigma_0| \geq \alpha^{-1/2}$ or $|\sigma_0| = 0$, but $|\sigma_i|$ oscillating rapidly between 0 and $\alpha^{-1/2}$ where $0 < |\sigma_0| < \alpha^{-1/2}$; the sets where $\sigma_i = 0$ constitutes a minimizing sequence for (3.1). It is not hard to construct examples where $0 < |\sigma_0| < \alpha^{-1/2}$ on a set of positive measure; for these problems an optimal geometry exists only in a generalized sense - and we see that the appropriate class of generalized structures is a sort of "fibred continuum."

The analogous plasticity problem is the optimal removal of area from Ω given a constraint on the limit multiplier of the load f. Making appropriate reductions and again ignoring a material constant, the limit multiplier of the load upon $\Omega \sim H$ is

$$L(\Omega\sim H) = \sup\{t: \exists\sigma\in X(f),\ \sigma = 0 \text{ a.e. on } H,\ |\sigma| \leq t^{-1}\}.$$

Therefore the geometry problem

$$(3.4) \qquad \inf_{\substack{H\subset\Omega \\ L(\Omega\sim H)\geq\lambda}} \text{Area } (\Omega\sim H)$$

reduces to the variational problem

$$(3.5) \qquad \inf_{\substack{\sigma\in X_f \\ |\sigma|\leq\lambda^{-1}}} \int_\Omega 1_{\{\sigma\neq 0\}}$$

Once again, the shape optimization problem can be solved by relaxing (3.5); this time the relaxed problem is

$$(3.6) \qquad \inf_{\substack{\sigma\in X_f \\ |\sigma|\leq\lambda^{-1}}} \lambda\int_\Omega |\sigma| \quad ,$$

and the appropriate class of generalized structures is, once again, the "fibred continua". (See figure 3).

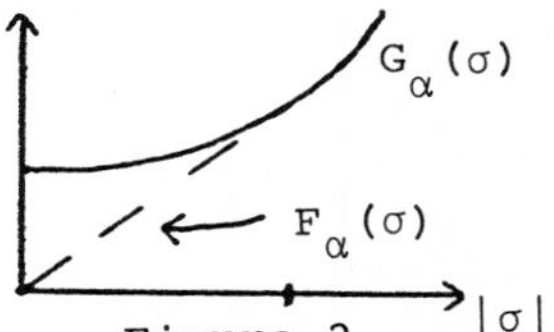

Figure 2.

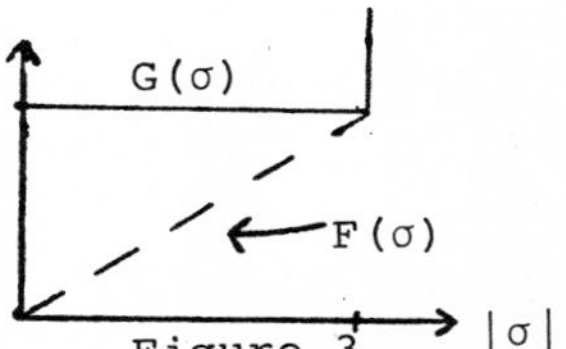

Figure 3.

One can say quite a bit more about the relaxed problems (3.3) and (3.6), using convex duality and other tools; proofs and further analysis, as well as a treatment of other model problems by a similar method, will be presented elsewhere. Here we focus instead on how to extend the method to a wider class of problems.

Our approach does seem to require that the constraints be closely related to the variational structure of the analysis problem; for the moment, it seems restricted (unlike the approach of section 2) to compliance problems in elasticity and to limit analysis problems in plasticity. There is, however, no difficulty in formulating analogues of (3.2) or (3.5) in a fully three-dimensional context, and in the presence of any finite number of loads. The obstacle to solving such problems lies in the computation of the correct relaxed problem. (We are convinced that it will not, in most cases, be the convex hull of the discontinuous integrand!) Here is a list of four specific versions of this problem.

1) (Antiplane shear of an elastic rod, as above, but under several different loads.) Let $f_1,\dots,f_N\colon \partial\Omega \to \mathbb{R}$ with $\int_{\partial\Omega} f_i = 0$ for each i, and fix $\{\alpha_i\}_{i=1}^N$, $\alpha_i \geq 0$. Find the relaxation of

$$\inf_{\sigma_i \in X(f_i)} \int_\Omega \chi(\sigma_1,\dots,\sigma_N) + \sum_{i=1}^N \alpha_i |\sigma_i|^2$$

where

$$\chi(\sigma_1,\dots,\sigma_N) = \begin{cases} 1 & \text{if } \sigma_i \neq 0 \text{ for some } i \\ 0 & \text{if } \sigma_1 = \dots = \sigma_N = 0. \end{cases}$$

2) (The same problem in a limit analysis formulation): for $\{f_i\}_{i=1}^N$ as above, find the relaxation of

$$\inf_{\substack{\sigma_i \in X(f_i) \\ |\sigma_i| \leq 1}} \int_\Omega \chi(\sigma_1,\dots,\sigma_N)$$

3) (Three-dimensional shape optimization, minimizing a linear combination of volume and compliance.) Let E denote the space of 3×3 symmetric tensors, and let $A\colon E \to E$ be an elastic stress-strain law, with $\|\sigma\|^2 = \langle\sigma, A^{-1}\sigma\rangle$ the corresponding "strain energy" norm on stresses. Let

$\Omega \subset \mathbb{R}^3$, $f:\partial\Omega \to \mathbb{R}^3$, and assume $\int_{\partial\Omega} f = 0$. For $\alpha > 0$, find the relaxation of

$$\inf_{\sigma \in X'(f)} \int_\Omega 1_{\{\sigma \neq 0\}} + \alpha\|\sigma\|^2 ,$$

where $X'(f) = \{\sigma \in L^2(\Omega;E): \text{div}\,\sigma = 0 \text{ on } \Omega, \sigma\cdot\nu = f \text{ on } \partial\Omega\}$.

4) (The analogue of 3, for plastic limit analysis.) Let K be a convex subset of E containing a neighborhood of the origin. Find the relaxation of

$$\inf_{\substack{\sigma \in X'(f) \\ \sigma(x)\in K \text{ a.e.}}} \int_\Omega 1_{\{\sigma \neq 0\}} ,$$

where Ω, f and $X'(f)$ are as in (3).

References

1) Lagrange, J.L., "Sur la figure des colonnes", Miscellanea Taurinensia V, 1770-1773, pg. 123.

2) Michell, A.G.M., "The limits of economy of material in frame structures", Phil. Mag. S6, Vol 8, No.47, pp 589-597.

3) Haug, E.J., and Cea, J., Proceedings of NATO ASI on optimization of distributed parameter structures, Iowa City, 1980. Sithoff and Nordhoff, to appear.

4) Sawzuk, A. and Mroz, Z., Optimization in structural design (Proceedings of 1973 IUTAM symposium in Warsaw, Poland), Springer-Verlag, 1975.

5) Prager, W., Introduction to structural optimization, International Centre for Mechanical Science, Udine, Courses and Lectures no. 212, Springer-Verlag, 1974.

6) Rozvany, G.I.N., Optimal design of flexural systems, Pergamon Press, 1976.

7) Cea, J. and Malanowski, K., "An example of a max-min problem in partial differential equations", SIAM J. Control vol 8, 1970, pp. 305-316.

8) Klosowicz, B. and Lurie, K.A., "On the optimal nonhomogeneity of a torsional elastic bar", Arch. of Mechanics, vol 24, 1971, pp 239-249.

9) Mroz, Z., "Limit analysis of plastic structures subject to boundary variations", Arch. Mech. Stos. vol 15, 1963, pp. 63-75.

10) Zavelani, A., Maier, G., and Binda, L., "Shape optimization of plastic structures by zero-one programming", in IUTAM Warsaw Symposium, 1973, see reference 3.

11) Lurie, K.A., Fedorov, A.V., and Cherkaev, A.V., "Regularization of optimal design problems for bars and plates and elimination of con-

tradictions within the necessary conditions of optimality", Journal of Opt. Th. and Appl., to appear, 1982

12) Cea, J., "Shape optimal design: problems and numerical methods", in Proc. of NATO-ASI on Optimization of Distributed Parameter Structure Iowa City, 1980

13) Olhoff, N., "Optimization of columns against buckling" in Proceeding of NATO-ASI on optimization of distributed parameter structures, 1980.

14) Cinquini, C. and Mercier, B., "Minimal cost and elastoplastic structures", Meccanica vol 11, no 4, 1976, pp 219-226.

15) Wheeler, L., "On the role of constant - stress surfaces in the problem of minimizing elastic stress concentration.", International J. of Solids and Structures vol 12, 1967, pp 779-789.

16) Olhoff, N. and Taylor, J.E., "On optimal structural remodeling", J. Opt. Theory and Applic. Vol 27, 1979, pp 571-582.

17) Fleury, C. and Schmit, L.A., "Primal and dual methods in structural optimization," J. Structural Div. ASCE vol 106, 1980, pp 1117-1133.

18) Cheng, K.T. and Olhoff, N., "An investigation concerning the optimal design of solid elastic plates", Int. J. of Solids and Structures, to appear.

19) Cheng, K.T. and Olhoff, N., "Regularized formulation for optimal design of axisymmetric plates", to appear.

20) Murat, F., "Contre-examples pour divers problemes ou le control intervient dans les coefficients", Annali di Mat. Pura ed Appl. Ser vol 112-113, 1977.

21) Simon, L., "On G-convergence of elliptic operators," Indiana Univ. Math. Journal vol 28, pp 587-594.

22) De Giorgi, E., "Convergence problems for functionals and operators", in Proc. of the International Meeting on Recent Methods in Nonlinear Analysis, Rome, 1978; De Giorgi, Magenes, & Mosco editors, Pitagora Editrice, Bologna, 1980.

23) Duvaut, G., "Comportement microscopique d'une plaque perforée périodiquement", to appear.

24) Cioranescu, D. and Saint Jean Paulin, J., "Homogenization dans des ouverts à cavités," C.R. Acad. Sci. Paris A, vol 284 (1977) pp 857-860.

25) Carbone, L., "Sur un problème d'homogénéisation avec des constraints sur le gradient", J. Math. Pures et Appl. 58, 1979, pp 275-297.

26) Strang, G. and Kohn, R., "Optimal design of cylinders in shear", to appear in proceedings of the 1981 MAFELAP Conference, Brunel University.

27) Kohn, R. and Strang, G., "Optimal design for torsional rigidity", to appear in proceedings of the conference on Mixed and Hybrid Methods in Finite Element Methods, Atlanta, 1981.

28) Cherepanov, G.P., "Inverse problems of the plane theory of elasticit

P.M.M. vol 38, no 6, 1974, pp 963-979.

29) Acker, A., "Interior free boundary problems for the Laplace equation", Arch. Rat. Mech. Anal. vol 75, 1981, pp 157-168.

30) Alt, H.W. and Caffarelli, L.A., "Existence and regularity for a minimum problem with free boundary", J. Riene Angew. Math. 325 (1981) pg 105.

31) Banichuk, N., Doklady Akad. Nauk USSR vol 242, pp 1042-1045, 1978.

32) Jouron, C., "Sur un problème d'optimisation où la constrainte porte sur la fréquence fondamentale", RAIRO Analyse Numerique vol 12, 1978, pp 349-376.

33) Murat, F., "Compacité par compensation", Ann. Scuola Norm.Sup Pisa vol 5, 1978, pp. 489-507.

34) Murat, F., "Compacite par compensation II", in proc. of the International Meeting on Recent Methods in Nonlinear Analysis, De Giorgi, Magenes, Mosco editors, Pitagora Editrice, Bologna, 1980.

35) Tartar, L., "Estimation de coefficients homogeneises", Springer Lect. Notes in Math. vol 704, pp 364-373.

36) Benedict, R.L., "Optimal design for elastic bodies in contact", in proceedings of the NATO-ASI on optimization of distributed parameter structures, (ref.3).

37) Banichuk, N.V.; Kartvelishvili, V.M.; and Mironov, A.A., "Optimization problems with local performance criteria in the theory of plate bending", Mechanics of Solids, 1978, no 1.

38) Ekeland, I. and Temam, R., Convex Analysis and Variational Problems, North-Holland, 1976.

39) Vogelius, M.; Kohn, R.; Papanicolaou, G., "Effective equations for plates and beams with rapidly varying thickness", to appear.

40) Olhoff, N.; Lurie, K.A.; Cherkaev, A.V.; Fedorov, A.V.; "Sliding Regimes and Anisotropy in Optimal Design of Vibrating Axisymmetric Plates", Int. J. Solids Structures, to appear.

41) Raitum, U.E., "On optimal control problems for linear elliptic equations", Soviet Math. Dokl. Vol 20, pp 129-132, 1979.

42) Raitum, U.E., "The extension of extremal problems connected with a linear elliptic equation", Soviet Math. Dokl. Vol 19, pp 1342-1345, 1978.

Coherent Medium Approach to Hopping Conduction

M. Lax
Department of Physics, The City College of the City University of New York, New York, N.Y. 10031 and Bell Laboratories, Murray Hill, N.J. 07974

T. Odagaki
Department of Physics, The City College of the City University of New York, New York, N.Y. 10031

Contents

1. Introduction

Stochastic transport in disordered or amorphous materials consists of hopping of carriers from one localized center to another with the assistance of other degrees of freedom such as phonons. An ac conductivity due to the hopping process was first observed in doped silicon by Pollak and Geballe[1] in 1961. The typical and a striking feature of the ac conductivity is that the observed data at various temperatures and frequencies for a given doping level can be fitted to a single curve by a relevant scaling of both the conductivity and the frequency, and the so-called ac part $(\mathrm{Real}\{\sigma(\omega)-\sigma(0)\})$ of the conductivity shows a power law dependence on the frequency at low frequencies. [See later Figs. 14.1 and 14.2.] The electric transport of this system is believed to be caused by electron hopping from a neutral donor to a donor ionized by a compensating acceptor. [See Fig. 1.1.]

A similar mechanism of conduction has been employed to explain low-frequency ac conductivity in many other systems such as doped VO_2,[2] Na-βAl_2O_3,[3] MoO_3,[4] Te-doped Se,[5] and spinel-type MnCoNiCu complex oxides[6], amorphous chalcogenides[5,7] and so on. In each case, carriers (electrons or ions) are assumed to move from one localized center to another by a hopping process. Environmental fluctuations and the assistance of other degrees of freedom makes the site-to-site motion of carriers stochastic.

Now, let us introduce a conditional probability $P(\mathbf{s},t\,|\,\mathbf{s}_0,0)$ of finding a moving carrier at a point $\mathbf{s}$ at time t if it was at a point $\mathbf{s}_0$ at t=0. It is natural to assume that $P(\mathbf{s},t\,|\,\mathbf{s}_0,0)$ obeys a random walk equation

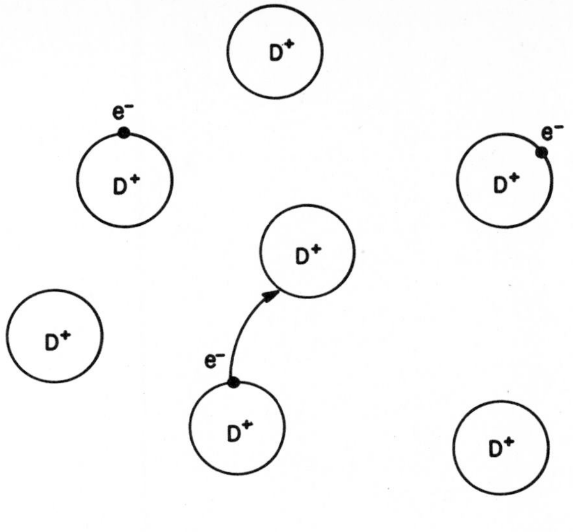

Fig. 1.1. Schematic illustration of hopping motion of an electron in a doped semiconductor. An electron jumps from a neutral donor to a donor ionized by a compensating acceptor.

$$\frac{\partial P(\mathbf{s},t)}{\partial t} = -\Gamma_{\mathbf{s}} P(\mathbf{s},t) + \sum_{\mathbf{s}'\neq\mathbf{s}} w_{\mathbf{ss}'} P(\mathbf{s}',t) \ , \tag{1.1}$$

where the total decay rate out of $\mathbf{s}$ is given by

$$\Gamma_{\mathbf{s}} = \sum_{\mathbf{s}'\neq\mathbf{s}} w_{\mathbf{s}'\mathbf{s}} \tag{1.2}$$

and for simplicity of notation we have omitted listing the initial conditions. The definition of $\Gamma_{\mathbf{s}}$ Eq. (1.2) insures conservation of probability $\sum_{\mathbf{s}} P(\mathbf{s},t\,|\,\mathbf{s}_0,0)=1$. We do not include any traps in the present paper. The jump rate $w_{\mathbf{s}'\mathbf{s}}$ from site $\mathbf{s}$ to $\mathbf{s}'$ is assumed to have an appropriate dependence on the distance $|\mathbf{s}'-\mathbf{s}|$ between two sites. As a simplified or tractable model for positionally random systems, hopping conduction on lattices will be studied, where the hopping sites $\{\mathbf{s}\}$ form a regular array of lattice points and a suitable distribution of jump rates is introduced to replace randomness in jump rates induced by positional disorder. In a sense, we study a random walk on a lattice in a random environment.

For later convenience, we introduce the Laplace transform of the probability $P(\mathbf{s},t\,|\,\mathbf{s}_0,0)$

$$\tilde{P}(\mathbf{s},u\,|\,\mathbf{s}_0) = \int_0^\infty e^{-ut} P(\mathbf{s},t\,|\,\mathbf{s}_0,0)\, dt \tag{1.3}$$

which obeys

$$(u+\Gamma_{\mathbf{s}})\tilde{P}(\mathbf{s},u\,|\,\mathbf{s}_0) - \sum_{\mathbf{s}'\neq\mathbf{s}} w_{\mathbf{ss}'}\, \tilde{P}(\mathbf{s}',u\,|\,\mathbf{s}_0) = \delta(\mathbf{s},\mathbf{s}_0) \ , \tag{1.4}$$

$\delta(\mathbf{s},\mathbf{s}_0)$ being a Kronecker δ-function.

A formal solution of Eq. (1.4) is readily given by a matrix element

$$\tilde{P}(\mathbf{s},u\,|\,\mathbf{s}_0) = \{(u\hat{1}-\hat{H})^{-1}\}_{\mathbf{s},\mathbf{s}_0} \ , \tag{1.5}$$

where a matrix $\hat{H}$ is defined by

$$\hat{H}_{\mathbf{s},\mathbf{s}'} = w_{\mathbf{ss}'} \quad \text{for } \mathbf{s} \neq \mathbf{s}' \tag{1.6}$$

and

$$\hat{H}_{\mathbf{s},\mathbf{s}} = -\Gamma_{\mathbf{s}} \ . \tag{1.7}$$

Equation (1.5) makes $\tilde{P}(\mathbf{s},u\,|\,\mathbf{s}_0)$ behave as the matrix element of a propagator and $\hat{H}$ plays the role of the corresponding Hamiltonian matrix.[8] Thus, we may call $(u\hat{1}-\hat{H})^{-1}$ a random walk propagator.

We are primarily interested in the ac conductivity due to the hopping motion of carriers described by Eq. (1.1). It has been shown[9,10] that when $kT \ll \hbar\omega$ the ac conductivity is reducible to the generalized Einstein relation

$$\sigma(\omega) = \frac{ne^2}{kT} D(\omega) \ , \tag{1.8}$$

and the frequency-dependent diffusion constant is written in terms of the second spatial moment of $\tilde{P}(\mathbf{s},i\omega\,|\,\mathbf{s}_0)$ as

$$D(\omega) = -\frac{\omega^2}{2d} \sum_{\mathbf{s}} < (\mathbf{s}-\mathbf{s}_0)^2\, \tilde{P}(\mathbf{s},i\omega\,|\,\mathbf{s}_0) > . \tag{1.9}$$

Here, d is the dimensionality of the system, n denotes the number density of carriers with charge e, k is the Boltzmann constant, T is the absolute temperature and $<\cdots>$ denotes the average over the ensemble of possible $w_{\mathbf{ss}'}$. Note that the average over the initial site $\mathbf{s}_0$ has been omitted in Eq. (1.9), because the ensemble average eliminates the dependence of the summand on $\mathbf{s}_0$. If the sites $\{\mathbf{s}\}$ form a regular lattice, $(\mathbf{s}-\mathbf{s}_0)^2$ can be taken outside of the ensemble average. (The randomness resides in the jump rates.) Eventually, in order to obtain the conductivity we must evaluate the ensemble average of Eq. (1.5). The dielectric constant $\epsilon(\omega)$ at low frequencies is expressible in terms of $\sigma(\omega)$.

Another interesting quantity is the decay of the carrier from its initial site which is simply described by $P(\mathbf{s}_0,t\,|\,\mathbf{s}_0,0)$ itself. Although there is extensive literature[11] on spectral diffusion and excitation transfer which are governed by the time decay function $P(\mathbf{s}_0,t\,|\,\mathbf{s}_0,0)$, we will not discuss $P(\mathbf{s}_0,t\,|\,\mathbf{s}_0,0)$ except for a few cases.

In passing, we define the dimensionless conductivity or diffusion constant by

$$\tilde{\sigma}(\omega) = \tilde{D}(\omega) \equiv \sigma(\omega)/(ne^2a^2w_0/kT) \ , \tag{1.10}$$

using scaling parameters a and w_0 of distance and frequency, respectively, which will appear in individual problems.

The present paper is organized as follows. In Sections 2 and 3 we present two exact expressions for the coherent medium which is defined through an average of the random walk propagator; one is the multiple scattering description by Lax[12,13] and another is a formal proof of the validity of the continuous time random walk (CTRW) description of Scher and Lax.[9] This was first given in t space by Klafter and Silbey,[14] using projection techniques developed by Zwanzig.[15] A shorter proof will be presented here in the Laplace transformed space. The multiple scattering method provides the starting point of the subsequent coherent medium approximation (CMA). The exact coherent medium is easily obtained at the high frequency limit, which is given in Section 4. In Section 5, we explain a generalized application of the coherent potential approximation[12,16] to obtain the averaged random walk propagator. In Section 6, we discuss other approximations proposed to treat the random walk equation with random jump rate. Sections 7-11 are devoted to the hopping conduction on one-dimensional chains. First, in Section 7 a bond-percolation model where a jump rate has a finite probability of vanishing is solved exactly to provide a standard against which approximate methods can be tested. Then, general results obtained by the coherent medium approximation are given in Section 8. The distribution of jump rate is classified into six categories according to the first and second moments of the inverse of the jump rate which determine the behavior of the ac conductivity in the vicinity of the static limit. An example is given for a case in which the dc conductivity vanishes even though a carrier can diffuse away infinitely from its initial position. The ac conductivity is evaluated in the present approximation for three types of the distribution of the jump rate, (1) binary jump rate [Section 9], (2) bond-percolation model [Section 10], and (3) a distribution which is derived to simulate a positionally random distribution of hopping sites [Section 11]. Applications to the three dimensional system are given in Sections 12-14. Section 12 treats the bond-percolation model and several critical behaviors at the percolation threshold are discussed. A comparison with computer simulation is given in Section 13. The method is also applied to reproduce the low-frequency conductivity in doped semiconductors (Section 14). A short summary is given in Section 15.

2. Exact coherent medium I - Multiple scattering formalism

An elegant multiple scattering formalism for wave propagation in random media was first given by Lax.[12,13] The idea can be employed to express the inverse matrix in Eq. (1.5) by an integral equation, even though there is no physical wave at all.

Let us define a matrix $\hat{G}$ (the random walk propagator) by

$$\hat{G} \equiv (u\hat{1} - \hat{H})^{-1} . \tag{2.1}$$

For any matrix $\hat{H}_1$, $\hat{G}$ can be expressed as

$$\hat{G} = \hat{G}_1 + \hat{G}_1 \hat{V} \hat{G} , \tag{2.2}$$

where

$$\hat{G}_1 = (u\hat{1} - \hat{H}_1)^{-1} , \tag{2.3}$$

and

$$\hat{V} = \hat{H} - \hat{H}_1 . \tag{2.4}$$

Now, suppose the difference matrix $\hat{V}$ can be written as a sum of submatrices.

$$\hat{V} = \sum_i \hat{V}_i . \tag{2.5}$$

Here, i stands for an index which represents the unit. For $\hat{H}$ defined by Eqs. (1.6) and (1.7), for example, the unit consists of a large matrix whose nonvanishing elements are those connected to a pair of sites. Then, Eq. (2.2) can be converted into the following equations of multiple scattering form:

$$\hat{G} = \hat{G}_1 + \sum_j \hat{G}_1 \hat{T}_j \hat{G}^j , \tag{2.6}$$

$$\hat{G}^j = \hat{G}_1 + \sum_{i \neq j} \hat{G}_1 \hat{T}_i \hat{G}^i , \tag{2.7}$$

where the transition matrix obeys

$$\hat{T}_i = \hat{V}_i + \hat{V}_i \hat{G}_1 \hat{T}_i . \tag{2.8}$$

The proof of these equations is found in Refs. 12 and 13. So far the choice of $\hat{H}_1$ is completely arbitrary and no approximation has been made. The best choice of $\hat{H}_1$ will be such that $\hat{H}_1$ describes the averaged properties of the system. The ensemble average of Eq. (2.6) leads to an exact definition of a coherent medium $\hat{\Sigma}$ through

$$< \hat{G} > = \hat{G}_c \equiv (u\hat{1} - \hat{\Sigma})^{-1} , \tag{2.9}$$

where $\hat{\Sigma}$ is chosen to satisfy

$$\Sigma_j < \hat{T}_j \hat{G}^j > = 0 . \tag{2.10}$$

Here, the matrix $\hat{H}_1$ in Eqs. (2.6) and (2.7) is replaced by the as yet unknown coherent matrix $\hat{\Sigma}$.

3. Exact coherent medium II - Projection operator method and the CTRW

Recently Klafter and Silbey[14] formally proved that the original random walk problem in a random environment can be rigorously reduced to a CTRW calculation of the Scher-Lax type. We shall supply a simpler version of the Klafter-Silbey derivation. Like their proof, the derivation is based on projection operator techniques. The simplicity is achieved by performing the derivation completely in the Laplace transform domain rather than working in the time domain and then transforming to the Laplace domain.

We introduce a projection operator **P** in the manner so heavily exploited by Zwanzig:[15]

$$\mathbf{P}(\text{anything}) = <\text{anything}> = \text{average of anything} . \tag{3.1}$$

Equation (2.1) can be written in the form

$$(u\hat{1} - \hat{H})\, \hat{G} = \hat{1} . \tag{3.2}$$

We are trying to evaluate $<\hat{G}> = \mathbf{P}\hat{G}$. If we also introduce the complementary projection operator

$$\mathbf{Q} = \mathbf{1} - \mathbf{P} , \tag{3.3}$$

$\mathbf{P}\hat{G}$ can be obtained using techniques previously introduced to deal with degenerate perturbation theory.[17] Multiplication of Eq. (3.2) by **P** yields

$$u\mathbf{P}\hat{G} - \mathbf{P}\hat{H}\hat{G} = \hat{1} \tag{3.4}$$

or

$$(u\hat{1} - \mathbf{P}\hat{H}\mathbf{P})\mathbf{P}\hat{G} = \hat{1} + \mathbf{P}\hat{H}\mathbf{Q}\hat{G} , \tag{3.5}$$

where the latter form was obtained by replacing $\hat{H}$ by $\hat{H}(\mathbf{P}+\mathbf{Q})$ and an extra **P** has been introduced at no extra charge since $\mathbf{P}^2 = \mathbf{P}$. Since we want to find $\mathbf{P}\hat{G} = <\hat{G}>$, we must eliminate $\mathbf{Q}\hat{G}$. To obtain $\mathbf{Q}\hat{G}$, multiply Eq. (3.2) by **Q** and use $\mathbf{Q}\hat{1} = (\mathbf{1}-\mathbf{P})\hat{1} = \hat{1}-\hat{1} = 0$ to obtain

$$(u\hat{1} - \mathbf{Q}\hat{H}\mathbf{Q})\mathbf{Q}\hat{G} = \mathbf{Q}\hat{H}\mathbf{P}\hat{G} . \tag{3.6}$$

Equation (3.6) may now be solved for $\mathbf{Q}\hat{G}$ and the result be inserted into Eq. (3.5) to obtain

$$[u\hat{1} - \hat{\Sigma}(u)]\, \mathbf{P}\hat{G} = \hat{1} , \tag{3.7}$$

where the "coherent medium" $\hat{\Sigma}(u)$ is given by

$$\hat{\Sigma}(u) = \mathbf{P}\hat{H}\mathbf{P} + \mathbf{P}\hat{H}\mathbf{Q}(u\hat{1}-\mathbf{Q}\hat{H}\mathbf{Q})^{-1}\mathbf{Q}\hat{H}\mathbf{P} , \tag{3.8}$$

and an extra **Q** has been inserted. Also, since $\mathbf{PQ} = 0$ we can write

$$\mathbf{P}<\hat{H}>\mathbf{Q}\hat{A}\mathbf{P} = 0 , \tag{3.9}$$

where $\hat{A}$ is any matrix. Thus we can write

$$\mathbf{P}\hat{H}\mathbf{Q}\hat{A}\mathbf{P} = \mathbf{P}\delta\hat{H}\mathbf{Q}\hat{A}\mathbf{P} = \mathbf{P}\delta\hat{H}\mathbf{Q}\delta\hat{A}\mathbf{P} , \tag{3.10}$$

where

$$\delta\hat{H} \equiv \hat{H} - <\hat{H}> ; \delta\hat{A} \equiv \hat{A} - <\hat{A}> . \tag{3.11}$$

Equation (3.9) can therefore be written in a notation similar to that used by Klafter and Silbey

$$\hat{\Sigma}(u) = \mathbf{P}\hat{H}\mathbf{P} + \mathbf{P}\delta\hat{H}(u\hat{1}-\mathbf{Q}\hat{H}\mathbf{Q})^{-1}\delta\hat{H}\mathbf{P} , \tag{3.12}$$

or

$$\hat{\Sigma}(u) = <\hat{H}> + <\delta\hat{H}(u\hat{1}-\mathbf{Q}\hat{H}\mathbf{Q})^{-1}\delta\hat{H}> . \tag{3.13}$$

The principal conclusion is that a "self-energy matrix", $\hat{\Sigma}(u)$ or coherent medium, exists with a formal expression, Eq. (3.8), for its evaluation.

Now, if we take the $\mathbf{s},\mathbf{s}_0$ matrix element of the algebraic identity

$$u(u\hat{1} - \hat{\Sigma})^{-1} = \hat{1} + \hat{\Sigma}(u\hat{1}-\hat{\Sigma})^{-1} , \tag{3.14}$$

we obtain the effective master equation

$$u<\tilde{P}(\mathbf{s},u\,|\,\mathbf{s}_0)> = \delta(\mathbf{s},\mathbf{s}_0) + \sum_{\mathbf{s}'} \hat{\Sigma}_{\mathbf{ss}'}<\tilde{P}(\mathbf{s}',u\,|\,\mathbf{s}_0)> \tag{3.15}$$

which is identical in form to Eq. (1.4) with the random $w_{\mathbf{ss}'}$ replaced by the non-random $\hat{\Sigma}_{\mathbf{ss}}$.

Particle conservation in the form

$$\sum_{\mathbf{s}} \tilde{P}(\mathbf{s},u\,|\,\mathbf{s}_0) = 1/u \tag{3.16}$$

must remain true for the averaged quantities. Thus,

$$\sum_{\mathbf{s}} \hat{\Sigma}_{\mathbf{ss}'}(u) = 0 , \tag{3.17}$$

just as $\sum_{\mathbf{s}} \hat{H}_{\mathbf{ss}'} = 0$.

Our coherent jump rate $\hat{\Sigma}_{\mathbf{ss}'}$ must display translational invariance

$$\hat{\Sigma}_{\mathbf{ss}'}(u) = \hat{\Sigma}_{\mathbf{s}-\mathbf{s}',0}(u) . \tag{3.18}$$

Therefore, Eq. (3.15) can be solved by introducing the Fourier transformed quantities:

$$<\tilde{P}(\mathbf{k},u)> = \sum_{\mathbf{s}} \exp\{-i\mathbf{k}\cdot(\mathbf{s}-\mathbf{s}_0)\}<\tilde{P}(\mathbf{s},u\,|\,\mathbf{s}_0)> , \tag{3.19}$$

$$\hat{\Sigma}(\mathbf{k},u) \doteq \sum_{\mathbf{s}\neq\mathbf{s}_0} \hat{\Sigma}_{\mathbf{ss}_0}(u)\exp\{-i\mathbf{k}\cdot(\mathbf{s}-\mathbf{s}_0)\} . \tag{3.20}$$

The use of the conservation law, Eq. (3.14), yields

$$\sum_{\mathbf{s}} \hat{\Sigma}_{\mathbf{ss}_0}(u)\exp\{-i\mathbf{k}\cdot(\mathbf{s}-\mathbf{s}_0)\} = \hat{\Sigma}(\mathbf{k},u)-\hat{\Sigma}(0,u) , \tag{3.21}$$

and

$$u<\tilde{P}(\mathbf{k},u)> = 1 + [\hat{\Sigma}(\mathbf{k},u)-\hat{\Sigma}(0,u)]<\tilde{P}(\mathbf{k},u)> . \tag{3.22}$$

The solution of the equation can be inverse Fourier transformed to yield

$$<\tilde{P}(\mathbf{s},u\,|\,\mathbf{s}_0)> = \frac{1}{N}\sum_{\mathbf{k}} \frac{\exp\{i\mathbf{k}\cdot(\mathbf{s}-\mathbf{s}_0)\}}{u+\hat{\Sigma}(0,u)-\hat{\Sigma}(\mathbf{k},u)} , \tag{3.23}$$

$$= \frac{1}{u+\hat{\Sigma}(0,u)} \frac{1}{N} \sum_{\mathbf{k}} \frac{\exp\{i\mathbf{k}\cdot(\mathbf{s}-\mathbf{s}_0)\}}{1-\{\hat{\Sigma}(\mathbf{k},u)/[u+\hat{\Sigma}(0,u)]\}} , \tag{3.24}$$

N being the total number of sites. This last result has precisely the same form as the CTRW result (Eq. 25) with Eq. (20) in Ref. 9(a) if one makes the identification

$$\Lambda(\mathbf{k},u) = \frac{\hat{\Sigma}(\mathbf{k},u)}{u+\hat{\Sigma}(0,u)} . \tag{3.25}$$

Note that the prefactor in Eq. (25) of Ref. 9(a) then reduces to

$$(1/u)\{1-\Lambda(0,u)\} = (1/u)u/[u+\hat{\Sigma}(0,u)] \tag{3.26}$$

in agreement with that in Eq. (3.24).

In particular, if $\hat{\Sigma}(\mathbf{k},u)$ is known, one can calculate $\tilde{\psi}(\mathbf{s},u)$ from

$$\tilde{\psi}(\mathbf{s},u) = \frac{1}{N} \sum_{\mathbf{k}} \Lambda(\mathbf{k},u) e^{i\mathbf{k}\cdot\mathbf{s}} \tag{3.27}$$

$$= \frac{\hat{\Sigma}_{\mathbf{s}0}}{u+\hat{\Sigma}(0,u)} . \tag{3.28}$$

Here, $\tilde{\psi}(\mathbf{s},u)$ is the Laplace transform of the probability density $\psi(\mathbf{s},t)$ that the time between hops is t and the displacement $\mathbf{s}$. Unfortunately, the expression Eq. (3.13) is purely formal, and Klafter and Silbey do not propose a better method of evaluating it than the procedure used by Scher and Lax of averaging over a distribution of nearest neighbor distances (Section 6).

4. A simple example of the exact coherent medium

A simple example of the exact coherent medium can be obtained from Eq. (3.13) if we take a limit of $u=\infty$ or the high frequency limit. In this limit, the coherent medium is given simply by

$$\hat{\Sigma}(\infty) = \langle \hat{H} \rangle . \tag{4.1}$$

For example, if the sites $\{\mathbf{s}\}$ form a regular array of lattice points and if $\hat{H}$ is given by

$$\hat{H}_{\mathbf{ss}'} = \begin{cases} w_{\mathbf{ss}'} & \text{if } \mathbf{s} \text{ and } \mathbf{s}' \text{ are nearest neighbors} \\ -\Gamma_{\mathbf{s}} = \sum\limits_{\mathbf{s}'(\text{n.n.})} w_{\mathbf{s}'\mathbf{s}} & \text{if } \mathbf{s}=\mathbf{s}' \\ 0 & \text{otherwise} \end{cases} , \tag{4.2}$$

then the exact coherent medium has the same structure as that of $\hat{H}$ with the random jump rate $w_{\mathbf{ss}}$ replaced by the coherent jump rate w_c, which is defined by

$$w_c = \langle w_{\mathbf{ss}'} \rangle . \tag{4.3}$$

Here, we have assumed that each jump rate is distributed independently.

5. The coherent medium approximation

If we neglect the difference between the "exciting" matrix $\hat{G}^j$ and the "average" matrix $\hat{G}_c$ in Eq. (2.6) and make the crude approximation

$$\hat{G}^j \approx \hat{G}_c \ , \tag{5.1}$$

the rigorous condition Eq. (2.10) reduces to the simpler condition

$$\sum_j < \hat{T}_j > = 0 \ . \tag{5.2}$$

If each sub-unit $\hat{V}_j$ in Eq. (2.5) has an independent distribution, Eq. (5.2) is further reducible to a single condition

$$< \hat{T}_j > = 0 \ , \tag{5.3}$$

which supplies a condition to determine the unknown coherent medium. This idea[12,13] has been widely used in the problems of the lattice vibration[18] and the electronic energy band[19] in alloys and in other areas,[16] and is referred to as the "coherent potential approximation" (CPA). We shall use the more general nomenclature, "coherent medium approximation" (CMA), because as we shall see in the following, the idea is useful even when there are no potentials in the problem at all.[10] [We treat the jump rate instead of the potential!]

For the sake of simplicity, we assume that the localized centers $\{\mathbf{s}\}$ form a regular lattice and that the jump rate $w_{\mathbf{ss}'}$ is zero unless $\mathbf{s}$ and $\mathbf{s}'$ are nearest neighbors. We expect to find a coherent medium $\hat{\Sigma}(u)$ which is described only by the nearest neighbor coherent jump rate $w_c(u)$:

$$\begin{aligned} \hat{\Sigma}_{\mathbf{ss}'} &= w_c(u) \quad (\mathbf{s},\mathbf{s}') \text{ nearest neighbors} \\ &= 0 \qquad\quad \text{otherwise} \end{aligned} \tag{5.4}$$

and

$$\hat{\Sigma}_{\mathbf{ss}} = -\, z w_c(u) \ , \tag{5.5}$$

where z is the coordination number of the lattice. If we replace $\hat{H}_1$ in Eqs. (2.2)-(2.4) by the coherent matrix, the difference matrix $\hat{V} \equiv \hat{H} - \hat{\Sigma}$ is written as

$$\hat{V} = \sum_{(\mathbf{s},\mathbf{s}')} \hat{V}(\mathbf{s},\mathbf{s}') \ , \tag{5.6}$$

where the summation runs all nearest neighbor pairs and the submatrix $\hat{V}(\mathbf{s},\mathbf{s}')$ of large dimension contains non-vanishing elements in the 2×2 submatrix $\overline{V}$, for example for a pair of sites 1 and 2,

$$\overline{V} = \begin{bmatrix} w_c(u) - w_{21} & w_{12} - w_c(u) \\ w_{21} - w_c(u) & w_c(u) - w_{12} \end{bmatrix} . \tag{5.7}$$

If we introduce the abbreviated notation

$$\overline{P}_{\mathbf{ss}'} \equiv \{(u\hat{1} - \hat{\Sigma})^{-1}\}_{\mathbf{ss}'} \equiv \overline{P}(\mathbf{s},u \mid \mathbf{s}') \tag{5.8}$$

and the 2×2 matrix

$$\overline{P} = \begin{bmatrix} P_{11} & P_{12} \\ P_{21} & P_{22} \end{bmatrix} , \tag{5.9}$$

then the condition Eq. (5.3) to determine $w_c(u)$ can be written more explicitly as

$$< \overline{V}(\overline{1} - \overline{P}\overline{V})^{-1} > = 0 \ , \tag{5.10}$$

where $\overline{1}$ is the unit 2×2 matrix, and the average is taken over the distribution for w_{12} and w_{21}.

For simplicity, we specialize to the case where $w_{\mathbf{ss}'} = w_{\mathbf{s}'\mathbf{s}}$. Then, it is easy to see that the four conditions contained in Eq. (5.10) reduces to a single condition

$$< \frac{w_c(u) - w_{12}}{1 - 2(\bar{P}_{11} - \bar{P}_{12})(w_c(u) - w_{12})} > = 0 \ , \tag{5.11}$$

or equivalently

$$\bar{P}_{11} - \bar{P}_{12} = \left\langle \frac{1}{(\bar{P}_{11} - \bar{P}_{12})^{-1} - 2w_c(u) + 2w_{12}} \right\rangle \ . \tag{5.12}$$

Because the coherent medium lattice is assume to be periodic, $\bar{P}$ can be evaluated in the usual way as we used in Section 3:

$$\bar{P}(\mathbf{s}, u \mid \mathbf{s}_0) = \frac{1}{N} \sum_{\mathbf{k}} \frac{\exp\{-i\mathbf{k}\cdot(\mathbf{s}-\mathbf{s}_0)\}}{u + zw_c(u)(1-f_{\mathbf{k}})} \ , \tag{5.13}$$

where

$$f_{\mathbf{k}} = \frac{1}{z} \sum_{\mathbf{s}}{}' \exp\{i\mathbf{k}\cdot(\mathbf{s}-\mathbf{s}_0)\} \tag{5.14}$$

and the primed sum is taken over the nearest neighbors of $\mathbf{s}_0$. Equation (5.13) can be used to obtain the relation

$$\bar{P}_{11} - \bar{P}_{12} = (1 - u\bar{P}_{11})/zw_c(u) \ , \tag{5.15}$$

so that only $\bar{P}_{11}$ needs to be evaluated.

By inserting Eq. (5.13) into Eq. (1.9), the diffusion constant of our coherent medium is simply given by

$$D(\omega) = a^2 w_c(i\omega) \ . \tag{5.16}$$

Here, a is the lattice constant (which may be modified for non-orthogonal lattices).

General results for limiting cases in the coherent medium approximation can be readily obtained. At the high frequency limit, our condition Eq. (5.11) yields

$$w_c(\infty) = < w_{12} > \tag{5.17}$$

in agreement with the exact result Eq. (4.3). That is, our CMA is exact at the high-frequency limit.

At the static limit u=0, Eq. (5.11) is reducible to

$$\frac{1}{zw_c(0)} = \left\langle \frac{1}{(z-2)\, w_c(0) + 2w_{12}} \right\rangle . \tag{5.18}$$

We may formally express the solution of Eq. (5.18) in terms of a continued fraction

$$w_c(0) = \cfrac{2}{z\left\langle \cfrac{1}{w_{12} + \cfrac{z-2}{z\left\langle \cfrac{1}{w_{12} + \cfrac{z-2}{\cdots}} \right\rangle}} \right\rangle} \ . \tag{5.19}$$

In particular, if z=2, that is for linear chains, the coherent jump rate at the dc limit is given by

$$\frac{1}{w_c(0)} = < \frac{1}{w_{12}} > \ , \tag{5.20}$$

which agrees with the result given by Bernasconi et al[20] on the basis of a scaling hypothesis.

The application of the CMA will be given in Sections 8-14.

6. Other approximations

A few approximate methods have been proposed to solve the random walk equation (1.1) or its Laplace transform, Eq. (1.4) with random jump rates. We summarize briefly these methods.

We start from Eq. (2.1) and separate the Hamiltonian matrix into the purely diagonal part $\hat{\Gamma}$ and the purely off-diagonal part $\hat{W}$:

$$\hat{H} = \hat{\Gamma} + \hat{W} \ , \tag{6.1}$$

where

$$\hat{\Gamma}_{\mathbf{ss}'} = -\hat{\Gamma}_{\mathbf{s}}\delta(\mathbf{s},\mathbf{s}') \ , \tag{6.2}$$

$$\hat{W}_{\mathbf{ss}'} = w_{\mathbf{ss}'}(1-\delta(\mathbf{s},\mathbf{s}')) \ . \tag{6.3}$$

Then, the random walk propagator $\hat{G}$ can be expanded into a series

$$\hat{G}=(u\hat{1}-\hat{\Gamma})^{-1}\{\hat{1}+\sum_{m=1}[\hat{W}(u\hat{1}-\hat{\Gamma})^{-1}]^m\} \ . \tag{6.4}$$

Scher and Lax[9,21] proposed to evaluate the average $<\hat{G}>$ in a Hartree approximation by

$$<\hat{G}> = <(u\hat{1}-\hat{\Gamma})^{-1}>\{\hat{1}+\sum_{m=1}[<\hat{W}><(u\hat{1}-\hat{\Gamma})^{-1}>]^m\} \ . \tag{6.5}$$

This procedure yields the coherent medium whose diagonal and off-diagonal matrix elements are given by a weighted average of $\Gamma_{\mathbf{s}}$ and $w_{\mathbf{ss}'}$, respectively,

$$\hat{\Sigma}_{\mathbf{ss}} = -\left\langle\frac{\Gamma_{\mathbf{s}}}{u+\Gamma_{\mathbf{s}}}\right\rangle / \left\langle\frac{1}{u+\Gamma_{\mathbf{s}}}\right\rangle , \tag{6.6}$$

and

$$\hat{\Sigma}_{\mathbf{ss}'} = \left\langle\frac{w_{\mathbf{ss}'}}{u+\Gamma_{\mathbf{s}'}}\right\rangle / \left\langle\frac{1}{u+\Gamma_{\mathbf{s}'}}\right\rangle \ . \tag{6.7}$$

The function $\tilde{\psi}(\mathbf{s},u)$ can be obtained using Eq. (3.28). This method has been successfully used to investigate many experiments.[6,9]

If we consider the random walk on a lattice and treat the case in which the jump rates are non-zero only between nearest neighbor pairs, we can express matrix elements of $\hat{G}$ as a renormalized perturbation series expansion.[22]

The off-diagonal element of $\hat{G}$ is written as

$$\hat{G}_{ij}=\hat{G}_{ii}w_{ij}\hat{G}_{jj}^{(i)}+\sum_{k\neq i,j}\hat{G}_{ii}w_{ik}\hat{G}_{kk}^{(i)}w_{kj}\hat{G}_{jj}^{(i,k)}+\cdots \ , \tag{6.8}$$

where the summation extends over all self-avoiding paths starting at i and ending at j, and the diagonal element is given by

$$\hat{G}_{ii} = (u - \Gamma_i - \Delta_i) \ , \tag{6.9}$$

and

$$\hat{G}_{kk}^{(i)} = (u - \Gamma_k - \Delta_k^{(i)}) \ , \tag{6.10}$$

and so on. The superscripts (i), (i,k), $\cdots$ denote that these sites must be excluded from the paths. Here, an abbreviation $i=\mathbf{s}_i$ is used. The "self-energy" Δ_i is given by

$$\Delta_i = \sum_k w_{ik}\hat{G}_{kk}^{(i)}w_{ki}+\sum_{k\neq l\neq i}w_{ik}\hat{G}_{kk}^{(i)}w_{kl}\hat{G}_{ll}^{(i,k)}w_{li}+\cdots \ , \tag{6.11}$$

where, again, the summations are over all closed self-avoiding walks starting at i and ending at i, and $\Delta_k^{(i)}$ is given by a similar series with an extra excluded site.

A simplification is achieved if one applies the expansion Eq. (6.8) to the Bethe lattice[23] where no closed loops occur in the lattice. Since an exclusion of a site in the Bethe lattice breaks the lattice into z semi-infinite branches (z being the coordination number), the self-energies are given by a finite sum

$$\Delta_i = \sum_j^{z}{}' w_{ij} \hat{G}_{jj}^{(i)} w_{ji} \ , \tag{6.12}$$

and

$$\Delta_k^{(i)} = \sum_{l \neq i}^{z-1}{}' w_{kl} \hat{G}_{ll}^{(k)} w_{lk} \ . \tag{6.13}$$

Here, use has been made of $\hat{G}_{ll}^{(i,k)} = \hat{G}_{ll}^{(k)}$ in the Bethe lattice. The primed sum again sums over nearest neighbors and in Eq. (6.13) one of the nearest neighbors of site k is excluded.

Now, let us introduce

$$g_i^{(j)} \equiv w_{ij} - w_{ij} \hat{G}_{jj}^{(i)} w_{ji} \ . \tag{6.14}$$

Then, the diagonal element of $\hat{G}$ is reduced to

$$\hat{G}_{ii} = \left(u + \sum_j^{z}{}' g_i^{(j)} \right)^{-1} , \tag{6.15}$$

and $g_i^{(j)}$ satisfies

$$\frac{1}{g_i^{(j)}} = \frac{1}{w_{ij}} + \frac{1}{u + \sum\limits_{k \neq i}^{z-1}{}' g_j^{(k)}} \ . \tag{6.16}$$

Here, we have assumed $w_{ij} = w_{ji}$. Equations (6.15) and (6.16) imply that $\hat{G}_{ii}$ is expressed in terms of a continued fraction:

$$\hat{G}_{ii} = \cfrac{1}{u + \sum\limits_j^{z}{}' \cfrac{1}{\cfrac{1}{w_{ij}} + \cfrac{1}{u + \sum\limits_{k \neq i}^{z-1}{}' \cfrac{1}{\cfrac{1}{w_{jk}} + \cfrac{1}{\cdots}}}}} \ . \tag{6.17}$$

Since a one-dimensional chain is a Bethe lattice with $z=2$, the expansions Eqs. (6.15)-(6.17) are valid for chains.

Bernasconi et al[24] examined the distribution function of the continued fraction $g_i^{(j)}$ for chains and proposed an approximate procedure to evaluate the effective continued fraction $g_c(u)$ by

$$g_c(u) = \left\langle \left(\frac{1}{w_{ij}} + \frac{1}{u + g_c(u)} \right)^{-1} \right\rangle \ . \tag{6.18}$$

From the effective continued fraction, we can obtain the coherent medium with the coherent jump rate $w_c(u)$ given by

$$w_c(u) = \{u + g_c(u)\} g_c(u)/u \ . \tag{6.19}$$

Movaghar[25] applied the same idea to Eq. (6.18) for the Bethe lattice with $z > 2$. However, the definition of the diffusion constant in the Bethe lattice is ambiguous. Movaghar assumed a completely random orientation of branches as in random branching high-polymers.

A comparison of these methods with an exact result will be given in Section 10.

7. Bond-percolation model in a chain I - Exact solution

The bond-percolation model is characterized by the following special type of distribution function $P(w_{ss'})$ for the nearest neighbor jump rate $w_{ss'}=w_{s's}$

$$P(w_{ss'}) = p\ \delta(w_{ss'}-w_0) + q\ \delta(w_{ss'}) \tag{7.1}$$

with $p+q=1$. Here, w_0 is a non-zero constant. This model is a realization of random interruptions in actual systems. The random distribution of the broken bonds decomposes the chain into a set of chunks. It is manifest that $\tilde{P}(\mathbf{s},u\,|\,\mathbf{s}_0)$ in the definition of $D(\omega)$ Eq. (1.9) is zero unless $\mathbf{s}$ belongs to the chunk which contains $\mathbf{s}_0$. An average treating all $\mathbf{s}_0$ as equally likely[26] yields

$$D(\omega) = \sum_{N=1}^{\infty} N(1-p)^2\, p^{N-1}\, D_N(\omega)\ , \tag{7.2}$$

where

$$D_N(\omega) = -\frac{\omega^2}{2N} \sum_{\mathbf{s},\mathbf{s}_0}^{N} (\mathbf{s}-\mathbf{s}_0)^2\, \tilde{P}(\mathbf{s},i\omega\,|\,\mathbf{s}_0) \tag{7.3}$$

is the diffusion constant of a chunk with N sites. The prefactor of the summand in Eq. (7.2)

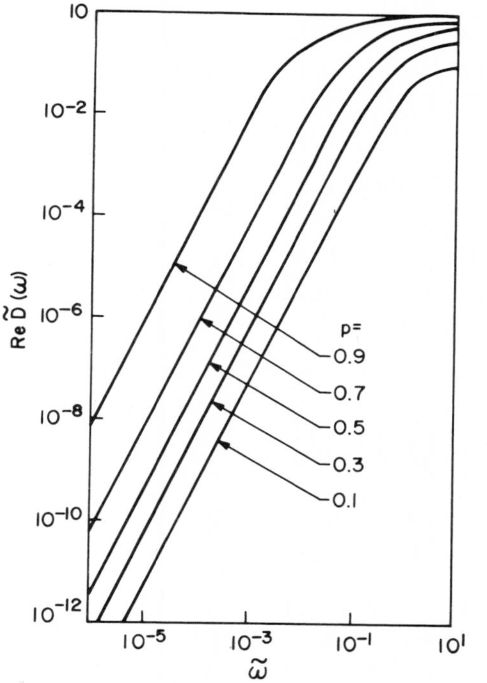

Fig. 7.1. The dependence of the real part of $\tilde{D}(\omega)$ on frequency for various values of p. The exact solution for the bond-percolation model in a chain. (Taken from ref. 27.)

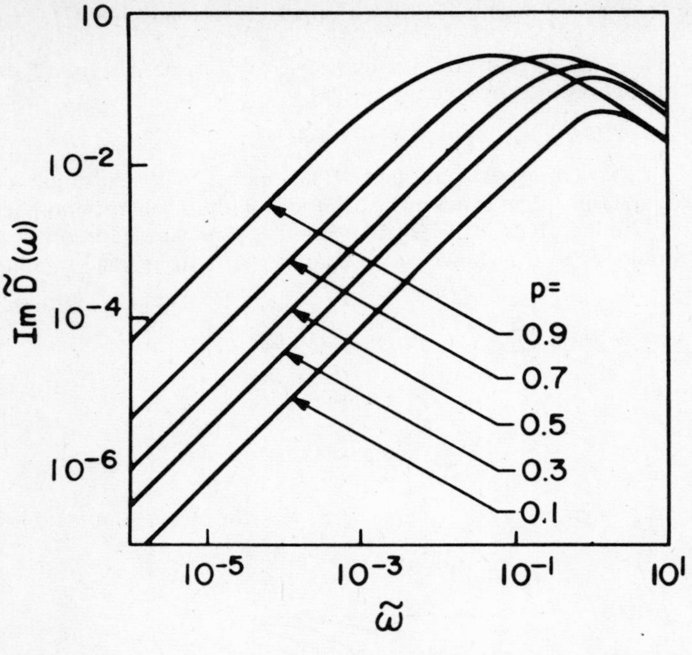

Fig. 7.2. The dependence of the imaginary part of $\tilde{D}(\omega)$ on frequency for various values of p. The exact solution for the bond-percolation model in a chain. (Taken from ref. 27.)

is the probability that a given site belongs to a chunk with N sites.

Now, the conductivity for the finite chunk Eq. (7.3) can be obtained by diagonalizing an $N \times N$ submatrix of $\hat{H}$ to yield[27]

$$D_N(\omega) = a^2 w_0 \left[1 + \frac{(1+4/i\tilde{\omega})^{1/2}}{N} \left\{ \frac{1}{z_+^{2N}+1} - \frac{1}{z_-^{2N}+1} \right\} \right], \tag{7.4}$$

where $\tilde{\omega} = \omega / w_0$ and $z_\pm = (\sqrt{i\tilde{\omega}} \pm \sqrt{4+i\tilde{\omega}})/2$. The frequency dependent diffusion constant can be easily evaluated from Eq. (7.2). Figures 7.1 and 7.2 show the frequency dependence of the real and imaginary part of the dimensionless diffusion constant $\tilde{D}(\omega) = D(\omega)/a^2 w_0$. We can show that the real and imaginary parts of $D(\omega)$ and hence of the conductivity vanish quadratically and linearly, respectively, with frequency. In fact, we have

$$A \equiv \lim_{\tilde{\omega} \to 0} \mathrm{Re}\tilde{D}(\omega)/\tilde{\omega}^2 = \frac{p(1+p)^2}{4(1-p)^4} \tag{7.5}$$

and

$$B \equiv \lim_{\tilde{\omega} \to 0} \mathrm{Im}\tilde{D}(\omega)/\tilde{\omega} = \frac{p}{2(1-p)^2}, \tag{7.6}$$

which are plotted in Fig. 7.3. The quantities A and B diverge at the percolation threshold $p=1$.

The exact diffusion constant may be used to determine an exact coherent jump rate between nearest neighbors through $w_c(\omega) = D(\omega)/a^2$.

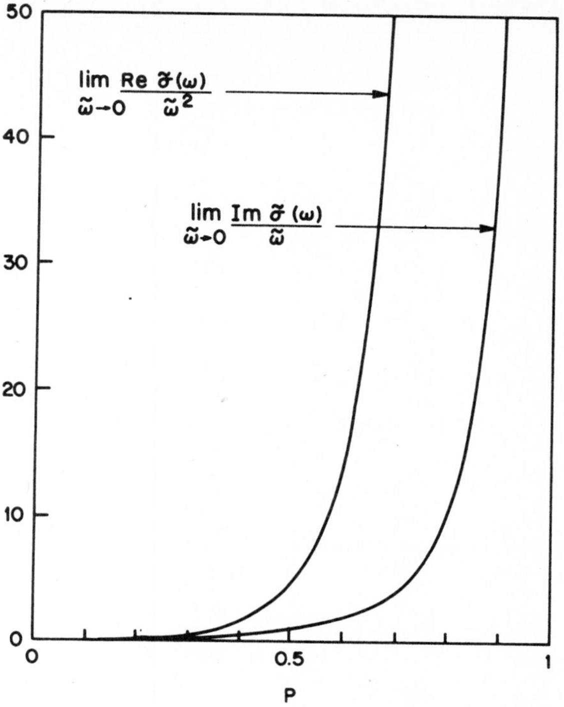

Fig. 7.3. Critical behavior of $\lim_{\tilde{\omega}\to 0} \mathrm{Re}\,\tilde{D}(\omega)/\tilde{\omega}^2$ and $\lim_{\tilde{\omega}\to 0} \mathrm{Im}\,\tilde{D}(\omega)/\tilde{\omega}$ at percolation threshold $p=1$. The exact solution for the bond-percolation model in a chain. Explicit forms are given in Eqs. (7.5) and (7.6).

8. The coherent medium approximation in one-dimensional chains

For one-dimensional chains, the diagonal element of the coherent random walk propagator $\bar{P}_{11}$ can be easily evaluated using Eq. (5.13) with $\mathbf{s}=\mathbf{s}_0$:

$$\bar{P}_{11} \equiv \bar{P}(\mathbf{s}_0,u\,|\,\mathbf{s}_0) = \{u(u+4w_c)\}^{-1/2} . \tag{8.1}$$

Noting Eq. (5.15), we can discuss general features of the ac conductivity on the basis of the CMA condition Eq. (5.12).

The high frequency behavior of the ac conductivity is determined by the first and second moment of the jump rate:

$$\tilde{\sigma}(\omega) \approx <w_{12}>/w_0 + 2i\{<w_{12}^2>-<w_{12}>^2\}/w_0^2\tilde{\omega} , \tag{8.2}$$

where w_0 is a relevant scaling parameter of the frequency and $\tilde{\omega} \equiv \omega/w_0$.

The low-frequency behavior of the conductivity is also obtained using an expansion for small u in Eq. (5.11) with Eqs. (5.15) and (8.1). We have found the following six regimes which are distinguished by the values of the first, $m_1 \equiv w_0<1/w_{12}>$, and second, $m_2 \equiv w_0^2<1/w_{12}^2>$, moments of the inverse of the jump rate[28]:

(1) $m_1, m_2 < \infty$:

$$\tilde{\sigma}(\omega) \approx m_1^{-1}+\{(m_2/m_1^2-1)/2m_1^{1/2}\}(i\tilde{\omega})^{1/2} . \tag{8.3}$$

(2) $m_1<\infty, m_2=\infty$ (logarithmic) with a distribution function $P(w_{12}) \propto w_{12}$:

$$\tilde{\sigma}(\omega) \approx m_1^{-1} - C_1 m_1^{-5/2}(i\tilde{\omega})^{1/2}\ln(i\tilde{\omega})^{1/2} . \tag{8.4}$$

(3) $m_1<\infty, m_2=\infty$ (non-logarithmic) with a distribution function $P(w_{12}) \propto (w_{12})^{\alpha}(0<\alpha<1)$:

$$\tilde{\sigma}(\omega) \approx m_1^{-1} + C_2 m_1^{-(4+\alpha)/2} .(i\tilde{\omega})^{\alpha/2} \tag{8.5}$$

(4) $m_1=\infty$ (logarithmic), $m_2=\infty$ with a distribution function $P(w_{12}) \propto$ constant:

$$\tilde{\sigma}(\omega) \approx -C_3/\ln(i\tilde{\omega}) . \tag{8.6}$$

(5) $m_1=\infty$, $m_2=\infty$ with a distribution function $P(w_{12}) \propto (w_{12})^{\alpha}$ $(-1<\alpha<0)$:

$$\tilde{\sigma}(\omega) \approx C_4(i\tilde{\omega})^{-\alpha/(2+\alpha)} . \tag{8.7}$$

(6) $m_1=\infty$, $m_2=\infty$ with a distribution function $P(w_{12}) \propto q\delta(w_{12}) + pF(w_{12})$ $(p + q=1)$, $\delta(w_{12})$ is a Dirac δ function and $F(w_{12})$ is any nonsingular function):

$$\tilde{\sigma}(\omega) \approx \{(1-q^2)/4q^2\}(i\tilde{\omega}) . \tag{8.8}$$

Here, $C_1,...,C_4$ are constants.

Using the approximation Eq. (6.18)[24,29] and a scaling argument on the distribution function of the continued fraction Eq. (6.16), Bernasconi et al[20] discussed general classes of the distribution function of the jump rate on the basis of the asymptotic behavior of $<P(\mathbf{s}_0,u\,|\,\mathbf{s}_0)>$ near $u=0$. Their results are summarized in a recent review article.[30] They classified Eq. (8.3)-(8.5) as class (a), and Eq. (8.6) as class (b) and Eq. (8.7) as class (c).

The conditional probability that a carrier remains at its initial site after an infinite time is given by

$$P_{00}(\infty) \equiv \lim_{t\to\infty} <P(\mathbf{s}_0,t\,|\,\mathbf{s}_0,0)> \tag{8.9}$$

$$= \lim_{u\to 0} u\,\bar{P}(\mathbf{s}_0,u\,|\,\mathbf{s}_0) . \tag{8.10}$$

Using the coherent jump rate in Eq. (8.1), we can easily verify that $P_{00}(\infty)=0$ for cases (1)-(5) and $P_{00}(\infty)=q$ for case (6). Therefore, cases (4) and (5) supply examples in which a carrier can diffuse away infinitely from its initial position even though the dc conductivity vanishes.[25,28,30]

9. Binary jump rate in a chain

Suppose a chain whose nearest neighbor jump rate is distributed according to a function

$$P(w_{12}) = p\ \delta(w_{12}-w_0) + q\ \delta(w_{12}-w_0\delta) , \tag{9.1}$$

where $p+q=1$, $w_0\neq 0$ and $0<\delta\leqslant 1$. ($\delta=0$ corresponds to the bond-percolation model, whose exact result has been given in Section 7. The CMA treatment of this case will be given in the next section.) It is clear that the first and the second moments of the inverse of the jump rate are finite:

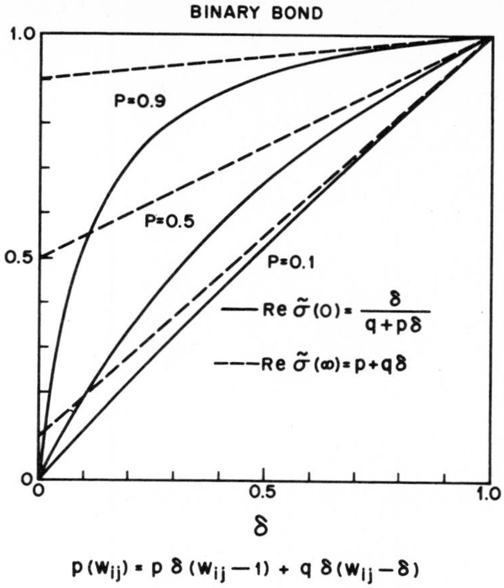

Fig. 9.1. The dependence on p and δ of $\tilde{\sigma}(\infty)$ and $\tilde{\sigma}(0)$ of a chain with a binary jump rate.

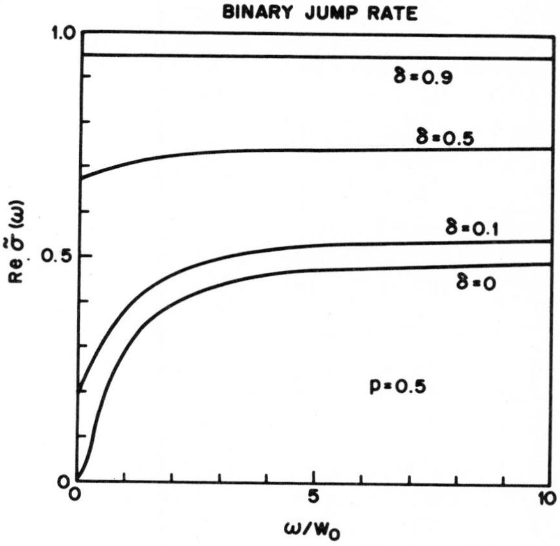

Fig. 9.2. The frequency dependence of the real part of $\tilde{\sigma}(\omega)$ for various values of p: binary jump in a chain. $\delta=0$ corresponds to the bond-percolation model.

$$m_1 = w_0 < 1/w_{12} > = (p\delta + q)/\delta \ , \tag{9.2}$$

and

$$m_2 = w_0^2 < 1/w_{12}^2 > = (p\delta^2 + q)/\delta^2 \ . \tag{9.3}$$

Therefore, this distribution belongs to case (1) in Section 8, and $\tilde{\sigma}(0)$ is given by $1/m_1 = \delta/(q+p\delta)$. Since the conductivity at $\omega = \infty$ is given by Eq. (8.2)

$$\tilde{\sigma}(\infty) = < w_{12} > / w_0 = p + q\delta \ , \tag{9.4}$$

the ac conductivity for fixed p and δ shows a dispersive frequency dependence between these limiting values $\tilde{\sigma}(0)$ and $\tilde{\sigma}(\infty)$ which are plotted in Fig. 9.1. Actual frequency dependence can be readily evaluated using Eqs. (5.11), (5.15), (8.1) and (5.16). Figure 9.2 shows the frequency dependence of the real part of $\tilde{\sigma}(\omega)$ for p=0.9, 0.5, 0.1. The ac part $\tilde{\sigma}(\omega) - \tilde{\sigma}(0)$ is given by

$$\tilde{\sigma}(\omega) - \tilde{\sigma}(0) = \frac{pq\sqrt{\delta}(1-\delta)^2}{2\sqrt{2}(q+p\delta)^{5/2}} (1+i)\omega^{1/2} \ , \tag{9.5}$$

near $\omega = 0$. The prefactor is shown in Fig. 9.3 as a function of δ for p=0.9, 0.5 and 0.1.

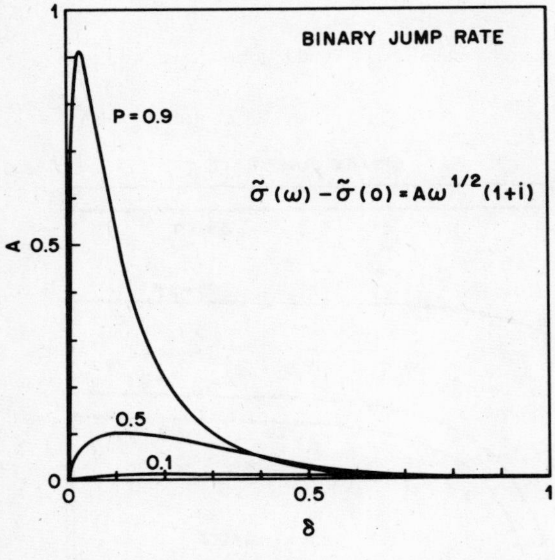

Fig. 9.3. Plots of the prefactor on the right-hand side of Eq. (9.5). The prefactor vanishes at $\delta=0$ and $\delta=1$ as it should be.

10. Bond-percolation model in a chain II. - Approximate treatment

We apply the approximate methods discussed in Section 5 and Section 6 to the bond-percolation model in a chain to provide a comparison of their relative merits.[26]

If we use the distribution function Eq. (7.1) in Eq. (5.11) for the CMA,[10] in Eq. (6.6) for

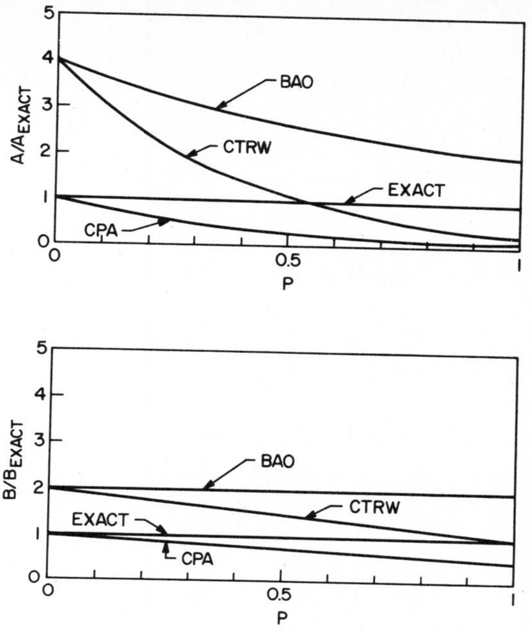

Fig. 10.1. A comparison of the relative merits of the various approximations in the low-frequency regime for the bond-percolation model in a chain: (1) Exact, (2) CMA [Eq. (5.11)], (3) SL [Eq. (6.6)], (4) BAO [Eq. (6.18)]. (Taken from ref. 27.)

the Scher-Lax approximation (SL)[9] and in Eq. (6.18) for the approximation due to Bernasconi et al (BAO),[24] we obtain the coherent jump rate for each method: for the CMA[10]

$$w_c(u)/w_0=[\tilde{u}+2(1-p)^2-(1-p)\{(\tilde{u}+2)^2+4p(p-2)\}^{1/2}]/\tilde{u} \ , \tag{10.1}$$

from Eq. (6.6)

$$w_c(u)/w_0=p(2-p+\tilde{u})\tilde{u}/[2(1-p)^2+(3-2p)\tilde{u}+\tilde{u}^2] \tag{10.2}$$

and from Eq. (6.18)

Table I. The low frequency behavior of $\tilde{\sigma}(\omega)$

Method	A	B
Exact	$p(1+p)^2/4(1-p)^4$	$p/2(1-p)^2$
CMA	$p(2-p)/8(1-p)^4$	$p(2-p)/4(1-p)^2$
SL	$p(4-3p)/4(1-p)^4$	$p(2-p)/2(1-p)^2$
BAO	$p(1+p)/(1-p)^4$	$p/(1-p)^2$

$$w_0(u)/w_0=[(1-p)^2+(1+p)u-(1-p)\{(1-p)^2+2(1+p)\tilde{u}+\tilde{u}^2\}^{1/2}]/2\tilde{u} \ . \tag{10.3}$$

Although we can compare the conductivity in the entire range of the frequency, we supply here only a comparison of coefficients A and B defined by Eqs. (7.5) and (7.6) in Table I and Fig. 10.1. In the latter, the ratio of A/A_{exact} and B/B_{exact} are plotted versus p. We would like to mention that Bernasconi et al believe their approximation Eq. (6.18) may not be adequate for the case in which completely broken jump rates are possible.[26,29,30]

11. Chains with positionally random localized centers

Let us consider a chain in which hopping centers are distributed randomly in space and the jump rate $w_{l,m}$ of a carrier between m and l hopping sites obeys

$$w_{l,m} = w_0 \exp(-|x_m - x_l|/R_d) \ . \tag{11.1}$$

Here, R_d is half an effective Bohr radius of the localized carrier and x_m denotes the position of hopping site m. We assume that $w_{l,m}$ is negligible unless m and l are adjacent hopping sites, because the carrier hopping will be most frequent between adjacent sites. As is well known, the probability density $N(x)$ that one finds an adjacent hopping site at a distance x from a given site in a uniform system obeys a Poisson distribution

$$N(x) = n_s \exp(-n_s x) \ , \tag{11.2}$$

where n_s is the number density of the hopping sites. Therefore, the jump rate w between adjacent centers is distributed according to

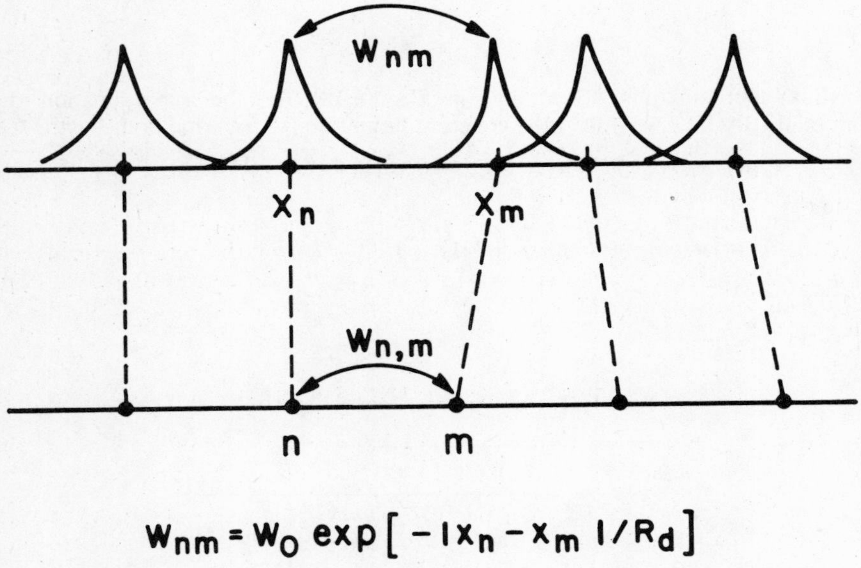

Fig. 11.1. Schematic illustration of mapping between the actual system and the lattice model. The jump rate between two adjacent centers is given by Eq. (11.1), which is mapped into w_{nm} on a regular chain. The latter obeys the distribution Eq. (11.3). (Taken from ref. 31.)

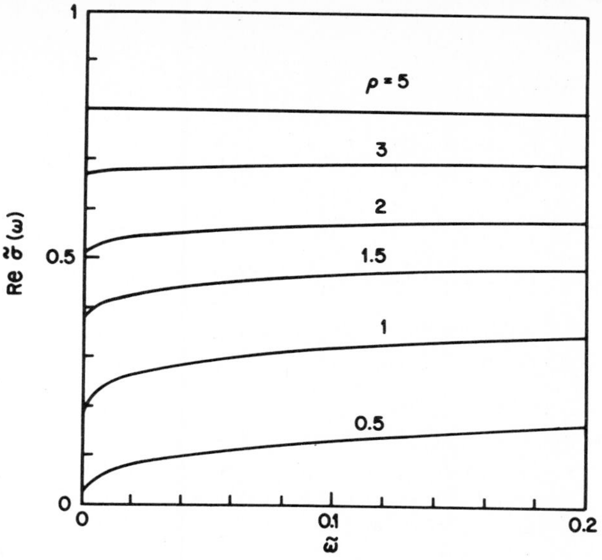

Fig. 11.2. The frequency dependence of the real part of $\tilde{\sigma}(\omega)$: a power law distribution in a chain. (Taken from ref. 31.)

$$P(w) = \begin{cases} \rho\, w^{\rho-1}/w_0^{\rho} & 0 \leqslant w \leqslant w_0 \\ 0 & \text{otherwise} \end{cases} , \tag{11.3}$$

where $\rho = n_s R_d$ is the dimensionless density of hopping sites.

Now, we map the actual positionally random system onto a regular lattice whose lattice constant is $a \equiv 1/n_s$ and the jump rate between nearest neighbors obeys the distribution Eq. (11.3). A schematic correspondence of the mapping is shown in Fig. 11.1.[31]

The distribution function Eq. (11.3) has already been used in Section 8 as a prototype function of the general distribution. Using the CMA approximation Eq. (5.11) with Eqs. (5.15) and (8.1), we obtain the frequency dependence of the ac conductivity as shown in Fig. 11.2.[31] As we decrease the density ρ from infinity to zero, the five regimes (1)-(5) given in Section 8 appear one by one. Explicitly, the low-frequency behavior of the conductivity is shown to have the following form:

(1) $\rho > 2$

$$\tilde{\sigma}(\omega) \approx \frac{\rho-1}{\rho} + \frac{1}{2\rho(\rho-2)} \left(\frac{\rho-1}{\rho} \right)^{1/2} (i\omega)^{1/2} , \tag{11.4}$$

(2) $\rho = 2$

$$\tilde{\sigma}(\omega) \approx \frac{1}{2} - \frac{1}{4\sqrt{2}} (i\tilde{\omega})^{1/2} \ln(i\tilde{\omega})^{1/2} , \tag{11.5}$$

(3) $1 < \rho < 2$

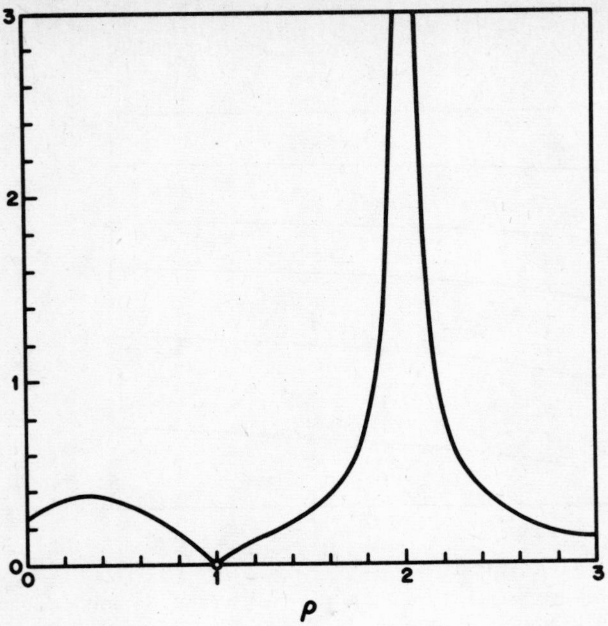

Fig. 11.3. The dependence of factor A on ρ. Open circle denotes a singular point. (Taken from ref. 31.)

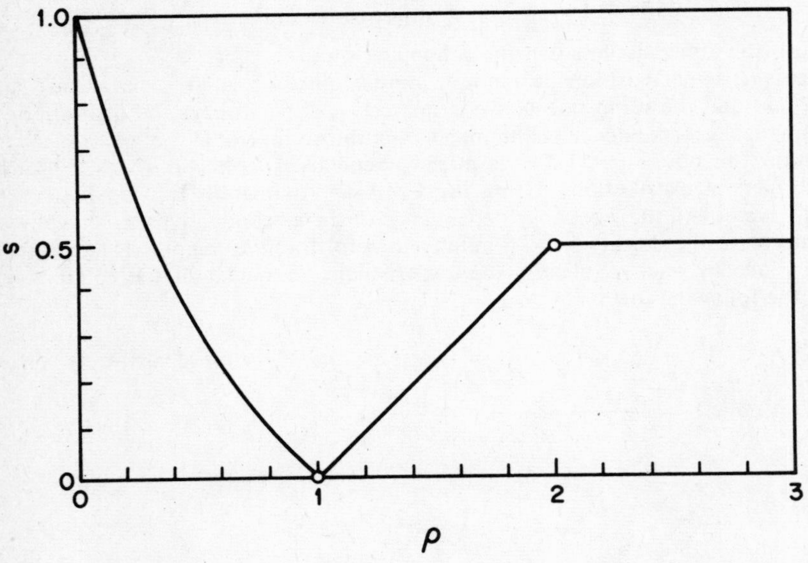

Fig. 11.4. The dependence of index s on ρ. Open circles denote singular points. (Taken from ref. 31.)

$$\tilde{\sigma}(\omega) \approx \frac{\rho-1}{\rho} + 2^{1-\rho}\left(\frac{\rho-1}{\rho}\right)^{(\rho+1)/2} \frac{(\rho-1)\pi}{\sin(\rho-1)\pi}(i\omega)^{(\rho-1)/2} , \tag{11.6}$$

(4) $\rho = 1$

$$\tilde{\sigma}(\omega) \approx -\frac{2}{\ln\{-i\tilde{\omega}/\ln(i\tilde{\omega})\}} , \tag{11.7}$$

(5) $0 < \rho < 1$

$$\tilde{\sigma}(\omega) \approx \left\{\frac{2^{1-\rho}\rho\pi}{\sin\rho\pi}\right\}^{-\frac{2}{1+\rho}} (i\omega)^{\frac{1-\rho}{1+\rho}} . \tag{11.8}$$

The ac part $\tilde{\sigma}(\omega)-\tilde{\sigma}(0)$ except for cases (2) and (4) has a form $A(i\omega)^s$. Figures 11.3 and 11.4 show the dependence of A and s on ρ, respectively. We can see that the ac part shows approximately symmetric behavior near $\rho=1$, which has been numerically observed by Richards and Renken.[32]

12. Bond-percolation model in a simple cubic lattice

We apply the CMA method to the bond percolation model in a simple cubic lattice. In order to obtain simple analytic results, we assume that $f_{\mathbf{k}}$ defined in Eq. (5.14) for the simple cublic lattice has a semi-elliptic density of states

$$\frac{1}{N}\sum_{\mathbf{k}}\delta(\epsilon-f_{\mathbf{k}}) = 2(1-\epsilon^2)^{1/2}/\pi , \tag{12.1}$$

with the correct behavior at the band edges. Using Eq. (12.1), the diagonal element of the random walk propagator Eq. (5.13) with $\mathbf{s}=\mathbf{s}_0$ is written as

$$\bar{P}(\mathbf{s},u\,|\,\mathbf{s}) = 2\{u+zw_c+\sqrt{u(u+2zw_c)}\}^{-1} . \tag{12.2}$$

The self-consistency equation (5.11) with Eqs. (7.1) and (12.2) can be solved analytically.[10] Figure 12.1 shows the frequency dependence of $\tilde{\sigma}(\omega)$. The dc conductivity vanishes at the percolation threshold p_c as

$$\tilde{\sigma}(0) = \begin{cases} \frac{3}{2}(p-p_c) & p \geqslant p_c \\ 0 & p \leqslant p_c \end{cases} , \tag{12.3}$$

where $p_c=1/3$. The ac part of the conductivity shows the following critical behavior at the percolation threshold:

$$\lim_{\tilde{\omega}\to 0}\frac{\mathrm{Re}\{\tilde{\sigma}(\omega)-\tilde{\sigma}(0)\}}{\tilde{\omega}^{3/2}} \sim (p-p_c)^{-3/2} \qquad (p>p_c) , \tag{12.4}$$

$$\lim_{\tilde{\omega}\to 0}\frac{\mathrm{Re}\tilde{\sigma}(\omega)}{\tilde{\omega}^{2}} \sim (p-p_c)^{-3} \qquad (p<p_c) \tag{12.5}$$

and

$$\lim_{\tilde{\omega}\to 0}\frac{\mathrm{Im}\tilde{\sigma}(\omega)}{\tilde{\omega}} \sim |p-p_c|^{-1} . \tag{12.6}$$

At $p=p_c$, $\tilde{\sigma}(\omega)$ vanishes as $(i\omega)^{1/2}$. Therefore, both $<P(\mathbf{s}_0,\infty\,|\,\mathbf{s}_0,0)>$ and $\tilde{\sigma}(0)$ vanish at

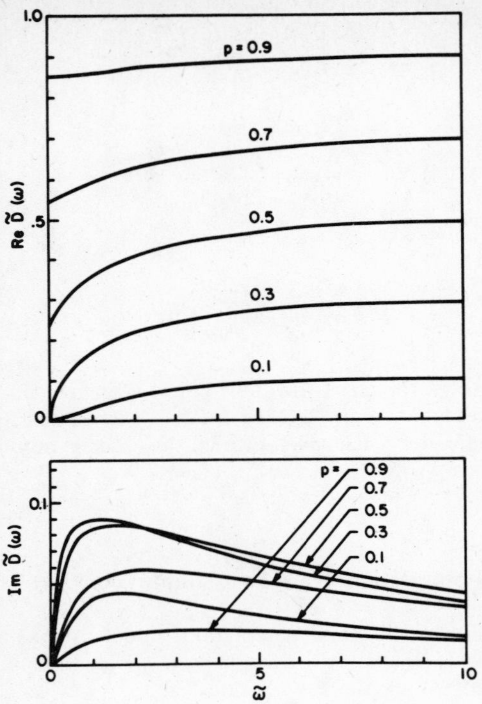

Fig. 12.1. The frequency and the probability dependence of the real and imaginary parts of the dimensionless ac conductivity $\tilde{\sigma}(\omega)$ for the bond-percolation model in the simple cubic lattice. (Taken from ref. 10.)

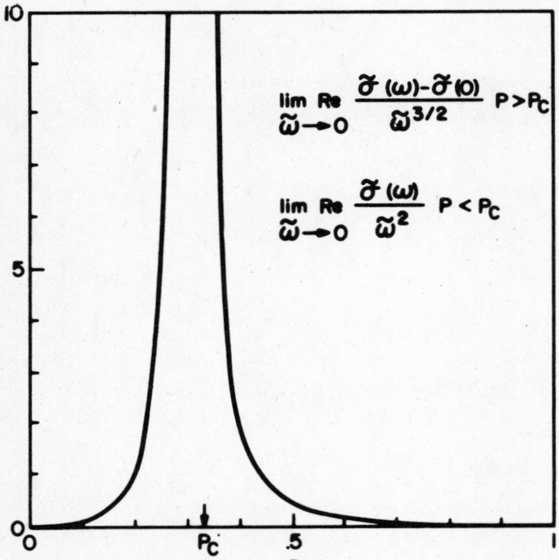

Fig. 12.2. The critical behavior of the real part of $\tilde{\sigma}(\omega)$ for the bond-percolation model in the simple cubic lattice at the static limit. [Equations (12.4) and (12.5).]

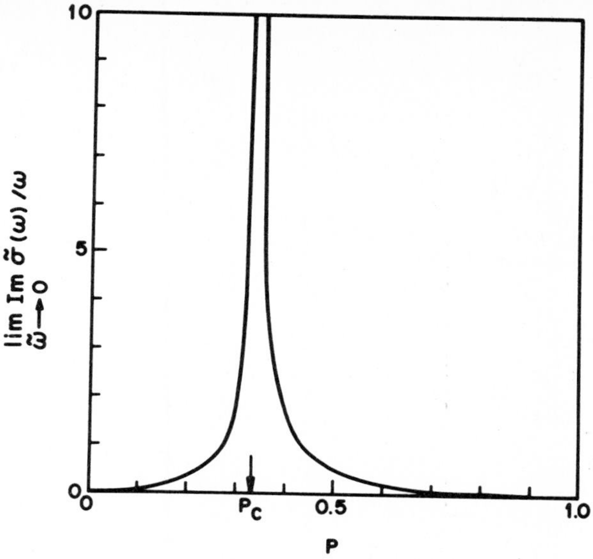

Fig. 12.3. The critical behavior of the imaginary part of $\tilde{\sigma}(\omega)$ for the bond-percolation model in the simple cubic lattice at the static limit. [(Eq. 12.6).]

the percolation threshold. The critical behavior of Eqs. (12.4)-(12.6) are depicted in Figs. 12.2 and 12.3. It should be remarked that the critical indices in Eqs. (12.3)-(12.6) are not changed even if we use the exact expression for $\bar{P}(\mathbf{s},u\,|\,\mathbf{s})$ instead of the approximation Eq. (12.2), although the proportional constants in Eqs. (12.4)-(12.6) are modified.[10]

13. Comparison with a computer simulation - Miller-Abrahams jump rate

In hopping between majority sites in doped semiconductors, Miller and Abrahams[33] derived the jump rate in the form

$$w_{ss'} \equiv w(r) = w_0(r/R_d)^{3/2} \exp(-r/R_d) , \tag{13.1}$$

where $r = |\mathbf{s}-\mathbf{s}'|$ and w_0 is a constant dependent on the temperature. Here, we have assumed that the activation energy is independent of r. Instead of treating the actual positionally disordered system we map the system onto a simple cubic lattice where the nearest neighbor jump rate obeys a distribution which is implied by the jump rate (13.1) and the distribution function $N(r)$ of the nearest neighbor distance r in a random system of uniform average density N_D. It is known that $N(r)$ is given by the Hertz distribution[34]:

$$N(r) = \frac{4\pi r^2 N_D}{z} \exp\{-\frac{4\pi N_D}{3z} r^3\} , \tag{13.2}$$

where $z(=6)$ is the coordination number of the simple cubic lattice and N_D is the density of hopping sites. The lattice constant a is chosen to be $(4\pi/3N_D)^{1/3}$. The CMA condition Eq. (5.11) reduces to

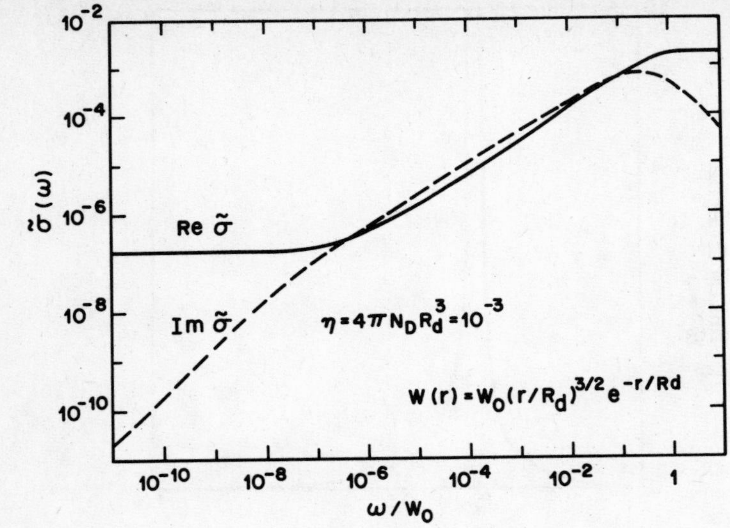

Fig. 13.1. Typical frequency dependence of the real and imaginary part of $\tilde{\sigma}(\omega)$ for the Miller-Abrahams jump rate (13.1). $\eta \equiv 4\pi N_D R_d^3 = 10^{-3}$.

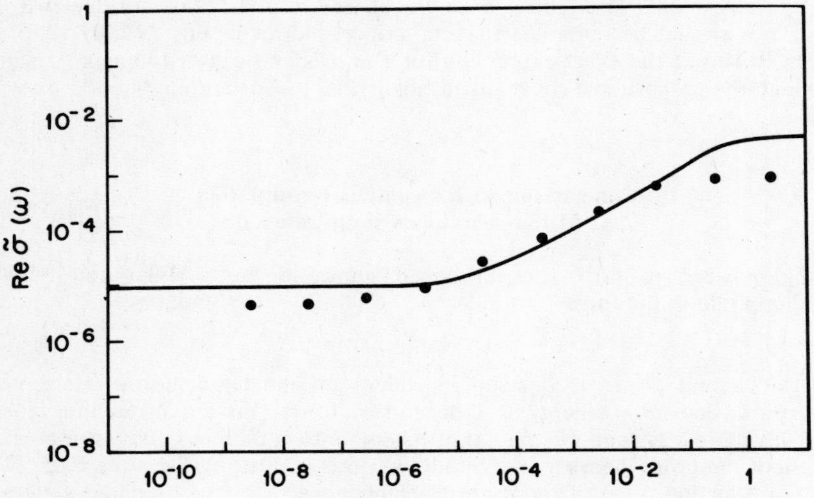

Fig. 13.2. A comparison of the frequency dependence of the real part of the dimensionless conductivity $\tilde{\sigma}(\omega)$ for the Miller-Abrahams jump rate. The solid line is the result given by the present coherent medium approximation and the solid circles are the result of the computer simulation given by McInnes et al (ref. 35). $\eta \equiv 4\pi N_D R_d^3 = 2.15 \times 10^{-3}$.

$$\int_0^\infty \frac{N(r)\{w_c - w(r)\}\, dr}{1 - 2(\bar{P}_{11} - \bar{P}_{12})\{w_c - w(r)\}} = 0 \; . \tag{13.3}$$

Typical results are shown in Fig. 13.1 for the frequency dependence of the real and imaginary part of the ac conductivity, where Eq. (12.2) has been used for $\bar{P}_{11}$ and the set of self-consistency equations (5.15), (12.2), (13.1), (13.2) and (13.3) has been solved numerically.

J. A. McInnes et al[35] performed a computer simulation on the positionally disordered system. A comparison is given in Fig. 13.2. The agreement between the CMA method and their computer simulation is good in the low-frequency region. The discrepancy at the high-frequency may be attributed to our use of the lattice model.

14. Impurity conduction

For impurity conduction which we have explained in Section 1, Scher and Lax[9] have shown that a simplified jump rate

$$w(r) = w_0 \exp(-r/R_d) \; , \tag{14.1}$$

describes experiments very well. We have used the same notation w_0 for the coefficient in Eq. (14.1), which takes a different value from w_0 in Eq. (13.1). Using the same method as in Section 13, we obtained the ac conductivity for doping levels $N_D = 2.7 \times 10^{17}\text{cm}^{-3}$ and $1.2 \times 10^{16}\text{cm}^{-3}$.[10] Figures 14.1 and 14.2 show a comparison with experiments given by Pollak

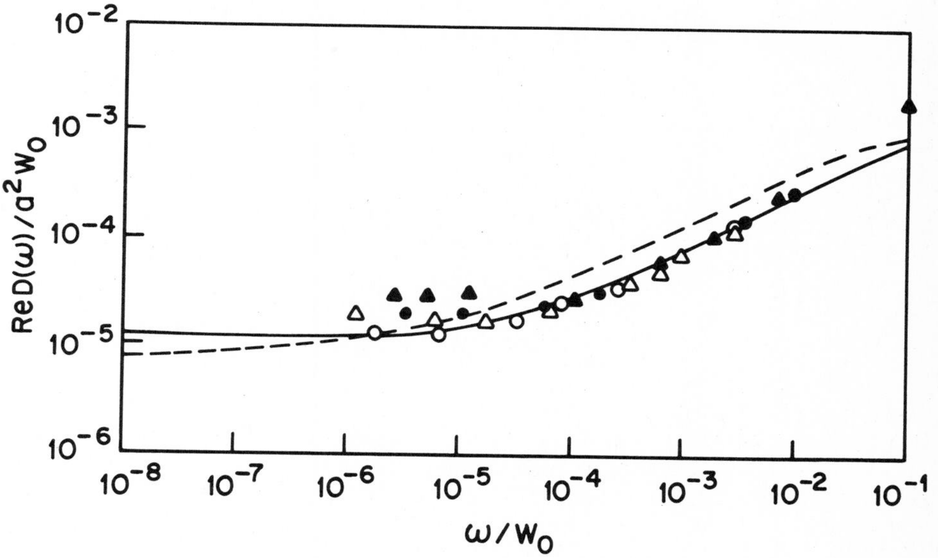

Fig. 14.1. Comparison of the theoretical and experimental values of $\mathrm{Re}\,\tilde{D}(\omega)$ for $N_D = 2.7 \times 10^{17}\text{cm}^{-3}$. The solid line is the present result and the broken line is the SL result. The latter and experimental data for $\mathrm{Re}\,\sigma(\omega)$ are taken from Ref. 9(b). The dimensionless ac conductivity or diffusion constant $\tilde{D}(\omega) = \sigma(\omega)/[(ne^2/kT)(a^2 w_0)]$ with $n = 0.8 \times 10^{15}\text{cm}^{-3}$ and $a = (4\pi N_D/3)^{-1/3}$ is shown. (Taken from ref. 10.)

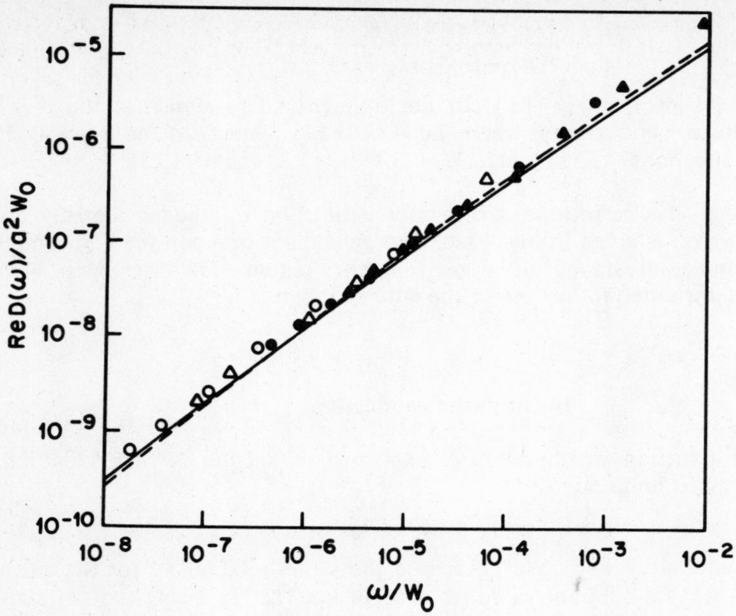

Fig. 14.2. Comparison of the theoretical and experimental values of Re $\tilde{D}(\omega)$ for $N_D=1.2\times10^{16}\text{cm}^{-3}$. Symbols are the same as in Fig. 14.1. According to SL, a is assumed to be $0.9\ (4\pi N_D/3)^{-1/3}$ in this plot. (Taken from ref. 10.)

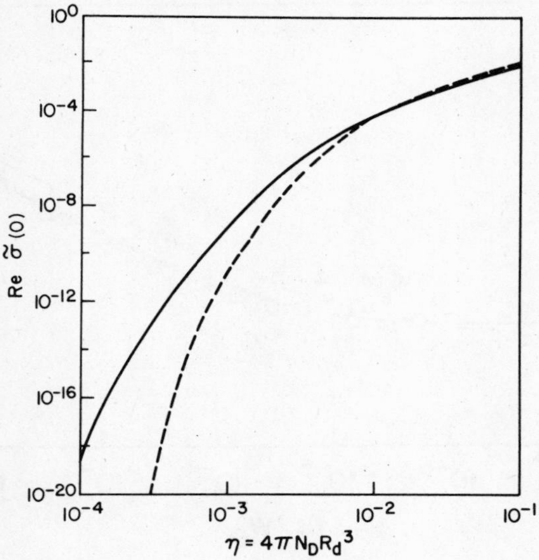

Fig. 14.3. Comparison of the present method (solid line) and the SL method (broken line) for the static dimensionless diffusion constant. Re $\sigma(0)/a^2w_0=\frac{1}{2}\exp\{-(18\ln 3/2)^{1/3}\eta^{-1/3}\}$ for the present method and $e^{\gamma}\eta^{1/4}\exp\{-2\eta^{-1/2}/3\}/6\sqrt{\pi}$ for the SL method with $\eta=4\pi N_D R_d^3$ and $\gamma=0.5772$. (Taken from ref. 10.)

and Geballe.[1] The results given by Scher and Lax[9] (SL) using the CTRW method are also shown in these figures. The ac part of our results shows the well-known ω^s dependence and the exponent s reads as 0.52 for three decades of $\tilde{\omega}$ in Fig. 14.1 and 0.77 for seven decades of $\tilde{\omega}$ in Fig. 14.2.

At low densities, the static conductivity can be shown to have the density dependence

$$\tilde{\sigma}(0) \approx \frac{1}{2} \exp\{-(18\ln 3/2)^{1/3} \eta^{-1/3}\} , \qquad (14.2)$$

where $\eta = 4\pi N_D R_d^3$. A comparison between the CTRW and the CMA methods for $\tilde{\sigma}(0)$ is given in Fig. 14.3.

15. Summary

We have presented the coherent medium approach to hopping conduction problems where the motion of carriers obey the usual random walk equation and discussed the existence of the coherent medium which is defined through an average of the random walk propagator. We have introduced the coherent medium approximation (CMA) to obtain an approximate but easily tractable coherent medium. The CMA is a generalized use of the coherent potential approximation (CPA) in the master equation. The CPA is one of the most fruitful methods in treating random systems and is widely used in electron[19] and phonon problems.[18] We have applied the CMA to various cases of hopping conduction in one- and three-dimensions and compared the results with experiment. A simple comparison of the CMA with other methods has also been given.

We have derived the CMA condition Eq. (5.11) from the multiple scattering formalism due to Lax.[12,13] An alternative derivation of Eq. (5.11) was given by Odagaki and Lax,[10] where a traditional idea of the effective medium approximation was used.[36] Namely, a random unit is subjected to an as yet unknown effective medium and the effective medium is determined to be such that the resulting extra perturbation is required to vanish on the average over all possibilities of the random unit. A similar condition to Eq. (5.11) has also been used in the problem of random resistor networks.[37]

We hope that further improvements of the coherent medium approximation will be developed including traps, asymmetric jump rates, cluster effects and a similar method will be applied to other problems, for example, optical properties and magnetic properties of random systems.

Acknowledgements

The work at City College was supported in part by a grant from the Army Research Office, the PSC-CUNY Research Award Program, Department of Energy, and in conjunction with the binational agreement between the National Science Foundation and the Japanese Society for the Promotion of Science, under contract No. INT-7918591.

References

[1]M. Pollak and T. H. Geballe, Phys. Rev. **122**, 1742 (1961).

[2]J. M. Reyes, M. Sayer, A. Mansingh and R. Chen, Can. J. Phys. **54**, 413 (1976).

[3]A. S. Barker Jr., J. A. Ditzenberger and J. P. Remeika, Phys. Rev. **B14**, 4254 (1976).

[4]M. Sayer, A. Mansingh, J. B. Webb and J. Noad, J. Phys. C: Solid State Phys. **11**, 315 (1978).

[5]R. M. Mehra, P. C. Mathur, A. K. Kathuria and R. Shyam, Phys. Rev. **B18**, 5620 (1978).

[6]M. Suzuki, J. Phys. Chem. Solids **41**, 1253 (1980).

[7]S. R. Elliott, Phil. Mag. **36**, 1291 (1977).

[8]Actually, $P(\mathbf{s},t\,|\,\mathbf{s}_0,0)$ can be written as an absolute square of a matrix element of $e^{-iHt/\hbar}$, where H is the total Hamiltonian of the underlying problem. See refs. 9 and 10.

[9]H. Scher and M. Lax,(a) Phys. Rev. **B7**, 4491 (1973);(b) ibid, 4502 (1973).

[10]T. Odagaki and M. Lax, Phys. Rev. B to be published.

[11]For example, T. Holstein, S. K. Lyo and R. Orbach, Phys. Rev. **B15**, 4693 (1977).

[12]M. Lax, Rev. Mod. Phys. **23**, 287 (1951); Phys. Rev. **85**, 621 (1952).

[13]M. Lax, "Wave Propagation and Conductivity in Random Media," in *Stochastic Differential Equations* SIAM-AMS (Soc. for Industrial and Applied Math. - American Mathematical Soc.) Proc. vol. **6**, 35-95 Amer. Math. Soc. Providence, R.I. (1973).

[14]J. Klafter and R. Silbey, Phys. Rev. Letts. **44**, 55 (1980).

[15]R. Zwanzig, J. Chem. Phys. **33**, 1338 (1960); See also *Lectures in Theoretical Physics* Vol. **III**, 106 (edited by W. E. Brittin), Interscience, New York (1961); and Phys. Rev. **124**, 983 (1961).

[16]See, for example, F. Yonezawa and K. Morigaki, Prog. Theor. Phys. Suppl. **53**, 1 (1973); R. J. Elliott, J. A. Krumhansl and P. L. Leath, Rev. Mod. Phys. **46**, 465 (1974).

[17]M. Lax, Phys. Rev. **79**, 200 (1950).

[18]D. W. Taylor, Phys. Rev. **156**, 1017 (1967).

[19]P. Soven, Phys. Rev. **156**, 809 (1967).

[20]J. Bernasconi, W. R. Schneider and W. Wyss, Z. Physik B **37**, 175 (1980).

[21]M. Lax and H. Scher, Phys. Rev. Letts. **39**, 781 (1977).

[22]E. Feenberg, Phys Rev. **74**, 206 (1948); E. N. Economou, "*Green's Functions in Quantum Physics*," (Springer-Verlag, Berlin, Heidelberg 1979).

[23]R. Abou-Chacra, P. W. Anderson, and D. J. Thouless, J. Phys. C: Solid State Phys. **6**, 1734 (1973).

[24]J. Bernasconi, S. Alexander and R. Orbach, Phys. Rev. Letts. **41**, 185 (1978).

[25]B. Movaghar, J. Phys. C: Solid State Phys. **13**, 4915 (1980).

[26]T. Odagaki and M. Lax, Phys. Rev. B to be published.

[27]T. Odagaki and M. Lax, Phys. Rev. Letts, **45**, 847 (1980).

[28]T. Odagaki and M. Lax, in preparation.

[29]J. Bernasconi, H. U. Beyeler, S. Strässler and S. Alexander, Phys. Rev. Letts. **42**, 819 (1979).

[30]S. Alexander, J. Bernasconi, W. R. Schneider and R. Orbach, Rev. Mod. Phys. **53**, 175 (1981).

[31]T. Odagaki and M. Lax, in preparation.

[32]P. M. Richards and R. L. Renken, Phys. Rev. **B21**, 3740 (1980).

[33]A. Miller and E. Abrahams, Phys. Rev. **120**, 745 (1960).

[34]S. Chandrasekhar, Rev. Mod. Phys. **15**, 1 (1943).

[35]J. A. McInnes, P. N. Butcher and J. D. Clark, Phil. Mag. **B41**, 1 (1980).

[36]R. Landauer, "Electrical Conductivity in Inhomogeneous Media," in *Proceedings of the First Conference on the Electrical Transport and Optical Properties of Inhomogeneous Media*, edited by J. C. Garland and D. B. Tanner, (A.I.P. New York, 1978), p 2.

[37]S. Kirkpatrick, Rev. Mod. Phys. **45**, 574 (1973).

NONLINEAR EVOLUTION EQUATIONS WITH RAPIDLY OSCILLATING INITIAL DATA

D. McLaughlin, G. Papanicolaou
Courant Institute of Mathematical Sciences

O. Pironneau
University of Paris XI

Abstract

We give a brief description of how one can analyze the behavior of solutions of nonlinear equations when the initial data oscillate very rapidly.

* * *

Let $u(t,x)$ and $p(t,x)$ be the velocity and pressure, respectively, of a viscous incompressible fluid at $x \in \mathbb{R}^3$ and at time $t \geq 0$. The Navier-Stokes equations

(1) $$u_t + u \cdot \nabla u + \nabla p = \nu \Delta u \ , \qquad \nabla \cdot u = 0$$

are satisfied with ν the kinematic viscosity. Suppose that at time $t = 0$ the velocity field is made up of a mean part and a fluctuating part. How does such an initial flow evolve?

To be more precise we suppose that

(2) $$u(o,x) = V(x) + W(\frac{x}{\varepsilon})$$

where $V(x)$ is a given smooth vector function and $W(y)$ is a periodic or stationary random vector function with mean zero. The small parameter ε is the ratio of the scales of variation of the two parts of the flow when the scale of variation of V is one. The Reynolds number $Re = LU/\nu$, where L is a length scale and U a velocity scale, can be calculated with $L = 1$ or $L = \varepsilon$ i.e., relative to the mean flow or relative to the fluctuating flow. To make the problem physically interesting the Reynolds number relative to the fluctuating flow must not be small. Therefore, we let ν be proportional to ε ($\nu \to \varepsilon\nu$ in (1)). Now the question is: how does $u(t,x)$, $t > 0$, behave as ε tends to zero?

The answer depends in general on two additional facts. First on whether $W(y)$ is periodic or random, which means on whether the power spectrum of W has zero as an isolated element or not. This is important because in the periodic case there is a clean separation of scales between the two parts of the flow which does not exist otherwise.

Second the behavior as $\varepsilon \to 0$ depends on the nature of the solutions of

$$(3) \qquad v_t + v \cdot \nabla v + \nabla P = 0 , \qquad \nabla \cdot v = 0 ,$$
$$v(o,x) = V(x)$$

which is the Euler equation for the mean flow. If v is smooth in some time interval we have one kind of behavior and if v has singularities another. Again, if the solution v is smooth there is only one scale of variation associated with it. If it has singularities there is scale mixing because v changes then rapidly and cannot be neatly distinguished from the fluctuating part of the flow.

In the simplest case: $v(t,x)$ smooth in $[o,T]$ and $W(y)$ periodic, we have the following result.

The solution $u^\varepsilon(t,x)$ of

$$(1') \qquad u^\varepsilon_t + u^\varepsilon \cdot \nabla u^\varepsilon + \nabla p^\varepsilon = \varepsilon \nu \Delta u , \qquad \nabla \cdot u^\varepsilon = 0$$
$$u^\varepsilon(o,x) = V(x) + W(\tfrac{x}{\varepsilon})$$

converges to the solution $v(t,x)$ of (3) exponentially fast as $\varepsilon \to 0$ and

$$(4) \qquad \lim_{\varepsilon \downarrow 0} \varepsilon \log |u^\varepsilon(t,x) - v(t,x)| = -S(t,x) , \qquad 0 \le t \le T ,$$

where $S(t,x)$ satisfies

$$(5) \qquad S_t + v \cdot \nabla S + \frac{\nu}{2} (\nabla S)^2 = \frac{\nu}{2} (\nabla \psi \cdot e)^2$$
$$S(o,x) = 0$$

$$(6) \qquad \psi_t + (v + \nu \nabla S) \cdot \nabla \psi = 0 , \qquad \psi(o,x) = x$$

Here e is a unit vector in one of the coordinate directions and it is assumed that in the interval $(o,T]$, the positive function $S(t,x)$ takes its minimum value for this choice of e.

The way one obtains (4) is by looking for a solution of (1') in the form

$$(7)\quad u^{\varepsilon}(t,x) = v(t,x) + \sum_{p} e^{-S^{(p)}(t,x)/\varepsilon} \left[W^{(p)}\left(\frac{\psi^{(p)}}{\varepsilon},x,t\right) + \varepsilon \hat{W}^{(p)}\left(\frac{\psi^{(p)}}{\varepsilon},x,t\right) + \ldots \right]$$

with $\psi^{(p)}(t,x)$ and $S^{(p)}(t,x)$ suitably determined in the expansion process.

We see therefore that in the simplest case of clean separation of scales there is very little energy transfer between the mean and fluctuating flow as one would expect.

How does one analyze situations where substantial energy transfer exists? This is a difficult question that can so far be answered only for model problems such as Burgers' equation [1],[2].

One can obtain some understanding of energy transfer by assuming that a fluctuating external force acts on the fluid. The type of fluctuating force one assumes is of course determined by what kind of energy transfer mechanism one wants to excite. The notion of an external turbulent force is not a primary physical one. One visualizes the interaction of fluid masses with different characteristic length scales by means of forces acting on bigger scales to stir them up or on smaller scales to keep them going. Therefore probing a fluid with a turbulent force can be viewed as an intermediate step in a grand self-consistent scheme that explains energy transfer between fluid elements with different characteristic length scales. The mathematical implementation of such a scheme is extremely difficult and almost nothing is known about it.

The force we choose is as follows. The Navier-Stokes equations are

$$(8)\quad u_t + u^{\varepsilon}\cdot\nabla u^{\varepsilon} = \nabla\cdot\tau + \frac{1}{\varepsilon} F\left(\frac{\theta^{-1}}{\varepsilon},x,t\right)\cdot\nabla\theta^{-1}$$

$$\nabla\cdot u = 0$$

$$u(o,x) \quad \text{given,}$$

and the stress tensor τ is

$$(9)\quad \tau_{ij} = -p^{\varepsilon}\delta_{ij} + \varepsilon\nu(u^{\varepsilon}_{i,j} + u^{\varepsilon}_{j,i}) , \qquad i,j = 1,2,3 .$$

The functions $\theta(t,x)$ and $\theta^{-1}(t,x)$ are defined below.

We look now for a solution of (8) in the form

(9) $$u_i(t,x) = v_i(t,x) + \theta_{i,K}(t,\theta^{-1}(t,x))\left[W_k\left(\frac{\theta^{-1}(t,x)}{\varepsilon},x,t\right) + \varepsilon\tilde{u}_k\left(\frac{\theta^{-1}(t,x)}{\varepsilon},x,t\right)\right]$$

(10) $$p^{\varepsilon}(t,x) = \pi\left(\frac{\theta^{-1}(t,x)}{\varepsilon},x,t\right) + P(t,x) + \varepsilon\tilde{p}^{\varepsilon}\left(\frac{\theta^{-1}(t,x)}{\varepsilon},x,t\right) .$$

We employ the summation convention and use the notation comma-subscript for derivative. Capital letter subscript means derivative with respect to the argument which is evaluated at θ^{-1}. When the expressions (9) and (10) are inserted in (8) and powers of ε are sorted out, we find that the fields W,π and v,P must satisfy the following equations.

(11) $$K_{in}K_{ik}W_m W_{k,m} + \pi_{,n} = \nu K_{ik}K_{in}K^{-1}_{mj}K^{-1}_{\ell j}W_{k,j\ell} + F_n$$

$$W_{k,k} = 0 ,$$

$$\langle\pi\rangle = 0 , \quad \langle F_n\rangle = 0 , \quad \langle W_m\rangle = 0$$

(12) $$v_t + v\cdot\nabla v = \nabla\cdot T , \qquad \nabla\cdot v = 0 , \quad v(o,x) = v_o(x) ,$$

(13) $$T_{ij} = -P\delta_{ij} - \theta_{j,p}\theta_{i,k} \langle W_p(\cdot,\nabla\theta)W_k(\cdot,\nabla\theta)\rangle$$

(14) $$\theta_t = v(t,\theta) , \quad \theta(o,x) = x .$$

In (11) $W_m = W_m(y;K)$ where K is a 3×3 nonsingular matrix on which W depends parametrically and $F_n(y)$ is the nth component of the given external force assumed to be stationary random process with mean zero. Mean values or expectations are denoted by $\langle\ \rangle$. One can show that for each K nonsingular and $\nu > 0$ (11) has a unique solution in a suitable space which is a stationary random field, is divergence free and has mean zero. Once $W(y;K)$ has been obtained one calculates

(15) $$C_{pk}(K) = \langle W_p(y;K)W_k(y;K)\rangle ,$$

which is independent of y by stationarity, and then the effective stress tensor is given by (13) or

(16) $$T_{ij} = -P\delta_{ij} - \theta_{j,p}\theta_{i,k}C_{pk}(\nabla\theta) .$$

The mean velocity $v(t,x)$ satisfies (12) which is the Euler equation with a new component in the stress tensor due to the external force. The mapping $x \to \theta(t,x)$ of $\mathbb{R}^3$ to $\mathbb{R}^3$ is just the orbit of the fluid particle at time t as (14) shows.

We see now the effect of the special external random force on the flow as follows. From (8) the force field is assumed to be transported along the characteristics of the mean flow. The fluctuation or microstructure field W is determined locally by its own steady Navier-Stokes equation (11) relative to a coordinate system adapted to the characteristics of the mean flow (parametric dependence on K). The microstructure flow W is affected by the mean flow through the slow changing of K $(K = \nabla\theta(t,x))$. The mean or macroscopic flow $v(t,x)$ satisfies the inviscid Euler equation (12) with effective stress tensor (13) or (16). The microstructure affects the macrostructure through the "Reynolds stress" term $\nabla\theta\nabla\theta\ C(\nabla\theta)$ in (16).

Depending on the type of force F one has, the new term in the effective stress in (16) may correspond to energy transfer to the mean flow or energy transfer out of it. It is very hard to say which forces produce a dissipative and which a pumping effect. Our main result is that if (12),(13),(14) have a smooth solution in some finite time interval then (11) is well defined for F small enough and the whole solution u^{ε} of (8) is well represented by (9),(10).

There are two issues that must be settled now in connection with the external force problem. The first one is the characterization of forces that lead to a dissipative mean stress tensor as mentioned above. The second and harder problem is the determination of a self-consistent scheme for energy transfer in a realistic situation in which each step in some iteration process is a problem of the form (8) with the force F determined through the flow at a previous step. This sounds like some kind of renormalization group analysis [3] but concrete results seem hard to get at present.

A number of other problems, mostly model problems that give rise to Reynolds stress terms, have been worked out [4] partly in connection with the theory of turbulent diffusion. Other nonlinear equations with rapidly oscillating initial data are considered in [5].

Note: This work was supported by the Air Force Office of Scientific Research, Grant AFOSR-80-0228.

REFERENCES

[1] J.M. Burgers, The Nonlinear Diffusion Equation, D. Reidel Publishing Co., 1974.

[2] S. Kida, Asymptotic properties of Burgers turbulence, J. Fluid Mech. (1979) 93, 2, pp. 337-377.

[3] J.-D. Fournier, Quelque methodes systematiques d'approximation en turbulence homogene, These, Université de Nice, 1977.

[4] D. McLaughlin, G. Papanicolaou and O. Pironneau, Selfconsistent advection of microstructure in viscous fluids. (To appear.)

[5] D. McLaughlin, G. Papanicolaou and L. Tartar, Weak limits of conservation laws with oscillating data, SIAM J. Appl. Math. (To appear.)

A COMPARISON OF TWO METHODS FOR DERIVING BOUNDS ON THE EFFECTIVE CONDUCTIVITY OF COMPOSITES

G. W. Milton
Laboratory of Atomic and Solid State Physics and
Materials Science Center, Clark Hall,
Cornell University, Ithaca, NY, 14853, U.S.A.
and
R. C. McPhedran
Department of Theoretical Physics,
The University of Sydney, Sydney, N.S.W., 2006, Australia

Abstract

An infinite set of bounds on the effective conductivity σ_e of a two-component composite has previously been derived using the analytic properties of σ_e as a function of σ_1 and σ_2 (the conductivities of the components). We show that this same set of bounds can alternatively be derived from variational principles. The bounds incorporate information about the microstructure of the composite in addition to the volume fractions of the components.

I. Introduction

There are two dominant methods for deriving rigorous bounds on the effective conductivity σ_e of a two-component composite. The first method, largely developed by Hashin and Shtrikman,[1] Brown[2] and Beran,[3] involves the use of variational principles. A second method, pioneered by Bergman,[4] makes use of the analytic properties of σ_e as a function of the conductivities σ_1 and σ_2 of the components.

Using the analytic method, an infinite set of bounds on σ_e has been derived by Milton.[5] The appropriate pair of bounds (one an upper bound and the other a lower bound) depend on the known structure and properties of the composite. Each bound-pair is characterized by an integer pair (L,M): L signifies that the bound-pair incorporates L structural parameters; M represents the number of known values of σ_e (for various values of σ_1 and σ_2) which are incorporated in the bounds. The set of bounds (1,0), (2,0) and (2,1) were originally derived using the analytic method by Bergman.[4]

In this paper we show that the entire set of bounds (L,O) where L is arbitrary can also be derived using variational principles. First we briefly describe these bounds.

II. Description of the (L,O) bounds

It is convenient to introduce a new variable

$$\rho = (\sigma_1 - \sigma_2)/(\sigma_1 + \sigma_2), \tag{1}$$

and to assume, without loss of generality that

$$\sigma_1 \geq \sigma_2 = 1. \tag{2}$$

The Lth order bound-pair incorporates the values of the structural parameters

$$Q_n = \left.\frac{d^n \sigma_e(e)}{d\rho^n}\right|_{\rho = 0}, \tag{3}$$

for n = 1 to n = L. It follows from the work of Brown[6] that for any three-dimensional isotropic composite material,

$$Q_1 = 2f_1 \tag{4}$$

$$Q_2 = 4f_1(1 + 2f_1)/3 \tag{5}$$

$$Q_3 = 16f_1^2 - 4f_1 + 16f_1f_2(2\zeta_1 + f_2)/3 \tag{6}$$

where $f_1 = 1 - f_2$ is the volume fraction occupied by component 1 and ζ_1 is a geometric parameter which can be written in terms of the three-point correlation function (not usually known) and can have any value in the range [0,1]. Figure 1 shows how ζ_1 depends on f_1 for periodic systems, in particular the simple cubic (SC), body centered cubic (BCC) and face centered cubic (FCC) lattices of spheres of uniform radius. This same parameter ζ_1 is also incorporated in bounds on the effective elastic moduli of two-component composites.

The values of the Q_n for higher values of n can in principle be determined from the n-point correlation function.[6] However, if $n \geq 2$, the n-point correlation function contains a vast amount of statistical information, not all of which is relevant to the determination of $\sigma_e(\sigma_1,\sigma_2)$. Indeed, composites with different geometrical structures may have the same characteristic function $\sigma_e(\sigma_1,\sigma_2)$. It is appropriate to think of the set of parameters Q_n as representing the relevant details of the microstructure of the composite.

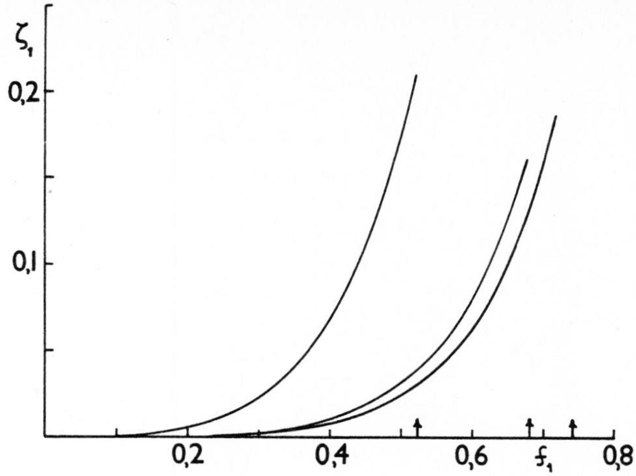

Fig. 1 The structural parameter ζ_1 as a function of the volume fraction f_1 of the inclusions for periodic lattices of spheres of uniform size. The curves from left to right correspond to SC, BCC and FCC lattices and each arrow denotes the volume fraction when the spheres touch. The graphs are plotted from the tabulated data of McPhedran and Milton.[7]

Numerical studies by McPhedran and Milton[7] have shown that provided σ_1/σ_2 is not zero or not infinite, the bounds converge rapidly to the exact value of σ_e as progressively more information is supplied, i.e., as L is increased. These studies have been carried out again for periodic systems, for SC, FCC and BCC lattices of spheres of uniform radius and for square and hexagonal arrays of cylinders. The convergence properties persist even for ratios of σ_1/σ_2 as large as 10,000:1 (Figs. 2,3). As yet we do not know of any analytic proof that the bounds always converge to the precise value of σ_e (provided $\sigma_1/\sigma_2 \neq 0$ or ∞), but it is conjectured that this is the case.

The prescription for caluclating the bounds depends on whether L is even or odd. In either situation the bounds can be expressed in terms of the following function:

$$g(\rho) = \left(1 + \sum_{h=1}^{N} \alpha_h \rho^h\right) / \left(1 + \sum_{h=1}^{N} \beta_h \rho^h\right). \tag{7}$$

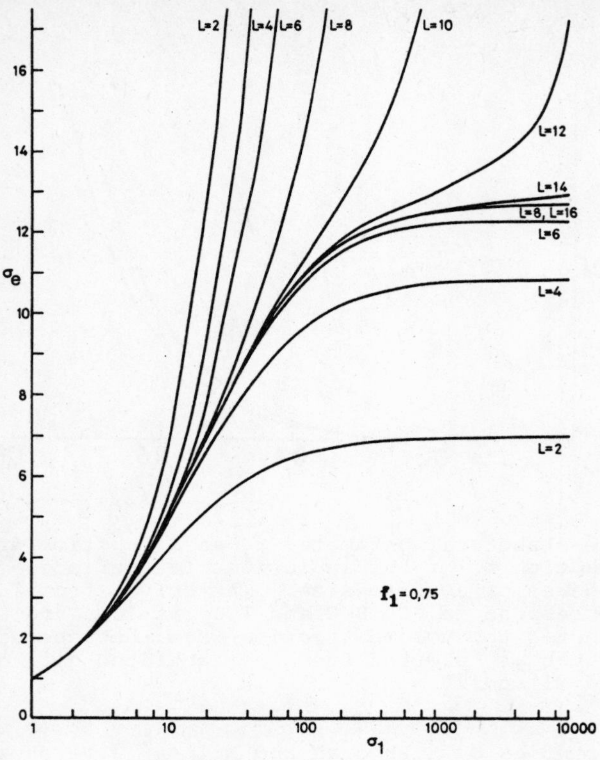

Fig. 2 The upper and lower Lth order bounds on the transverse conductivity of a square array of circular cylinders of uniform diameter, conductivity σ_1, in a matrix of unit conductivity. The cylinders occupy 75% of the composite and so are close to touching. All the bounds, for even L, are attainable by geometries consisting of multicoated cylinders.[5] The graph is taken from McPhedran and Milton.[7]

Here the 2N real coefficients α_h and β_h are such that

$$\left.\frac{d^n g}{d\rho^n}\right|_{\rho=0} = Q_n \quad (n=1,2...,L). \tag{8}$$

If L is even and 2N = L, then the α_h and β_h are determined by the 2N constraints (8). In this case $g(\rho)$ is a <u>lower bound</u> on σ_e. Alternatively if 2N = L + 2 and we impose the additional constraints

$$g(1) = \infty, \tag{9}$$

$$g(-1) = 0, \tag{10}$$

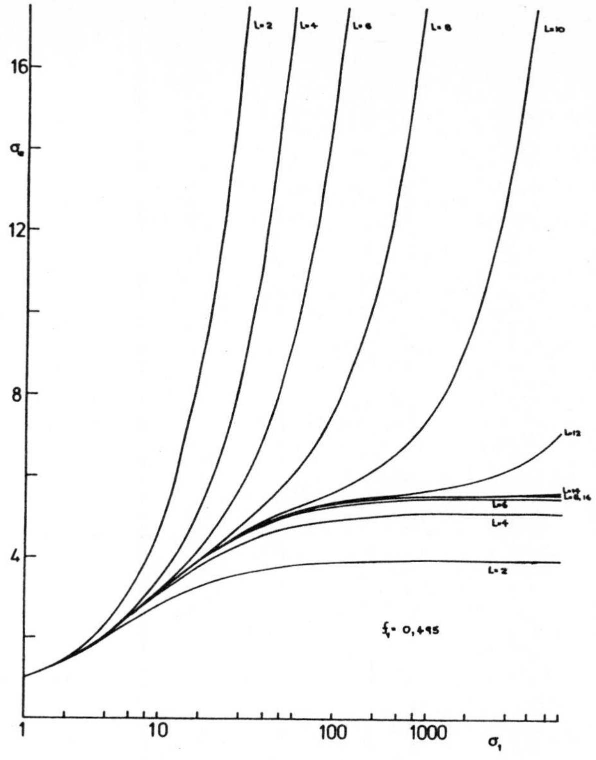

Fig. 3 As for Fig. 2, but for a cubic array of spheres occupying 49.5% of the composite. It is not known if these bounds are attainable. The graph is taken from McPhedran and Milton.[7]

then $g(\rho)$, which is again uniquely determined, is an <u>upper bound</u> on σ_e.

If L is odd and $2N = L + 1$, then $g(\rho)$ is an <u>upper or lower bound</u> on σ_e, depending on whether we impose the addition constraint (9) or the constraint (10).

As an example of the use of these bounds, let us take $L = 1$. We find that

$$\frac{1 + \rho}{1 - (f_1 - f_2)\rho} \leq \sigma_e \leq \frac{1 + (f_1 - f_2)\rho}{1 - \rho} , \tag{11}$$

or equivalently

$$[f_1/\sigma_1 + f_2]^{-1} \leq \sigma_e \leq f_1\sigma_1 + f_2 . \tag{12}$$

The bounds (12) are precisely the same as those derived by Wiener.[8] Similarly the L = 2, L = 3, L = 4, and L = 5 bounds correspond exactly with the bounds derived by Hashin and Shtrikman,[1] Beran,[3] Phan-Thien and Milton,[9] and Elsayed.[10] These bounds were derived using variational principles.

We will now show that the entire set of bounds, for arbitrary L, can in fact be derived from variational principles.

III. Derivation of bounds using classical variational principles

According to one of the classical variational principles, if we choose any trial electric field $\underset{\sim}{E}^T$ such that

$$\langle \underset{\sim}{E}^T \rangle = \hat{\underset{\sim}{x}}, \tag{13}$$

$$\underset{\sim}{\nabla} \times \underset{\sim}{E}^T = 0, \tag{14}$$

where $\hat{\underset{\sim}{x}}$ is a unit vector in the x direction, and the brackets $\langle\ \rangle$ denote a volume average, then the quantity

$$\sigma_{eU} = \langle \sigma \underset{\sim}{E}^T \underset{\sim}{E}^T \rangle, \tag{15}$$

is an upper bound on σ_e. The proof is straightforward. If $\underset{\sim}{E}$ denotes the actual electric field (with $\langle \underset{\sim}{E} \rangle = \hat{\underset{\sim}{x}}$) and $\underset{\sim}{E}^* = \underset{\sim}{E}^T - \underset{\sim}{E}$, then

$$\sigma_{eU} = \langle \sigma \underset{\sim}{E}\underset{\sim}{E} \rangle - 2\langle \sigma \underset{\sim}{E}^* \underset{\sim}{E} \rangle + \langle \sigma \underset{\sim}{E}^* \underset{\sim}{E}^* \rangle. \tag{16}$$

By applying Green's theorem we find

$$\langle \sigma \underset{\sim}{E}\underset{\sim}{E} \rangle = \sigma_e, \tag{17}$$

$$\langle \sigma \underset{\sim}{E}^* \underset{\sim}{E} \rangle = 0, \tag{18}$$

and hence

$$\sigma_{eU} = \sigma_e + \langle \sigma \underset{\sim}{E}^* \underset{\sim}{E}^* \rangle, \tag{19}$$

which is clearly greater than σ_e.

The actual electric field $\underset{\sim}{E}$ will be an analytic function of ρ, and so if ρ is small we can write

$$\underset{\sim}{E}(\rho) = \sum_{n=0}^{\infty} \rho^n \underset{\sim}{E}_n \tag{20}$$

where the fields $\underset{\sim}{E}_n$ are independent of ρ, and $\underset{\sim}{E}_0 = \hat{\underset{\sim}{x}}$. Following the method outlined by Beran,[3] we take as our trial field

$$\underset{\sim}{E}^T = \sum_{n=0}^{M} a_n \underset{\sim}{E}_n \tag{21}$$

where, from (13), we require that $a_0 = 1$. The other coefficients a_n will be chosen to minimize σ_{eU}. Observe that

$$\sigma = 1 + \Omega(\sigma_1 - 1) = [2\rho\Omega + (1 - \rho)]/(1 - \rho) \tag{22}$$

where $\Omega(\underset{\sim}{z})$ is one if $\underset{\sim}{z}$ is in component 1 and zero otherwise.

Substituting (21) in (15) we find

$$\sigma_{eU} = [1 + \rho(f_1 - f_2) + H_{ij}a_ia_j + 2L_ia_i]/(1 - \rho), \tag{23}$$

where the summation over i and j extends from 1 to M and

$$H_{ij} = 2\rho \langle\Omega \underset{\sim}{E}_i\underset{\sim}{E}_j\rangle + (1 - \rho) \langle\underset{\sim}{E}_i\underset{\sim}{E}_j\rangle \tag{24}$$

$$L_i = 2\rho \langle\Omega\underset{\sim}{E}_i\hat{\underset{\sim}{x}}\rangle + (1 - \rho) \langle\underset{\sim}{E}_i\hat{\underset{\sim}{x}}\rangle \tag{25}$$

The best possible choice of $\underset{\sim}{a} = (a_1, a_2 \ldots a_m)$ is found by setting $\partial\sigma_{eU}/\partial\underset{\sim}{a} = 0$ giving

$$\underset{\sim}{a} = -\underset{\approx}{H}^{-1}\underset{\sim}{L} \tag{26}$$

and substituting this in (23),

$$\sigma_{eU} = [1 + \rho(f_1 - f_2) - \underset{\sim}{L}^T\underset{\approx}{H}^{-1}\underset{\sim}{L}]/(1 - \rho), \tag{27}$$

where $\underset{\sim}{L}^T$ is the transpose of $\underset{\sim}{L}$.

Note that all the elements of the matrix $\underset{\approx}{H}$ and the column vector $\underset{\sim}{L}$ are linear functions of ρ. It follows that $\sigma_{eU}(\rho)$, given by (27), can be expressed in the form (7) where N = M + 1. Also, from (27),

$$\sigma_{eU}(1) = \infty. \tag{28}$$

If instead of choosing the optimum values, we had taken $a_h = \rho^h$ then E^* would be of order ρ^{M+1} and consequently from (19) $\sigma_{eU} - \sigma_e$ would be of order $\rho^{2(M+1)}$. The bounds (27) being more restrictive must similarly coincide with the exact value of σ_e up to terms of order

ρ^{2M+1}. Consequently

$$\left.\frac{d^n \sigma_{eU}}{d\rho^n}\right|_{\rho=0} = Q_n \quad (n=1,2\ldots,2M+1). \tag{29}$$

Since $\sigma_{eU}(\rho)$ can be expressed in the form (7) where N = M + 1, the 2(M + 1) constraints (28) and (29) are sufficient to determine the function $\sigma_{eU}(\rho)$. Thus if we know Q_n, for n = 1 up to n = 2M + 1, we can calculate the bound σ_{eU}. We do not need to evaluate the quantities $\langle \Omega \underset{\sim}{E}_i \underset{\sim}{E}_j \rangle$, $\langle \underset{\sim}{E}_i \underset{\sim}{E}_j \rangle$, $\langle \Omega E_i \hat{\underset{\sim}{x}} \rangle$ and $\langle \underset{\sim}{E}_i \hat{\underset{\sim}{x}} \rangle$. The bound $\sigma_{eU}(\rho)$ is identical to the Lth order upper bound where L = 2M + 1.

The other classical variational principle states that if we choose any trial current field $\underset{\sim}{J}^T$ such that

$$\langle \underset{\sim}{J}^T \rangle = \hat{\underset{\sim}{x}}, \tag{30}$$

$$\underset{\sim}{\nabla} \cdot \underset{\sim}{J}^T = 0, \tag{31}$$

then

$$\sigma_{eL} = 1/\langle \underset{\sim}{J}^T \underset{\sim}{J}^T/\sigma \rangle \tag{32}$$

is a lower bound on σ_e. Following Beran[3] we express the true current field $\underset{\sim}{J}$ as a power series in ρ

$$\underset{\sim}{J} = \sum_{n=0}^{\infty} \underset{\sim}{J}_n \rho^n, \tag{33}$$

where $\underset{\sim}{J}_0 = \hat{\underset{\sim}{x}}$. We take as our trial field

$$\underset{\sim}{J}^T = \sum_{n=0}^{M} b_n \underset{\sim}{J}_n, \tag{34}$$

where from (30) we require $b_0 = 1$. Provided $\underset{\sim}{b} = (b_1, b_2 \ldots, b_M)$ is chosen to maximize σ_{eL} we find that $\sigma_{eL}(\rho)$ coincides with the Lth order lower bound, where L = 2M + 1.

IV. Bounds derived from the Hashin-Shtrikman Variational Principle

According to the Hashin-Shtrikman variational principle,[1] if we take any trial "polarization field" $\underset{\sim}{P}^T$ and let $\underset{\sim}{A}^T$ denote the associated field which is determined by the constraints

$$\sigma_0 \underset{\sim}{\nabla} \cdot \underset{\sim}{A}^T + \underset{\sim}{\nabla} \cdot \underset{\sim}{P}^T = 0, \tag{35}$$

$$\underset{\sim}{\nabla} \times \underset{\sim}{A}^T = 0, \tag{36}$$

$$\langle \underset{\sim}{A}^T \rangle = 0, \tag{37}$$

where σ_0 is a constant, then

$$\sigma_{eb} = \langle \sigma_0 - \frac{\underset{\sim}{P}^T \cdot \underset{\sim}{P}^T}{(\sigma - \sigma_0)} + 2\underset{\sim}{P}^T \cdot \hat{\underset{\sim}{x}} + \underset{\sim}{P}^T \cdot \underset{\sim}{A}^T \rangle \tag{38}$$

is an upper bound on σ_e when $\sigma_0 \geq \sigma_1$ and a lower bound when $\sigma_0 \leq \sigma_2 (=1)$.

Let us take $\sigma_0 = \sigma_2 = 1$ and choose as our trial field

$$\underset{\sim}{P}^T = (\sigma - \sigma_0)\underset{\sim}{E}^T = (\sigma_1 - 1) \left(\sum_{n=0}^{M} a_n \Omega \underset{\sim}{E}_n \right), \tag{39}$$

in which case

$$\underset{\sim}{A}^T = \tfrac{1}{2}(\sigma_1 - 1)[a_0 \hat{\underset{\sim}{x}} + \sum_{n=0}^{M} a_n (\underset{\sim}{E}_{n+1} - \underset{\sim}{E}_n)]. \tag{40}$$

Substituting (39) and (40) in (38) and selecting the optimum value of $\underset{\sim}{a} = (a_0, a_1 \ldots, a_m)$ we find that

$$\sigma_{eb} = 1 - 2\rho(\underset{\sim}{G}^T \underset{\sim}{F}^{-1} \underset{\sim}{G}), \tag{41}$$

where

$$G_i = \langle \Omega \underset{\sim}{E}_i \hat{\underset{\sim}{x}} \rangle, \tag{42}$$

$$F_{ij} = \rho(\delta_{i0} \langle \Omega \underset{\sim}{E}_j \hat{\underset{\sim}{x}} \rangle + \langle \Omega \underset{\sim}{E}_i \underset{\sim}{E}_{j+1} \rangle) - \langle \Omega \underset{\sim}{E}_i \underset{\sim}{E}_j \rangle. \tag{43}$$

Since i and j range from 0 to M and all the matrix elements of F_{ij} are linear in ρ, it follows that σ_{eb} can be expressed in the form (7) where $N = M + 1$.

Now denote $\underset{\sim}{P}^* = \underset{\sim}{P}^T - \underset{\sim}{P}$ and $\underset{\sim}{A}^* = \underset{\sim}{A}^T - \underset{\sim}{A}$ where $\underset{\sim}{P} = (\sigma - \sigma_0)\underset{\sim}{E}$ and $\underset{\sim}{A} = \underset{\sim}{E} - \hat{\underset{\sim}{x}}$. As Hashin and Shtrikman[1] prove

$$\sigma_{eb} - \sigma_e = \langle \underset{\sim}{A}^* \cdot \underset{\sim}{P}^* - \frac{(\underset{\sim}{P}^* \cdot \underset{\sim}{P}^*)}{(\sigma - \sigma_0)} \rangle . \tag{44}$$

If we had taken $a_n = \rho^n$ then $\underset{\sim}{P}^*$ would be of order ρ^{M+2}, $\underset{\sim}{A}^*$ of order ρ^{M+1} and $\sigma_{eb} - \sigma_e$ of order ρ^{2M+3}. The bound (41) being more restrictive must also coincide with σ_e up to terms of order ρ^{2M+2}.

Thus

$$\left.\frac{d^n \sigma_{eb}}{d\rho^n}\right|_{\rho=0} = Q_n \quad (n=1,2...,2M+2). \tag{45}$$

The 2(M + 1) constraints (45) are sufficient to determine the function σ_{eb}, since it can be expressed in the form (7) where N = M + 1.

We have established that the bound σ_{eb} coincides with the Lth order lower bound, where L = 2(M + 1). Similarly, if $\sigma_0 = \sigma_1$, the bound σ_{eb} is identical with the Lth order upper bound, where L = 2(M + 1).

V. Conclusion

In summary, if L is odd then the Lth order upper and lower bounds can be derived using the classical variational principles. If L is even then the Lth order upper and lower bounds can be derived using the Hashin-Shtrikman variational principles. If L is either even or odd then the bounds can be derived using the analytic method.

Thus the analytic method and the variational method are equally powerful when applied to the problem of finding bounds on σ_e, for real σ_1 and σ_2, given the values of Q_n for n = 1 up to n = L. However, each method has its own advantages: the analytic method seems most suitable for deriving bounds on σ_e when σ_1 and σ_2 are complex;[11,12] the variational method appears most appropriate for deriving bounds on σ_e for multi-component materials.

Achnowledgments

This work was supported in part by the Department of Energy through a grant (Grant #XH-9-8158-1) from the Solar Energy Research Institute. It was undertaken while one of the authors (G.W.M.) was a recipient of a Sydney University Traveling Scholarship.

References

1. Z. Hashin and S. Shtrikman, J. Appl. Phys. 33, 3125 (1962).
2. W. F. Brown, Trans. Soc. Rheol. 9, 357 (1965).
3. M. Beran, Nuovo Cimento 38, 771 (1965).
4. D. J. Bergman, Phys. Rep. 43, 377 (1978).
5. G. W. Milton, J. Appl. Phys., in press.
6. W. F. Brown, J. Chem. Phys. 23, 1514 (1955).
7. R. C. McPhedran and G. W. Milton, submitted to Appl. Phys.

8. O. Wiener, Abh. Math. Phys. Kl. Königl, Saechs. Ges. 32, 509 (1912).
9. N. Phan-Thien and G. W. Milton, submitted to Proc. Roy. Soc. Lond.
10. M. A. Elsayed, J. Math. Phys. 15, 2001 (1974).
11. D. J. Bergman, Phys. Rev. Lett. 44, 1285 (1980).
12. G. W. Milton, Appl. Phys. Lett. 37, 300 (1980).

FLUCTUATION CORRECTIONS TO THE MEAN FIELD

DESCRIPTION OF A NONUNIFORM FLUID

J. K. Percus*
Courant Institute of Mathematical Sciences
and Department of Physics
New York University
New York, N. Y. 10012

The interaction of a fluid of particles is separated into a core-core repulsion and a communally generated attractive field. Existence of a self-maintained density profile between liquid and gas phases is investigated in the Van der Waals model, and then from a microscopic point of view under the assumptions of local homogeneity and fluctuationless mean field. The particle degrees of freedom are rigorously eliminated, and the pure field system analyzed first under the previous assumptions and then in the presence of leading order field fluctuations. Divergent broadening of the density profile in two or three dimensional space is found, and related to the capillary wave picture.

1. Introduction

I would like to discuss the propagation of a very special field through a special disordered medium. The medium can be imagined as a gas of (perhaps soft) microscopic billiard balls. Except to the expert, it is not a very interesting fluid. As a gas, it occupies an arbitrarily large volume under sufficiently low pressure. Increasing its density requires proportionally increasing the pressure, at first, and then increasing it more and more rapidly as the available volume gets used up. But suppose now that each billiard ball is also the source of an attractive force field, so that each ball or core is immersed in the communal force field of the others. Then as the density is increased at fixed temperature, each core finds itself eventually sitting in an increasingly deep, somewhat erratic, potential well from which it escapes only occasionally -- into another well -- if its thermal energy is not too large. For this trapped or liquid state, the system pressure can be quite low, and so there is the possibility of a fluid at a given temperature and pressure possessing either a low density disordered gas phase, a high density more or less ordered liquid phase, or both, in spatially separate enclaves.

The spatial transition between two phases, as a self-supported rapidly changing density, must have some unique properties, and it is these that we want to focus on. The possibility of maintaining a density gradient in the absence of external forces can be assessed quite easily in a preliminary way. A unit volume of billiard balls at $\underset{\sim}{r}$ will be subject to an average force $-\nabla P_0(n(\underset{\sim}{r}))$, where $P_0(n)$ is the equation of state and $n(\underset{\sim}{r})$ the spatially varying density. The average potential, due to the attractive force field, felt by a particle in this volume is a linear functional of the density, and hence (if intrinsically isotropic) can be expanded as $-an(\underset{\sim}{r}) - b\nabla^2 n(\underset{\sim}{r}) + \cdots$. Thus, the average forces will be balanced if $-\nabla P_0(n(\underset{\sim}{r})) + n(\underset{\sim}{r})\nabla(an(\underset{\sim}{r}) + b\nabla^2 n(\underset{\sim}{r})) = 0$, which we shall rewrite as [1]

$$b\nabla^2 n(\underset{\sim}{r}) + an(\underset{\sim}{r}) - \mu_0(n(\underset{\sim}{r})) = -\mu \tag{1.1}$$

where $\mu_0'(n) = P_0'(n)/n$.

μ is the constant of integration, and $\mu_0(n)$ -- called the chemical potential -- starts logarithmically with n and eventually increases in the same fashion as $P_0(n)$.

Eq. (1.1) is consistent with two different uniform densities n_1 (liquid) and n_2 (gas) only if

$$\mu(n_1) = \mu = \mu(n_2) \tag{1.2}$$

where $\mu(n) \equiv \mu_0(n) - an$.

If a system is to have simultaneously $n(\underset{\sim}{r}) = n_1$, say as $z \to \infty$, and $n(\underset{\sim}{r}) = n_2$, say as $z \to -\infty$, (1.2) is not sufficient. Take $\partial/\partial z$ of (1.1) and multiply by $n(\underset{\sim}{r})$:

$$bn(\underset{\sim}{r})\nabla^2 \frac{\partial}{\partial z} n(\underset{\sim}{r}) = \frac{\partial}{\partial z} P(n(\underset{\sim}{r})) \tag{1.3}$$

where $P(n) \equiv P_0(n) - \frac{1}{2} an^2$.

Since $n\nabla^2 \partial n/\partial z = \frac{1}{2}\nabla\cdot(n\nabla\, \partial n/\partial z - \nabla n\, \partial n/\partial z) + \frac{1}{2}\partial(n\nabla^2 n)/\partial z$, integration over a large rectangular parallelopiped yields as well the requirement

$$P(n_1) = P(n_2). \tag{1.4}$$

Indeed $\mu'(n) = P'(n)/n$, and (1.2, 1.4) are readily identified with equality of pressure and chemical potential for bulk gas and liquid.

If (1.1, 1.2, 1.4) are satisfied, a spatial two-phase separation does occur, whose form we shall examine later. Our objective now will be two-fold: 1) to put the mean field approximation implicit in (1.1) on a somewhat firmer basis, and 2) to examine the nature of the field fluctuations which have been neglected in (1.1), and their effect on observable quantities.

2. Mean Field Approximation

The conceptual division referred to above -- particle cores immersed in a self-generated force field -- can be developed, although not uniquely, into an effective computational tool. One possibility is to describe the system jointly by the locations $\{\underset{\sim}{r}_i\}$ of its particles -- coupled by some "billiard-ball" interaction potential $\phi_0(\underset{\sim}{r}_i-\underset{\sim}{r}_j)$ -- and the potential field $v(\underset{\sim}{r})$ resulting from the additional attractive forces. Thus for the dynamics of the particles,

$$m\ddot{\underset{\sim}{r}}_i = -\sum_j \nabla\phi_0(\underset{\sim}{r}_i-\underset{\sim}{r}_j) - \nabla v(\underset{\sim}{r}_i) , \tag{2.1}$$

with the convention $\nabla\phi_0(0) = 0$. If the attractive potential giving rise to the field is written as $-\phi_1(\underset{\sim}{r}_i-\underset{\sim}{r}_j)$ for a given particle pair, then of course

$$v(\underset{\sim}{r}) = -\sum_j \phi_1(\underset{\sim}{r}-\underset{\sim}{r}_j) , \tag{2.2}$$

where now $\nabla\phi_1(0) = 0$. Suppose that $\phi_1(\underset{\sim}{r}-\underset{\sim}{r}')$, as the kernel of an integral operator, is invertible. Then we can rewrite (2.2) -- including a propagation delay mechanism which vanishes as the field mass density $M \to 0$ -- as

$$M\ddot{v}(\underset{\sim}{r}) = -\int \phi_1^{-1}(\underset{\sim}{r}-\underset{\sim}{r}')v(\underset{\sim}{r}')d\underset{\sim}{r}' - \sum_j \delta(\underset{\sim}{r}-\underset{\sim}{r}_j) . \tag{2.3}$$

The reason for doing this is that then both (2.1) and (2.3) are consequences of the equations of motion for the Hamiltonian [2] ($\underset{\sim}{p}_i$ = particle momentum, $\pi(\underset{\sim}{r})$ = field momentum)

$$\begin{aligned} H = \sum_i p_i^2/2m + \frac{1}{2}\sum_{i,j} \phi_0(\underset{\sim}{r}_i-\underset{\sim}{r}_j) + \sum_i v(\underset{\sim}{r}_i) \\ + \frac{1}{2}\iint \phi_1^{-1}(\underset{\sim}{r}-\underset{\sim}{r}')v(\underset{\sim}{r})v(\underset{\sim}{r}')d\underset{\sim}{r}d\underset{\sim}{r}' + \int \pi(\underset{\sim}{r})^2/2Md\underset{\sim}{r} , \end{aligned} \tag{2.4}$$

as is readily verified. Furthermore, if the total energy is rewritten by substituting v from (2.3) in (2.4), there results

$$E = \sum_i p_i^2/2m + \int \pi(\underset{\sim}{r})^2/2M\, d\underset{\sim}{r} + \frac{1}{2} \iint \dot{\pi}(\underset{\sim}{r})\phi_1(\underset{\sim}{r}-\underset{\sim}{r}')\dot{\pi}(\underset{\sim}{r}')d\underset{\sim}{r}d\underset{\sim}{r}'$$

$$+ \frac{1}{2} \sum_{i,j} \left(\phi_0(\underset{\sim}{r}_i-\underset{\sim}{r}_j) - \phi_1(\underset{\sim}{r}_i-\underset{\sim}{r}_j)\right) , \qquad (2.5)$$

the coordinate part of which is the full particle interaction energy.

In this paper, we will restrict attention to time-average equilibrium properties. Introducing the microscopic particle density

$$\rho(\underset{\sim}{r}) \equiv \Sigma\delta(\underset{\sim}{r} - \underset{\sim}{r}_i) , \qquad (2.6)$$

the average particle density and particle pair density defined by

$$\begin{aligned} n(r) &= \langle\rho(\underset{\sim}{r})\rangle \\ n_2(\underset{\sim}{r},\underset{\sim}{r}') &= \langle \sum_{i\neq j} \delta(\underset{\sim}{r}-\underset{\sim}{r}_i)\delta(\underset{\sim}{r}'-\underset{\sim}{r}_j)\rangle \\ &= \langle\rho(\underset{\sim}{r})\rho(\underset{\sim}{r}') - \rho(\underset{\sim}{r})\delta(\underset{\sim}{r}-\underset{\sim}{r}')\rangle \end{aligned} \qquad (2.7)$$

will be of principal interest, as well as the average field $\bar{v}(\underset{\sim}{r}) \equiv \langle v(\underset{\sim}{r})\rangle$. For a uniform simple fluid of density n, the pair density is translation-invariant and isotropic, so that one has

$$\begin{aligned} n(\underset{\sim}{r}) &= n \\ n_2(\underset{\sim}{r},\underset{\sim}{r}') &= n^2 g(|\underset{\sim}{r}-\underset{\sim}{r}'|) , \end{aligned} \qquad (2.8)$$

where g is the radial distribution function.

On averaging (2.3), we have to start with the exact

$$\int \phi_1^{-1}(\underset{\sim}{r}-\underset{\sim}{r}')\bar{v}(\underset{\sim}{r}')d\underset{\sim}{r}' = -n(\underset{\sim}{r}) . \qquad (2.9)$$

The approximations take place in (2.1), which we first multiply by $\delta(\underset{\sim}{r}-\underset{\sim}{r}_i)$ and average, obtaining (by use of $\langle\dot{A}B\rangle = -\langle A\dot{B}\rangle$)

$$m \sum_i \langle\dot{\underset{\sim}{r}}_i\dot{\underset{\sim}{r}}_i\cdot\nabla\delta(\underset{\sim}{r}-\underset{\sim}{r}_i)\rangle = -\int \nabla\phi_0(\underset{\sim}{r}-\underset{\sim}{r}') \langle\rho(\underset{\sim}{r})\rho(\underset{\sim}{r}')\rangle d\underset{\sim}{r}'$$

$$- \langle\nabla v(\underset{\sim}{r})\rho(\underset{\sim}{r})\rangle . \qquad (2.10)$$

Since, at reciprocal temperature β, $\langle\dot{r}_{i\mu}\dot{r}_{i\nu}f(\underset{\sim}{r}_i)\rangle = (\delta_{\mu\nu}/m\beta)\langle f(\underset{\sim}{r}_i)\rangle$, this becomes, after a minor transformation,

$$\langle\nabla\rho(r)\rangle/\beta - \frac{1}{2} \int \nabla\phi_0(\underset{\sim}{r}')\langle\rho(\underset{\sim}{r}+\underset{\sim}{r}')\rho(\underset{\sim}{r}) - \rho(\underset{\sim}{r})\rho(\underset{\sim}{r}-\underset{\sim}{r}')\rangle d\underset{\sim}{r}'$$

$$= -\langle\nabla v(\underset{\sim}{r})\rho(\underset{\sim}{r})\rangle . \qquad (2.11)$$

Two approximations are now required. The first one, the local thermodynamics approximation, has only to do with the properties of the underlying core fluid under the assumption that the density is slowly varying in space on the scale of the range of particle-particle correlations. It avers that under these circumstances, the pair density -- as well as other intensive quantities -- can be taken as that of the uniform system at the local density. The assumed slow variation thus means that (2.11) can be written as

$$-\langle v(\underset{\sim}{r})\rho(\underset{\sim}{r})\rangle = \nabla n(\underset{\sim}{r})/\beta - \frac{1}{2}\nabla\cdot\int \underset{\sim}{r}'\nabla\phi_0(\underset{\sim}{r}')\langle\rho(\underset{\sim}{r}+\underset{\sim}{r}')(\rho(\underset{\sim}{r})-\delta(\underset{\sim}{r}'))\rangle d\underset{\sim}{r}'$$

$$= \nabla[n(\underset{\sim}{r})/\beta - \frac{1}{6}\,n(\underset{\sim}{r})^2\int \underset{\sim}{r}'\cdot\nabla\phi_0(\underset{\sim}{r}')g_0(\underset{\sim}{r}';n(\underset{\sim}{r}))d\underset{\sim}{r}'] \;, \tag{2.12}$$

but the bracketed expression is just the well-known virial form for the pressure, and so

$$\nabla P_0(n(\underset{\sim}{r})) = -\langle\nabla v(\underset{\sim}{r})\rho(\underset{\sim}{r})\rangle \;. \tag{2.13}$$

The second approximation, the mean field approximation, is that the variations of the field $v(\underset{\sim}{r})$ are negligible (or at least uncorrelated with $\rho(\underset{\sim}{r})$), so that $v(\underset{\sim}{r})$ can be replaced by its mean in (2.13):

$$\nabla P_0(n(\underset{\sim}{r})) = -n(\underset{\sim}{r})\nabla\bar{v}(\underset{\sim}{r}) \;. \tag{2.14}$$

From (1.1), this is equivalent to

$$\mu_0(n(\underset{\sim}{r})) + \bar{v}(\underset{\sim}{r}) = \mu \tag{2.15}$$

for a suitable constant μ. Hence if the inverse of $\mu_0(n)$ is designated by $n_0(\mu)$, (2.9) and (2.15) combine to give

$$\int\phi_1^{-1}(\underset{\sim}{r}-\underset{\sim}{r}')(\mu_0(\underset{\sim}{r}') - \mu)d\underset{\sim}{r}' = n_0(\mu_0(\underset{\sim}{r})) \tag{2.16}$$

which can be solved at leisure. A model ϕ_1 makes the job easier and more explicit, and we shall choose the "Yukawa potential"

$$\phi_1(\underset{\sim}{r}-\underset{\sim}{r}') = Ke^{-\gamma|\underset{\sim}{r}-\underset{\sim}{r}'|}\,/\,|\underset{\sim}{r}-\underset{\sim}{r}'| \tag{2.17}$$

with the convenient property that the inverse operator is simply

$$\phi_1^{-1} = \frac{1}{4\pi K}\,(\lambda^2 - \nabla^2) \;. \tag{2.18}$$

Eq. (2.16) now becomes

$$\frac{1}{4\pi K} \nabla^2 \mu_0 + n_0(\mu_0) - \frac{\lambda^2}{4\pi K} (\mu_0 - \mu) = 0 . \tag{2.19}$$

Let us look at the possibility of a planar interface. Thus (2.19) has only a z-dependence, and it can be reduced in differential order by multiplying by $\mu_0'(z)$ and integrating. This yields

$$\frac{1}{8\pi K} \mu_0'(z)^2 + P_0(\mu_0(z)) - \frac{\lambda^2}{8\pi K} (\mu_0(z) - \mu)^2 = P \tag{2.20}$$

for suitable constant P. Of course, we have simply found the "energy integral" for (2.19), and the resulting profile is determined by the "potential"

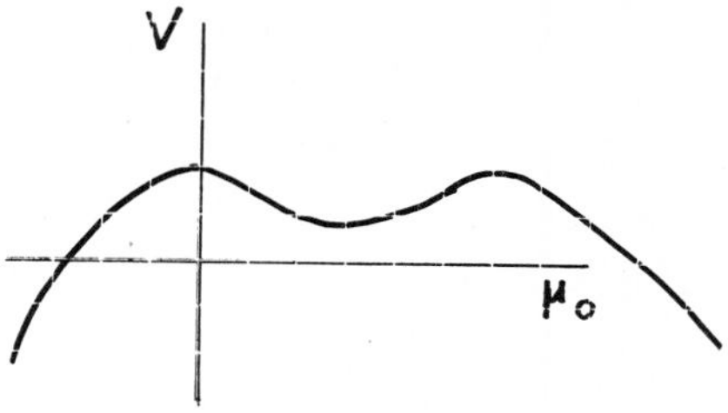

$$V(\mu_0) = P_0(\mu_0) - \frac{\lambda^2}{8\pi K} (\mu_0 - \mu)^2 \tag{2.21}$$

In particular, asymptotically constant values of $\mu_0(\infty) = \mu_1$, $\mu_0(-\infty) = \mu_2$ are obtained providing that $V(\mu_1) = V(\mu_2)$ and $V'(\mu_1) = 0 = V'(\mu_2)$, which we write as

$$P_0(\mu_1) - \frac{\lambda^2}{8\pi K} (\mu_1 - \mu)^2 = P = P_0(\mu_2) - \frac{\lambda^2}{8\pi K} (\mu_2 - \mu)^2$$

$$\mu_1 - \frac{4\pi K}{\lambda^2} n_0(\mu_1) = \mu = \mu_2 - \frac{4\pi K}{\lambda^2} n_0(\mu_2) . \tag{2.22}$$

P and μ can be identified with the common pressure and chemical potential of the system with full interaction $\phi_0 - \phi_1$.

A prototypical example due to Van der Waals [1] (but in the context of (1.1), which is equivalent to (2.16) when $\phi_1 = a + b\nabla^2$) is that in which $P = \mu = 0$, $\mu_2 = n_0(\mu_2) = 0$, and $P_0(\mu_0) = (\lambda/\mu_1)^2 \mu_0^3 (2\mu_1 - \mu_0) \ / \ 8\pi K$, so that

$$V(\mu_0) = \frac{-1}{8\pi K} \left(\frac{\lambda}{\mu_1}\right)^2 \mu_0^2 (\mu_1 - \mu_0)^2 . \tag{2.23}$$

From (2.20), then, $\lambda \mu_1 dz/d\mu_0 = (1/\mu_0 + 1/(\mu_1 - \mu_0))$, producing the profile

$$\mu_0(z) = \frac{\mu_1}{1 + e^{-\lambda z}} = \frac{\mu_1}{2} (1 + \tanh \tfrac{1}{2}\lambda z) . \tag{2.24}$$

As for the density, (2.16) now reads $n = (\lambda^2 - d^2/dz^2)\mu_0/4\pi K$, or

$$n(z) = \frac{n_1}{1+e^{-\lambda z}} \frac{1+3e^{-\lambda z}}{(1+e^{-\lambda z})^2}$$

$$n_1 = \lambda^2 \mu_1 / 4\pi K \ . \tag{2.25}$$

3. Introduction of Field Fluctuations

Since we are restricting our attention to thermal equilibrium, it behooves us to make use of the well developed formalism of equilibrium statistical mechanics. According to this, momenta contribute only trivially to thermodynamic quantities, and the probability distribution in coordinate space for N particles at reciprocal temperature β is given by

$$\rho(\mathbf{r}_1,\ldots,\mathbf{r}_N) = \exp - \beta(\Phi(\mathbf{r}_1,\ldots,\mathbf{r}_N) - F)$$

$$\text{where} \qquad e^{-\beta F} = \int\cdots\int e^{-\beta\Phi(\mathbf{r}_1,\ldots,\mathbf{r}_N)} d\mathbf{r}_1\cdots d\mathbf{r}_N \ . \tag{3.1}$$

ϕ is the full coordinate space potential and F is then the Helmholtz free energy. If the number of particles, N, is not fixed but is rather determined by a reservoir of chemical potential μ , then instead

$$\rho_N(\mathbf{r}_1,\ldots,\mathbf{r}_N) = \frac{1}{N!} \exp \beta(\Phi(\mathbf{r}_1,\ldots,\mathbf{r}_N) - N\mu - \Omega)$$

$$\text{where} \quad e^{-\beta\Omega} = \sum_N \frac{1}{N!} e^{\beta N\mu} \int\cdots\int e^{-\beta\Phi(\mathbf{r}_1,\ldots,\mathbf{r}_N)} d\mathbf{r}_1\cdots d\mathbf{r}_N \ . \tag{3.2}$$

Ω is the grand potential, with the value $-PV$ for a uniform system of volume V. Finally, if the degrees of freedom include a field $\{v(\mathbf{r})\}$, (3.2) must be generalized to

$$\rho_N(\mathbf{r}_1,\ldots,\mathbf{r}_N;v(\mathbf{r})) = \frac{1}{N!} \exp - \beta(\Phi(\mathbf{r}_1,\ldots,\mathbf{r}_N;v(\mathbf{r})) - N\mu - \Omega)$$

$$\text{where} \quad e^{-\beta\Omega} = \sum_N \frac{1}{N!} e^{\beta N\mu} \iint \int\cdots\int e^{-\beta\Phi(\mathbf{r}_1,\ldots,\mathbf{r}_N;v(\mathbf{r}))} d\mathbf{r}_1\cdots d\mathbf{r}_N Dv \ . \tag{3.3}$$

Here $\iint Dv$ denotes a functional integral over $\{v(\mathbf{r})\}$; the only property that we will (implicitly) use is that it is translation-invariant, but in fact it will always be associated in our discussion with an integrand that permits immediate interpretation as a Wiener integral.

We will concentrate on the grand potential Ω , since it serves as generating function for all expectations by virtue of the obvious relation

$$\delta\Omega = <\delta\Phi(\underset{\sim}{r}_1,\ldots,\underset{\sim}{r}_N;v(\underset{\sim}{r})) \quad . \tag{3.4}$$

If the argument leading to (2.4) is to be believed, our particle system with core interaction ϕ_0 , attractive tail $-\phi_1$ (with the convention $\phi_0(0) = \phi_1(0) = 0$) , and external potential field u, can be represented by [5,6]

$$e^{-\beta\Omega[u]} = \sum_N \frac{1}{N!} \iint \int\cdots\int e^{-\frac{1}{2}\beta\Sigma\phi_0(\underset{\sim}{r}_i-\underset{\sim}{r}_j)} e^{-\beta\Sigma(u(\underset{\sim}{r}_i)+v(\underset{\sim}{r}_i)-\mu)} e^{-\frac{1}{2}\beta\int v(\underset{\sim}{r})\phi_1^{-1}v(\underset{\sim}{r})d\underset{\sim}{r}} d\underset{\sim}{r}_1,\ldots,d\underset{\sim}{r}_N Dv \tag{3.5}$$

(which can be defined providing that ϕ_1 is positive definite). To verify this, we perform the translation $v(\underset{\sim}{r}) \rightarrow v(\underset{\sim}{r}) - \phi_1\rho(\underset{\sim}{r})$, $\rho(\underset{\sim}{r})$ as in (2.6), converting (3.5) to

$$e^{-\beta\Omega[u]} = I_1 \sum_N \frac{1}{N!} \int\cdots\int e^{-\frac{1}{2}\beta\Sigma(\phi_0(\underset{\sim}{r}_i-\underset{\sim}{r}_j)-\phi_1(\underset{\sim}{r}_i-\underset{\sim}{r}_j))} e^{-\beta\Sigma(u(\underset{\sim}{r}_i)-\mu)} d\underset{\sim}{r}_1\ldots d\underset{\sim}{r}_N \tag{3.6}$$

where $I_1 = \iint e^{-\frac{1}{2}\beta\int v(\underset{\sim}{r})\phi_1^{-1}v(\underset{\sim}{r})d\underset{\sim}{r}} Dv$.

Thus, except for an irrelevant constant, (3.6) is precisely the required expression for the system of interacting particles.

Our strategy however will be to carry out the integration over $\underset{\sim}{r}_1,\ldots,\underset{\sim}{r}_N$ and summation over N instead. Doing so, (3.5) becomes

$$e^{-\beta\Omega[u]} = \iint e^{-\beta\Omega_0[u+v]} e^{-\frac{1}{2}\beta\int v(\underset{\sim}{r})\phi_1^{-1}v(\underset{\sim}{r})d\underset{\sim}{r}} Dv , \tag{3.7}$$

which will serve as our basic expression. We have in fact set up a probability density in v-space

$$\mathcal{P}_v[u] = e^{-\beta(\mathcal{H}_v[u]-\Omega[u])}$$

where

$$\mathcal{H}_v[u] = \Omega_0(u+v) + \frac{1}{2}\int v(\underset{\sim}{r})\phi_1^{-1}v(\underset{\sim}{r})d\underset{\sim}{r}$$

$$\text{and} \quad e^{-\beta\Omega[u]} = \iint e^{-\beta\mathcal{H}_v[u]}Dv \quad . \tag{3.8}$$

From (3.4), it readily follows by choosing $\delta\Phi = Q$ that any expectation $\langle Q[u]\rangle$ in the presence of the external field u can be written as

$$\langle Q[u]\rangle = \iint \langle Q[u+v]\rangle_0 \mathcal{P}_v[u]Dv \ , \tag{3.9}$$

and in particular

$$n(\underset{\sim}{r}) = \iint n_0(\underset{\sim}{r}|u+v)\mathcal{P}_v[u]Dv \ . \tag{3.10}$$

However, operations on the field can be used for the same purpose. For instance, applying $\delta/\delta v(\underset{\sim}{r})$ to the integrand of (3.5) causes the integral to vanish, thereby providing the relation

$$n(\underset{\sim}{r}) = -\phi_1^{-1}\bar{v}(\underset{\sim}{r}) \tag{3.11}$$

(directly derivable from (3.8) as well), precisely as in (2.16).

To make use of (3.8), we have to evaluate $\Omega_0[u+v]$ and carry out the requisite integrations over v. Each task will be undertaken by essentially a transcription of the approximations of Section 2. To start with, we have observed that for a uniform system, $\Omega = -PV = -\int P d\underset{\sim}{r}$. This suggests that if the reference system is not varying rapidly in space, so that the local pressure would be expected to depend upon the local excess chemical potential,

$$\Omega_0[\mu] = -\int P_0(\mu-u(\underset{\sim}{r}))d\underset{\sim}{r} \ . \tag{3.12}$$

Indeed, since (from (3.4)) $n(\underset{\sim}{r}) = \delta\Omega/\delta u(\underset{\sim}{r})$, this implies

$$n_0(\underset{\sim}{r}) = \partial P_0(\mu-u(\underset{\sim}{r}))/\partial\mu \ , \tag{3.13}$$

just a statement of the local thermodynamics approximation. We thus have

$$\mathcal{H}_v[\mu] = \int [\tfrac{1}{2}v(\underset{\sim}{r})\phi_1^{-1}v(\underset{\sim}{r}) - P_0(\mu-u(\underset{\sim}{r})-v(\underset{\sim}{r}))]d\underset{\sim}{r} \ , \tag{3.14}$$

an expression that we will retain.

We proceed next to approximations on $\mathcal{P}_v[u]$ that allow the integration Dv to be performed. Simplest is the no-fluctuation assumption: the distribution of v is concentrated at the minimum of

$\mathcal{H}_v[u]$. Thus $\bar{v}$ minimizes $\mathcal{H}_v[u]$, and from $\delta\mathcal{H}_v[u]/\delta v(\underset{\sim}{r}) = 0$, we have

$$\phi_1^{-1}\bar{v}(\underset{\sim}{r}) = -n_0(\mu-u(\underset{\sim}{r})-\bar{v}(\underset{\sim}{r})) \ , \tag{3.15}$$

subsequent to which (3.10) or (3.11) tells us that

$$n(\underset{\sim}{r}) = n_0(\mu-u(\underset{\sim}{r})-\bar{v}(\underset{\sim}{r})) \ . \tag{3.16}$$

Eq. (3.15), when $u(\underset{\sim}{r}) = 0$, is just the mean field approximation (2.16), the consequences of which we have analyzed above.

We can now sequentially improve the mean field approximation by expanding $\mathcal{H}_v[u]$ about the minimizing field $\bar{v}$. The next non-vanishing order is the second. From (3.14), we have then

$$\begin{aligned}\mathcal{H}_v[u] = &\int [\tfrac{1}{2}\ \bar{v}(\underset{\sim}{r})\phi_1^{-1}\bar{v}(\underset{\sim}{r}) - P_0(\mu-u(\underset{\sim}{r})-\bar{v}(\underset{\sim}{r}))]d\underset{\sim}{r} \\ &+ \tfrac{1}{2}\int \delta v(\underset{\sim}{r})[\phi_1^{-1}-n_0'(\mu-u(\underset{\sim}{r})-v(\underset{\sim}{r}))]\delta v(\underset{\sim}{r})d\underset{\sim}{r} \ ,\end{aligned} \tag{3.17}$$

where $\delta v(r) = v(\underset{\sim}{r})-\bar{v}(\underset{\sim}{r})$, converting $\mathcal{P}_v[u]$ to a Gaussian. Now from (3.10) and (3.13),

$$n(r) = \iint n_0(\mu-u(\underset{\sim}{r})-\bar{v}(\underset{\sim}{r})-\delta v(\underset{\sim}{r}))\mathcal{P}_v[u]Dv \ , \tag{3.18}$$

for which only the distribution of the single quantity $\delta v(\underset{\sim}{r})$ is needed. But this is Gaussian, $<\delta v(\underset{\sim}{r})> = 0$, and so we need only

$$\zeta(\underset{\sim}{r})^2 \equiv <\delta v(\underset{\sim}{r})^2> = \frac{1}{\beta}\ [\phi_1^{-1}-n_0'(\mu-u(\underset{\sim}{r})-\bar{v}(\underset{\sim}{r}))]^{-1}(\underset{\sim}{r},\underset{\sim}{r}) \ , \tag{3.19}$$

as a consequence of which

$$\mu(\underset{\sim}{r}) = \int n_0(\mu-u(\underset{\sim}{r})-\bar{v}(\underset{\sim}{r})-\Delta)e^{-\Delta^2/2\zeta(\underset{\sim}{r})^2}d\Delta/\sqrt{2\pi\zeta(r)^2} \ . \tag{3.20}$$

In other words, the full effect of fluctuations in v at this level is that the profile that would have resulted in the absence of Δ is now broadened via a chemical potential broadening with an amplitude of $\zeta(\underset{\sim}{r})$. It is then the structure of $\zeta(\underset{\sim}{r})$ that we must examine in detail.

4. Capillary Wave Picture

With the phase interface problem in mind, let us set $u(\underset{\sim}{r}) = 0$. This however raises a problem. Taking the gradient of (3.15), with

$u(\underset{\sim}{r}) = 0$, we have

$$(\phi_1^{-1} - n_0'(n-\bar{v}(\underset{\sim}{r}))\nabla\bar{v}(\underset{\sim}{r}) = 0 \quad , \tag{4.1}$$

informing us that the inverse operator of (3.19) does not exist. Since the particle number is no longer fixed, a self-supported phase interface can be translated arbitrarily in the system with no change of controlling parameters. The null eigenvector $\nabla\bar{v}(\underset{\sim}{r})$ represents precisely such an infinitesimal translation. It would disappear under a weak applied field $u(\underset{\sim}{r})$ or under fixing of particle number, and so we shall simply imagine that (3.19) represents a principal part or generalized inverse, with the null eigenspace projected out.

Let us also anticipate a problem that will arise from the local thermodynamics approximation which, in effect, says that the correlation range of the core fluid is zero. This means that there is no built-in limitation on the rapidity with which the density can change and thus allows excitation of arbitrarily large wave vector to propagate unchecked. A semblance of reality can be recovered by instituting a cutoff K_M on wave vector space, and this is what we shall do whenever necessary.

With these caveats out of the way, what is the nature of the dispersion $\zeta(\underset{\sim}{r})$? We once more have in mind the situation of a plane symmetric interface, with tail potential modeled by (2.18). Hence it is the operator

$$\mathcal{L} = \frac{1}{4\pi K}\left(\lambda^2 - \frac{\partial^2}{\partial z^2}\right) - n_0'(\mu-\bar{v}(z)) - \frac{1}{4\pi K}\nabla_T^2$$

$$\nabla_T^2 = \frac{\partial^2}{\partial x^2} + \frac{\partial^2}{\partial y^2} \tag{4.2}$$

that we must analyze. In the special case of (2.23), the analysis can be carried out in complete detail [7]. But this is not necessary. Let us observe instead that if the system is infinite in the z-direction, but a periodic box $L \times L$ in cross-section, then $\mathcal{L}$ has the spectrum

$$\ell_{\alpha\underset{\sim}{k}} = \ell_\alpha + \frac{k^2}{4\pi K} , \qquad \underset{\sim}{k} = \frac{2\pi}{L}(N_1, N_2)$$

where

$$-\frac{1}{4\pi K}\frac{\partial^2\psi_\alpha}{\partial z^2} + \left(\frac{\lambda^2}{4\pi K} - n_0'(\mu-\bar{v}(z)\right)\psi_\alpha = \ell_\alpha\psi_\alpha . \tag{4.3}$$

The lowest eigenvalue in z-space is $\ell_0 = 0$, corresponding to the nodeless state $\psi_0 = \bar{v}'(z)$, and $-n_0'(\mu-v(z))$ is asymptotically 0 , with a negative well in the vicinity of the origin, so that ℓ_1 is at a finite separation above ℓ_0. Thus, there is a sequence of states

$$\psi_{o\underset{\sim}{k}}(\underset{\sim}{r}) = \frac{C}{L^2}\,\bar{v}'(z)e^{i\underset{\sim}{k}\cdot\underset{\sim}{x}}$$

$$C = \left[\int (\bar{v}'(z))^2 dz\right]^{-\frac{1}{2}} \tag{4.4}$$

$$\ell_{ok} = k^2/4\pi K \,,$$

whose eigenvalues have 0 as limit point, which can make a divergent contribution to (3.19) (for $L \to \infty$):

$$\begin{aligned} \zeta(\underset{\sim}{r})^2 &= \frac{1}{\beta} \sum_{(\alpha,k)\neq(0,0)} |\psi_{\alpha\underset{\sim}{k}}(\underset{\sim}{r})|^2/\ell_{\alpha\underset{\sim}{k}} \\ &= \frac{4\pi K}{\beta}\,\frac{C^2}{L^{D-1}}\,\bar{v}'(z)^2 \sum_{|k|>0}^{K_M} \frac{1}{k^2} + \dots \,, \end{aligned} \tag{4.5}$$

where we have generalized from 3 to D-dimensional space.

The behavior of (4.5) under increase of the cross-section diameter L is a sensitive function of the dimensionality D of the space in which the system lives. Since $\underset{\sim}{k}$ is one dimension lower, we see that (note that K changes its dimension as D changes).

$$\zeta(\underset{\sim}{r})^2 = \begin{cases} \dfrac{4K}{\pi\beta}\, C^2\,\bar{v}'(z)^2 L + \dots & D = 2 \\[2ex] \dfrac{K}{\pi\beta}\, C^2\,\bar{v}'(z)^2 \ln(K_M L) + \dots & D = 3 \\[2ex] \dfrac{2K}{\pi\beta}\, C^2\,\bar{v}'(z)^2 K_M + \dots & D = 4 \end{cases} \tag{4.6}$$

Hence the interface of a 2-dimensional fluid rapidly broadens — as $L^{\frac{1}{2}}$ — with increasing interfacial area; for a 3-dimensional fluid, the broadening is much slower, and it is first in 4 dimensions that a density profile independent of L can be achieved in the usual thermodynamic limit of infinite volume.

This conclusion, that fluctuations — field fluctuations in our picture — prohibit the establishment of a volume-independent density profile in two or three dimensions, is not new. It has usually been obtained in the context of capillary waves [8], bulk motions of the interfacial region. Transcription to this picture is not difficult.

We return to (3.18) and note that if

$$n_0(\mu-\bar{v}(z)) = \bar{n}(z) \tag{4.7}$$

is the mean field approximation profile, then to first order in Δ in the argument,

$$\bar{n}(z-\Delta) = n_0(\mu - \bar{v}(z) + \bar{v}'(z)\Delta) . \tag{4.8}$$

Thus, (3.18) can be rewritten

$$n(\underset{\sim}{r}) = \iint \bar{n}(z-\Delta)\mathcal{P}_{\bar{v}-\bar{v}'(z)\Delta}[u]Dv . \tag{4.9}$$

If the distribution is now taken as Gaussian in Δ, (2.20) changes to

$$n(\underset{\sim}{r}) = \int \bar{n}(z-\Delta)e^{-\Delta^2/2\xi(z)^2} d\Delta/\sqrt{2\pi\xi(z)^2}$$

where

$$\xi(z)^2 = \left\langle\left(\frac{\delta v(z)}{\bar{v}'(z)}\right)^2\right\rangle = \frac{1}{\beta}[\phi_1^{-1} - n_0'(\mu-\bar{v}(z))]^{-1}(\underset{\sim}{r},\underset{\sim}{r})/\bar{v}'(z)^2 . \tag{4.10}$$

But then according to (4.6), $\bar{v}'(z)^2$ cancels completely in (4.10), e.g. in three dimensions

$$\xi(\underset{\sim}{r})^2 = \frac{KC^2}{\pi\beta} \ell n(K_M L) + - \tag{4.11}$$

is independent of z. It thus represents a bodily motion of the whole interface at each location in the transverse plane — in other words, a capillary motion. However, we are left with the feeling that the notion of a force field tethered to a disordered gas is more basic, more controllable, and more amenable to meaningful approximation.

*Supported in part by the Department of Energy, Contract DE-AC02-76ER03077 and National Science Foundation, Grant CHE-80011285.

REFERENCES

1. J.D. Van der Waals, Z.Phy. Chem. 13, 657 (1894).
2. J.K. Percus, Courant Institute Course Notes, "Many-Body Theory," 1965 (unpublished).
3. J.K. Percus, Trans. N.Y. Acad. Sci. 26, 1062 (1964).
4. N.G. Van Kampen, Phys. Rev. 135, A 362 (1964).
5. M. Kac, in "Applied Probability," ed. L.A. MacColl (McGraw-Hill, New York 1957).
6. A.J.F. Siegert, in "Statistical Physics," 1962 Brandeis Lecture (Benjamin, New York, 1963).
7. J.K. Percus, in "Studies in Statistical Mechanics," ed. J.L. Lebowitz and E.W. Montroll (North-Holland, Amsterdam 1981).
8. Buff, Lovett, and Stillinger, Phys. Rev. Lett. 15, 621 (1965).

FINGERING IN POROUS MEDIA

P. G. Saffman
Department of Applied Mathematics
California Institute of Technology
Pasadena, CA 91125

Microscopic Fingering

Fingering is the phenomena which occurs when an attempt is made to displace a fluid saturating a porous medium by injecting another immiscible fluid of smaller viscosity. In this case, it is well known that the injected fluid tends to advance by sending out fingers and a fraction of the original fluid is left behind in the pores and interstices of the medium. A problem of major interest is to understand how the amount left behind and its spatial distribution depends upon the properties of the fluids and the medium.

One difficulty is that the fingering occurs on two scales, the microscopic and the macroscopic. The microscopic scale describes the motion of the interface in the pores and crevices. Unless the speeds are high, the viscosity small or the porosity large, the viscous forces dominate and the problem is essentially that of low Reynolds number (Stokes) flow of two immiscible fluids in a region of irregular geometry. The real problem is impossibly hard, but the fluid dynamical processes can be exemplified by motion in a capillary tube, where it is imagined the more viscous fluid wets the walls and that a finger of the less viscous fluid advances along the axis of the tube displacing the original more viscous fluid but leaving a film of it adhering to the wall. This flow is not too difficult to set up experimentally and experiments were carried out by Fairbrother & Stubbs (1935) and more extensively by Taylor (1961) and Cox (1962). On dimensional grounds, it is expected that the fraction of viscous fluid left behind, m say, is a function only of $\mu u/T$ and μ/μ_1, where u is the velocity of the tip of the finger, μ is the viscosity of the displaced fluid, μ_1 is the viscosity of the displacing fluid, and T is the interfacial tension. Since the inertia forces are assumed negligible and the flow is governed by a balance between viscous and surface tension forces, the size of the capillary does not enter and

$$m = F\left(\frac{\mu u}{T}, \frac{\mu_1}{\mu}\right) . \tag{1}$$

The experiments were carried out with $\mu_1/\mu << 1$ and the form of F for this case measured. In the limits of small and large values of $\mu u/T$,

$$m \doteq \left(\frac{\mu u}{T}\right)^{\frac{1}{2}} \quad \text{for} \quad \frac{\mu u}{T} < 0.1, \tag{2}$$

and

$$m \doteq 0.6 \quad \text{for} \quad \frac{\mu u}{T} > 2. \tag{3}$$

The calculation of F from first principles using the Stokes (creeping flow) equations is a problem of considerable difficulty. Bretherton (1961) attempted to tackle the $\mu u/T << 1$ limit, in which the thickness of the film of viscous fluid is small, by matching a lubrication flow solution for the film shape with the approximately constant curvature shape over the front. However, the analysis gives

$$m \propto \left(\frac{\mu u}{T}\right)^{2/3}, \tag{4}$$

and the conflict between the experiments and the asymptotic calculation has still not apparently been resolved.

An attempt is currently under way at the California Institute of Technology to solve the problem for finite $\mu u/T$ by numerical solution of the governing equations. There is a serious question about the uniqueness of the mathematical problem, since recent work has uncovered numerous examples of non-uniqueness of flows with free surfaces (e.g. Chen & Saffman 1980), and it is hoped that the work may give some information about this matter. In the event that non-unique solutions are obtained, we shall be faced with the serious problem of deciding what determines the actual value of m. This investigation is, however, still in a preliminary stage and no results are yet available.

If the porous medium were just a bundle of straight parallel capillary tubes, then the fingering would be entirely microscopic and the amount left behind described by (1), but of course this is a gross oversimplification. An attempt was made by Saffman (1963, 1969) to incorporate (1) into a more realistic (but still oversimplified) model of a porous medium as a random collection of capillary tubes, inclined at random directions with equal probability. Then the amount left behind by microscopic fingering in a capillary inclined at an angle θ to the direction of the mean flow is

$$F\left(\frac{3\mu U \cos\theta}{T}, \frac{\mu_1}{\mu}\right) \tag{5}$$

where U is the 'average velocity' of the interfaces or the velocity of the 'average interface'. However, (5) is predicated on the assumption that the finger goes all the way from one end of the capillary to the other. In fact, fluid will be trapped in a capillary because the less viscous fluid gets to the far end through some other capillary before the finger gets there. In this way, a fraction m_t will be trapped in the capillaries by purely kinematic effects, independent of the average speed U, and if $\mu_1 << \mu$ this fluid will in fact be trapped permanently. Making the ad hoc but reasonable guess that the proportion trapped in a capillary is on average proportional to $(1 - \cos\theta)$, one obtains the expression for the total fraction left behind

$$m = m_t + \frac{T}{3\mu U} \int_0^{\frac{3\mu U}{T}} F\left(\xi, \frac{\mu_1}{\mu}\right) \left(1 - 2m_t + \frac{2m_t T\xi}{3\mu U}\right) d\xi, \tag{6}$$

where m_t is a function only of the geometry, and $F(\xi,\mu_1/\mu)$ can be determined from experiments in straight capillaries and also, it is to be hoped, eventually from numerical solutions. For small values of $\mu U/T$,

$$m \doteqdot m_t + \frac{2}{\sqrt{3}}\left(1 - \frac{4m_t}{5}\right)\left(\frac{\mu U}{T}\right)^{\frac{1}{2}} \tag{7}$$

whereas for large values of $\mu U/T$,

$$m \doteqdot 0 \cdot 6 + 0 \cdot 4\, m_t \,. \tag{8}$$

The extent to which this result models the real phenomena is not known, as suitable experiments with which to compare are not known to the author. One reason why experiment is not simple and it may be difficult to interpret the results unambiguously is that the distribution of fluid may also be affected by macroscopic fingering on a scale large compared to the pore size.

Macroscopic Fingering

The above description of microscopic fingering gives a picture of the flow divided into two regions, one occupied by a single phase of viscous fluid and the other a mixture of the two phases. It is reasonable to suppose that under certain conditions the transition region, in which variations of microscopic finger are dominant, is thin compared with some overall macroscopic dimension of the flow and the flow can be modelled as composed of two macroscopic domains, separated by an interface, in each of which an appropriate form of Darcy's law is satisfied. The fraction m of original fluid in the region behind the interface is then regarded, in the spirit of Darcy's law, as a continuous function of position and time, with a discontinuity at the interface

$$m = \bar{F}\left(\frac{\mu U_n}{T}, \frac{\mu_1}{\mu}\right) \qquad (9)$$

where U_n is the normal velocity of the interface and $\bar{F}$ depends also on the porosity, permeability, etc., and also on the wetting properties of the fluids.

An important question is under what conditions is it reasonable to assume a macroscopic **interface**. The author can at present offer no satisfactory answer to this question, other then the speculation that $s = \mu U k^{\frac{1}{2}}/L T$, where k is the permeability and L the macroscopic length scale, measures the fuzziness of the interface produced by microscopic fingering and $s << 1$ is sufficient. A suitable name for s is the sharpness parameter. Note that it is of course possible to have $\mu U/T = O(1)$ and $s << 1$, if the permeability is sufficiently small.

Macroscopic fingering can be thought of as the finite amplitude stage of the so-called Saffman-Taylor instability of the plane interface in a porous medium separating two fluids with the less viscous driving the more viscous fluid. As shown by Saffman & Taylor (1958), the case in which m = constant is mathematically equivalent to the case $m = 0$, and for theoretical calculations it is therefore sufficient to suppose that one fluid completely expels the other if the dependence of m on U is negligible. Exact solutions for two-dimensional flow, based on the further assumption that the pressure drop across the interface is constant, can be obtained which describe both the growth of infinitesimal disturbances to finite size (Saffman 1959) and the shapes of steady fingers (Saffman & Taylor 1958). Although, these two-dimensional solutions are unlikely to be directly relevant to porous media, where it is expected that the fingers will be macroscopically three-dimensional, they can be tested experimentally in the device known as the Hele-Shaw cell, which is the space between two parallel, closely spaced flat plates. The average velocity in the gap is

$$u = -\frac{b^2}{12\mu}\frac{\partial p}{\partial x}, \qquad v = -\frac{b^2}{12\mu}\frac{\partial p}{\partial x} \tag{10}$$

where b is the distance between the plates, and the equations are those of a two-dimensional porous medium with permeability $b^2/12$ obeying Darcy's law.

Experiments in Hele-Shaw cells show good qualitative agreement with the predictions of the theory (see plates 2 and 3 of Saffman & Taylor 1958) and excellent quantitative agreement between observed and measured shapes is found for the faster bubbles (for which the assumption of constant interfacial pressure drop is most reasonable) provided the finger width is taken from the data. For some recent observations on radial fingering associated with injection or extraction from a point source in a Hele-Shaw cell, see Paterson (1981). There are, however, serious problems which arise from a further development of the theory and the attempt to predict the width of the finger, which indicate that although it may be reasonable under certain (but not yet well established) conditions to suppose that the microscopic fingering is restricted to a mathematically sharp interface, yet the boundary conditions on the macroscopic (Darcy) flow depend on the structure of the interface and plausible assumptions about the jump conditions lead to results in contradiction with experiment. In view of the several recent and proposed numerical calculations of unsteady macroscopic interfaces in porous media (e.g. Meng & Thompson 1978, Glimm et al 1980) it cannot be stressed too strongly that the question of the boundary conditions needs to be addressed and settled before the calculations can be regarded as more than an academic exercise.

The Hele-Shaw cell is an excellent phenomenon to study in this connection, as it contains a great deal of the physics without the complication of the irregular or random geometry. Suppose we consider a finger moving in a channel in a Hele-Shaw cell. The parallel plates are the surfaces $z = \pm\frac{1}{2}b$, and the edges of the channel are $y = \pm a$. For simplicity, suppose that $\mu_1 = 0$ (e.g. a finger of air displaces a heavy oil). In the region external to the finger, the average velocity $\underset{\sim}{u}_E(x,y,t)$ satisfies

$$\underset{\sim}{u}_E(x,y,t) = -\frac{b^2}{12\mu}\nabla p_E, \quad \text{div}\ \underset{\sim}{u}_E = 0 . \tag{14}$$

In the region occupied by the finger, there is a film of viscous fluid of total thickness mb, whose average velocity satisfies the equation

$$\underset{\sim}{u}_I\ (x,y,t) = -\frac{m^2 b^2}{12\,\mu}\nabla p_I , \tag{15}$$

together with the equation of continuity

$$\frac{\partial m}{\partial t} + \text{div}(m\ \underset{\sim}{u}_I) = 0 . \tag{16}$$

In general, $m = m(x,y,t)$. The remaining equation is

$$\frac{1}{2} b\, T\ \nabla^2 m + p_I = 0 \tag{17}$$

which expresses continuity of force across the surfaces of the finger parallel to the plates.

The boundary conditions at the edge of the finger are

$$(\underset{\sim}{U} - \underset{\sim}{u}_E) \cdot \underset{\sim}{n} = m\ (\underset{\sim}{U} - \underset{\sim}{u}_I) \cdot \underset{\sim}{n} \tag{18}$$

where $\underset{\sim}{U}$ is the velocity of the finger and $\underset{\sim}{n}$ is the normal direction,

$$m = F\ (\mu\, \underset{\sim}{U}.\underset{\sim}{n}/T) \tag{19}$$

and

$$p_E = p_I + \frac{T}{b}\, f\left(\frac{\mu \underset{\sim}{U}\cdot\underset{\sim}{n}}{T}, \frac{b}{R}\right) \tag{20}$$

where R is the radius of curvature of the edge in the x-y plane. The functions F and f are at present unknown and are to be determined by the 'microscopic' conditions at the edge. In the porous medium case, they would result from studies of the microscopic fingering process. In the Hele-Shaw flow, they can in principle be found from solutions of the Stokes equations in the $\underset{\sim}{n}-z$ plane. (Actually, Glimm et al 1980 assume for their calculations that $F = 0$, i.e. that the saturation m is continuous across the interface. This is equivalent to supposing that there is no difference between microscopic and macroscopic fingering, other than a change of scale, and that the variations on the microscopic scale can be described adequately by macroscopic equations. For porous media in which there is a large and continuous gradation of pore sizes, this approach may be a good approximation, but it does not seem to be a satisfactory representation of interfacial motion in a Hele-Shaw cell and is probably not good if the distribution of pore sizes in the medium is sharp).

Steady solutions of these equations and their linear stability are currently under study with plausible forms assumed for F and f. They have been solved (McLean & Saffman 1981) for the case

$$F = \text{const.}, \quad f = f_o + c\,\frac{b}{R}\,. \tag{21}$$

The analysis of Saffman & Taylor (1958) is the special case $c = 0$. The steady solutions are then arbitrary and the asymptotic width $2\lambda a$ of the finger could not be determined. However, taking $\lambda = \frac{1}{2}$ gave shapes which were in excellent agreement with the observed shapes for faster fingers. Slower fingers were observed to have values of $\lambda > \frac{1}{2}$, which depend on $\mu U/T$, and shapes which do not conform to those calculated for these values of λ. The inclusion of the term involving c removes the arbitrariness and gives λ as a unique function of $\mu UT\, b^2/a^2$, which agrees reasonably with the Saffman-Taylor experiments, and those of Pitts (1980) if $c \approx 3\text{-}4$. The observed shapes are in excellent agreement with calculation when the asymptotic widths are fixed equal.

However, the apparent success of the theory employing (21) is probably illusory. Apart from the fact that the value of c is somewhat large, a value of 1 being much more plausible, a study of the stability of the fingers reveals that they are exponentially unstable to infinitesimal disturbances. The observations show that they are extremely stable, without a trace of unsteadiness, and the conclution must be that the boundary condition (21) are wrong, at least for unsteady motion. A further test would be to carry out experiments for different values of a/b.

We conclude that microscopic and macroscopic fingering are strongly coupled, and although it is too much to hope for a complete theoretical solution in the foreseeable future, research should at least uncover the empirical dependence.

Acknowledgement

This work was supported by the Department of Energy (Office of Basic Energy Sciences). Thanks are also due to Control Data Corporation for the granting of time on the CYBER 203 at the C.D.C. Service Center, Arden Hills, Minnesota.

References

Bretherton, F. P. 1961. The motion of long bubbles in tubes. J. Fluid Mech. 10, 166-188.

Chen, B. & Saffman, P. G. 1980. Steady gravity-capillary waves on deep water 2. Numerical results for finite amplitude. Studies in App. Math. 62, 95-112.

Cox, B. G. 1962. On driving a viscous fluid out of a tube. J. Fluid Mech. 14, 81-96.

Fairbrother, F. & Stubbs, A. E. 1935. The 'bubble tube' method of measurement. J. Chem. Soc. 1, 527-530.

Glimm, J., Marchesin, D. & McBryan, O. 1980. Statistical fluid dynamics: unstable fingers. Commun. Math. Phys. 74, 1-13.

Meng, J. C. S. & Thomson, J. A. L. 1978. Numerical studies of some nonlinear hydrodynamic problems by discrete vortex element methods. J. Fluid Mech. 84, 433-453.

Paterson, L. 1981. Radial fingering in a Hele-Shaw cell. J. Fluid Mech. (to appear).

Pitts, E. 1981. Penetration of fluid into a Hele-Shaw cell. J. Fluid Mech. 97, 53-64.

Saffman, P. G. 1959. Exact solutions for the growth of fingers from a flat interface between two fluids in a porous medium or Hele-Shaw cell. Qu. J. Mech. App. Math. 12, 146-150.

Saffman, P. G. 1963. The displacement of a viscous fluid from a porous medium. Proceedings 1963 Heat Transfer and Fluid Mechanics Institute p176-182. Stanford U. Press.

Saffman, P. G. 1969. A mathematical treatment of dispersion in flow through a branching tree. Proc. CIBA Symposium on Circulatory and Respiratory Mass Transport. (Ed. Wolstenholme & Knight). Churchill.

Saffman, P. G. & Taylor, G. I. 1958. The penetration of a fluid into a porous medium or Hele-Shaw cell containing a more viscous liquid. Proc. Roy. Soc. A 245, 312-329.

Taylor, G. I. 1961. Deposition of a viscous fluid on the wall of a tube. J. Fluid Mech. 10, 161-165.

ON THE EFFECTIVE THERMAL CONDUCTIVITY AND PERMEABILITY OF REGULAR ARRAYS OF SPHERES

A.S. Sangani and A. Acrivos
Department of Chemical Engineering
Stanford University
Stanford, CA 94305

Abstract

We consider two-phase materials in which the discrete phase consists of equal-sized spheres fixed in a periodic array and summarize the results for the effective thermal conductivity and the permeability of such materials. The results are given for the whole range of volume fractions of spheres for three cubic arrays: simple, body-centered and face-centered.

1. Introduction

The problem of determining the effective parameters, e.g. conductivity, permeability, etc. of macroscopically homogeneous two phase materials has received a great deal of attention owing to its importance in many branches of mathematical physics. Here we shall be concerned with one very special aspect of this field, viz. when the discrete phase consists of equal-sized spheres fixed in a regular array. This renders the problem deterministic and therefore amenable to a unique solution. In spite of their specialized nature, the solutions to be presented provide valuable information which, at least qualitatively speaking, is applicable to more general systems. Also the method of solution which we have employed is of interest in its own right and might be found useful in many other applications.

We shall consider separately the two problems of thermal conduction and of flow through a periodic array.

2. The Effective Thermal Conductivity

Consider a composite material consisting of spherical particles of radius a and conductivity α embedded in a homogeneous matrix of unit conductivity. Rayleigh was the first to analyze the case when the spheres are arranged in a simple cubic array. He developed a solution by replacing the spheres by dipoles and higher order multipoles and obtained for the effective thermal conductivity k^*:

$$k^* = 1 - 3c\left[\frac{2+\alpha}{1-\alpha} + c - \frac{(1-\alpha)d}{(4+3\alpha)}\,c^{10/3} + O(c^{14/3})\right]^{-1} \qquad (1)$$

where c is the volume fraction of the solids. The value $d = 4.95$ given by Rayleigh was later corrected by Runge [9] to 1.57. Meredith and Tobias [7] extended Rayleigh's analysis and calculated the coefficient of the $O(c^{14/3})$ term. The validity of Rayleigh's method was questioned in recent years (see for example, Levine [4]; Jeffrey [3]) because it involved the summation of a non-absolutely convergent series. O'Brien [8] and McPhedran and McKenzie [6] have, however, modified Rayleigh's method in order to overcome this difficulty, hence, this modified Rayleigh's method has now a sound theoretical basis. McPhedran and McKenzie also gave an expression for k^* to $O(c^{35/3})$. Unfortunately, their calculated coefficients are somewhat incorrect because they were obtained using an axially symmetric potential, rather than a potential possessing the four-fold symmetry of the cubic array.

Recently Zuzovski and Brenner [13] resolved this problem using the method of generalized functions, which avoids the difficulty encountered in Rayleigh's original treatment, and found that the coefficient of the $O(c^{14/3})$ term reported by Meredith and Tobias [7] was in error. Unfortunately, owing to a numerical slip, the "corrected coefficient" given by Zuzovski and Brenner is itself incorrect.

We have modified the approach taken by Zuzovski and Brenner [13] and have derived expressions for k^* to $O(c^9)$ for any α for simple, body-centered, and face-centered cubic arrays. We have also calculated k^* over the complete range of volume fractions of solids for the two special cases: $\alpha = \infty$ and $\alpha = 0$. Our numerical results for $\alpha = \infty$ are indistinguishable from McPhedran and McKenzie's [6] for simple cubic arrays, and from McKenzie, McPhedran and Derrick's [5] for body-centered and face-centered cubic arrays.

At the other extreme, i.e. for closely packed arrays, Batchelor and O'Brien [1] used lubrication theory and obtained, for perfectly conducting solids $(\alpha = \infty)$, the expression

$$k^* \sim -K_1 \ln(1 - \chi) - K_2 \qquad c \longrightarrow c_{max}, \tag{2}$$

where $\chi = (c/c_{max})^{1/3}$ with c_{max} being the volume fraction of the particles when they are touching each other. The constant K_1 equals $\pi/2$, $\sqrt{3}\pi/2$ and $\sqrt{2}\pi$ for, respectively, simple, body-centered, and face-centered cubic arrays for which the corresponding values for c_{max} are $\pi/6$, $\sqrt{3}\pi/8$, and $\sqrt{2}\pi/6$. On the other hand, the $O(1)$ constant K_2 in (2) cannot be determined from lubrication theory alone. By extending the series for k^* to $O(c^{15})$ for $\alpha = \infty$ we have been able to compute K_2 for the three cubic arrays which we found to equal 0.7, 2.4 and 7.2 respectively.

The method of solution, which is described in more detail elsewhere [10] is as follows:

Let the centers of the spherical particles be located at

$$\underline{r}_{\underline{n}} = n_1\underline{a}_{(1)} + n_2\underline{a}_{(2)} + n_3\underline{a}_{(3)} \qquad (n_1, n_2, n_3 = 0, \pm 1, \pm 2, \cdots), \tag{3}$$

where $\underline{a}_{(1)}$, $\underline{a}_{(2)}$, and $\underline{a}_{(3)}$ are the basic vectors determining the unit cell of the array. Further suppose that a temperature gradient in the x_1-direction of unit magnitude is applied by some external means to this composite material. We are interested in calculating along x_1 the average heat flux which gives us directly k^*.

We recall that the temperature T is a continuous harmonic function everywhere within the composite except at the surface of the particles where its normal derivative suffers a jump discontinuity if α differs from unity. This jump in the derivative is such that the heat flux is continuous everywhere. The gradient of the temperature is a periodic function and since a temperature gradient of unit magnitude is externally applied in the x_1-direction, the temperature can be expressed as

$$T = x_1 + \tilde{T} \quad , \tag{4}$$

where $\tilde{T}$ is a spatially periodic function

$$\tilde{T}(\underline{r}) = \tilde{T}(\underline{r} + \underline{r}_n) \tag{5}$$

with $\underline{r}_n$ given by (3). In order to solve for $\tilde{T}$, we first obtain the periodic fundamental singular solution to Laplace's equation which satisfies

$$\Delta S_1(\underline{r}) = \frac{4\pi}{\tau_o} - 4\pi \sum_{\underline{n}} \delta(\underline{r} - \underline{r}_n) \qquad \left(\Delta = \frac{\partial^2}{\partial x_1^2} + \frac{\partial^2}{\partial x_2^2} + \frac{\partial^2}{\partial x_3^2} \right) , \tag{6}$$

where τ_o is the volume of a unit cell and $\delta(\underline{r} - \underline{r}_n)$ is the Dirac's delta function defined by

$$\int_\tau \delta(\underline{r} - \underline{r}_n) d\underline{r} = \begin{cases} 1 & \text{when } \underline{r}_n \in \tau \\ 0 & \text{when } \underline{r}_n \notin \tau \end{cases} \tag{7}$$

and

$$\delta(\underline{r} - \underline{r}_n) = 0 \qquad \text{for } \underline{r} \neq \underline{r}_n . \tag{8}$$

The constant $4\pi/\tau_o$ in (6), which represents the strength of uniformly distributed sinks, must be added to the right-hand side of the equation in order to ensure that the total heat generated within a unit cell is zero.

As shown by Hasimoto [2] a solution to (6) is

$$S_1 = \frac{1}{\pi\tau_o} \sum_{\underline{k}_n \neq \underline{0}} \frac{e^{-2\pi i(\underline{k}_n \cdot \underline{r})}}{k_n^2} \tag{9}$$

where

$$\underline{k}_{\underline{n}} = n_1\, \underline{b}_1 + n_2\, \underline{b}_2 + n_3\, \underline{b}_3 \tag{10}$$

are the reciprocal lattice vectors given by

$$\underline{k}_{\underline{n}} \cdot \underline{a}_{(j)} = n_{(j)} \qquad (j = 1, 2, 3). \tag{11}$$

The basic vectors $\underline{b}_{(1)}$, $\underline{b}_{(2)}$, and $\underline{b}_{(3)}$ in the reciprocal lattice are related to the basic vectors in physical space by

$$\underline{b}_1 = \frac{\underline{a}_{(2)} \cdot \underline{a}_{(3)}}{\tau_o}, \quad \underline{b}_{(2)} = \frac{\underline{a}_{(3)} \cdot \underline{a}_{(1)}}{\tau_o}, \quad \underline{b}_{(3)} = \frac{\underline{a}_{(1)} \cdot \underline{a}_{(2)}}{\tau_o}. \tag{12}$$

The temperature field outside any sphere can now be expressed as a sum of the successive derivatives of S_1 multiplied by some unknown coefficients, i.e.

$$T = x_1 + \underline{G}\, S_1 \tag{13}$$

where $\underline{G}$ is the differential operator:

$$\underline{G} = \sum_{m+n+p=1}^{\infty} A_{mnp} \frac{\partial^{m+n+p}}{\partial x_1^m\, \partial x_2^n\, \partial x_3^p}. \tag{14}$$

This formal solution (13,14) is identical to that derived by Zuzovski and Brenner [13] using generalized functions. The above expression for T is singular at $\underline{r} = \underline{r}_{\underline{n}}$ and therefore is valid only outside the particles. Within the latter, a different expression for T is needed which is regular everywhere. The requirements of continuity of temperature and flux at the surface of the particles then lead to a relation for determining these unknown coefficients A.

On using certain properties of the spherical harmonics and rearranging (14) into a more convenient form we have been able to derive a set of linear equations for the unknowns A in (14) whose coefficients we determined via an explicit expression. This set of linear equations was solved either by successive approximations (for $\chi \ll 1$) as a series in χ or by the method of "direct substitution" [10] for selected values of χ.

Table 1 gives k*, computed via the direct substitution method, as a function of χ for perfectly conducting spheres $(\alpha = \infty)$ for each of the three cubic arrays. Also, Figure 1 depicts k* versus χ for simple cubic arrays with $\alpha = 0$ and $\alpha = \infty$. For $\alpha = \infty$, it was found that when $\chi < 0.9$ both methods converged to the same answer, but that for $0.9 < \chi < 1.0$ the series method appeared superior in that K_2 in (2) rapidly approached a constant value as the number of terms in the series increased. For simple, body-centered and face-centered cubic arrays K_2 thus calculated equaled 0.7, 2.4 and 7.2, respectively. In contrast, for $\alpha < 40$, the direct substitution method converged more quickly and the results for the case of touching spheres ($\chi = 1$) are shown in Table 2. Of particular interest are the values of k^* for $\alpha = 0$ ($\chi = 1$) which equals 0.344, 0.220, and 0.164, respectively, for the above three cubic arrays. It should be noted that these values are within 15% of the Haskin-Shrikman upper bounds for such systems.

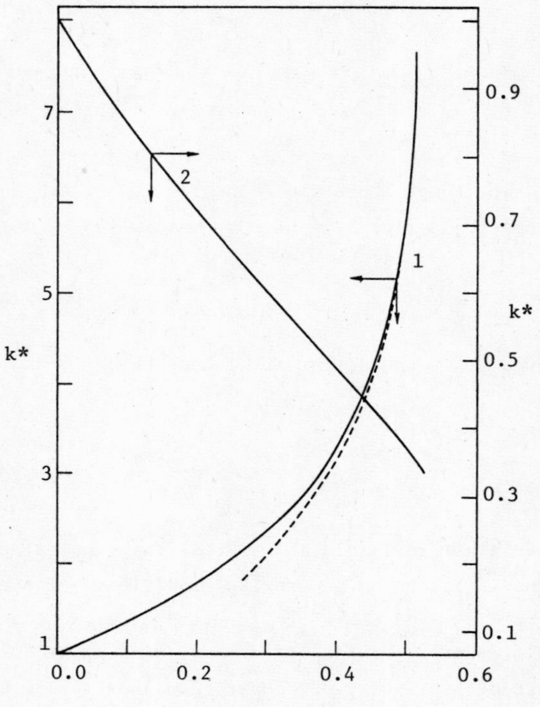

Figure 1.
The effective conductivity k^* as a function of volume fraction c for a simple cubic array of perfectly conducting (1) and perfectly insulating (2) spheres (—— computed values, ---- Eq. (2) with $K_1 = \pi/2$ and $K_2 = 0.7$) [10].

3. The Effective Permeability

Here we consider the corresponding problem of slow flow past a periodic array of spheres and wish to determine the permeability or more specifically the drag force F exerted by the fluid moving with average speed U on a representative sphere in the assembly as a function of c, the volume fraction of the spheres. As before, the centers of the spheres are located at $\underline{r}_n$ - c.f. (3).

Hasimoto [2] showed that in case of simple cubic array

$$K^{-1} = 1 - 1.7601\, c^{1/3} + c - 1.5593\, c^2 + O(c^{8/3}) \tag{15}$$

TABLE 1

k* as a function of α for the three cubic arrays and for $\alpha = \infty$ [10]

χ	SC	BCC	FCC
0.1	1.00157	1.00204	1.00222
0.2	1.0216	1.0164	1.0179
0.3	1.0430	1.0561	1.0612
0.4	1.104	1.1365	1.1492
0.5	1.210	1.2788	1.3060
0.6	1.383	1.5168	1.5713
0.7	1.660	1.915	2.023
0.8	2.125	2.618	2.848
0.85	2.491	3.199	3.556
0.90	3.045	4.11	4.706
0.95	4.06	5.8	6.99
0.97	4.84	-	8.74
0.99	6.55	-	-

TABLE 2

k* as a function of α for various closely packed arrays (χ=1) [10]

α	SC	BCC	FCC
0	0.344	0.220	0.164
1	1.0	1.0	1.0
2	1.4585	1.6196	1.6874
5	2.422	3.026	3.338
10	3.474	4.60	5.29
20	4.81	6.45	7.68
30	-	7.6	9.1
40	6.41	8.2	10.0

where

$$K \equiv \frac{F}{6\pi\mu Ua}$$

with μ being the viscosity of the fluid and a, as before, the radius of the spheres. Clearly, (15) is meaningless for c beyond approximately 0.2 when K becomes negative.

Here we modify Hasimoto's analysis and calculate K over the complete range of c for all three of the cubic arrays. Our method of solution, which is described in more detail elsewhere [11] is as follows

A periodic fundamental solution to the creeping flow equations can be obtained by solving

$$\mu\Delta v_i = \frac{\partial q}{\partial x_i} + \delta_{i1} \sum_{\underline{n}} \delta(\underline{r} - \underline{r}_{\underline{n}}) , \qquad \frac{\partial v_i}{\partial x_i} = 0 . \tag{16}$$

As shown by Hasimoto (1959)

$$v_i = v_{\underline{0}}\delta_{i1} - \frac{1}{4\pi\mu}\left\{ S_1\delta_{i1} - \frac{\partial^2 S_2}{\partial x_1 \partial x_i} \right\} \tag{17}$$

$$\frac{\partial q}{\partial x_i} = - \frac{\delta_{i1}}{\tau_o} + \frac{1}{4\pi} \frac{\partial^2 S_1}{\partial x_1 \partial x_1} , \tag{18}$$

where τ_o is, as before, the volume of the unit cell, S_1 is given by (9) and

$$S_2 = \frac{-1}{4\pi^3 \tau_o} \sum_{\underline{k}_{\underline{n}} \neq \underline{0}} \frac{e^{-2\pi i(\underline{k}_{\underline{n}} \cdot \underline{r})}}{k_{\underline{n}}^4} . \tag{19}$$

S_1 satisfies (6) while

$$\Delta S_2 = S_1 . \tag{20}$$

To obtain a completely general solution for the velocity u_i and for the pressure p, we add to the fundamental solution the derivatives of v_i and S_1 multiplied by some unknown coefficients, but omit from the sum those terms which do not satisfy the symmetry conditions of the system. We thus arrive at:

$$u_1 = U_0 - \frac{1}{4\pi\mu}\left\{\underline{G}\left(S_1 - \frac{\partial^2 S_2}{\partial x_1^2}\right) + \underline{H}\,\frac{\partial^2 S_1}{\partial x_1^2} - \underline{L}\left(\frac{\partial^4}{\partial x_1^4} - 6\,\frac{\partial^4}{\partial x_2^2 \partial x_3^2} + \frac{\partial^4}{\partial x_3^4}\right) S_1\right\} \tag{21}$$

$$u_2 = \frac{1}{4\pi\mu}\left\{G\,\frac{\partial^2 S_2}{\partial x_1 \partial x_2} - \underline{H}\,\frac{\partial^2 S_1}{\partial x_1 \partial x_2} - \underline{L}\,\frac{\partial}{\partial x_1}\left(\frac{\partial^3}{\partial x_2^3} - 3\,\frac{\partial^3}{\partial x_3^2 \partial x_2}\right)\right\} S_1 \tag{22}$$

$$u_3 = \frac{1}{4\pi\mu}\left\{\underline{G}\,\frac{\partial^2 S_2}{\partial x_1 \partial x_3} - \underline{H}\,\frac{\partial^2 S_1}{\partial x_1 \partial x_3} - \underline{L}\,\frac{\partial}{\partial x_1}\left(\frac{\partial^3}{\partial x_3^3} - 3\,\frac{\partial^3}{\partial x_2^2 \partial x_3}\right) S_1\right\} \tag{23}$$

$$\frac{\partial p}{\partial x_i} = \frac{-F}{\tau_o}\,\delta_{i1} + \frac{1}{4\pi}\,\underline{G}\,\frac{\partial^2 S_1}{\partial x_1 \partial x_i}\ , \tag{24}$$

where $\underline{G}$, $\underline{H}$, and $\underline{L}$ are the differential operators

$$\left\{\begin{matrix}\underline{G}\\ \underline{H}\\ \underline{L}\end{matrix}\right\} = \sum_{M=0}^{\infty}\ \sum_{m=0}^{m \le \frac{1}{2}M} \left\{\begin{matrix}A_{nm}\\ B_{nm}\\ C_{nm}\end{matrix}\right\} \left\{\frac{\partial^{2n}}{\partial x_1^{2n}}\left[\left(\frac{\partial}{\partial \xi}\right)^{4m} + \left(\frac{\partial}{\partial \eta}\right)^{4m}\right]\right\} (M = n+2m) \tag{25}$$

with $\xi = x_2 + ix_3$, $\eta = x_2 - ix_3$, (26)

and where the unknown coefficients A_{nm}, B_{nm}, and C_{nm} are to be determined by applying the no-slip boundary condition at the surface of the spheres. The above expressions for the components of the velocity differ from those given by Hasimoto [2] in two important aspects. First, the terms containing the differential operator $\underline{L}$ are absent in Hasimoto's solution, and second, the differential operators defined by (26) are a special form of those given by Hasimoto and are particularly convenient when dealing with problems involving cubic arrays of spheres.

Since it has been shown by Hasimoto [2] that

$$A_{oo} = 3\pi\mu UaK \qquad U_o = U + \frac{2\,B_{oo}}{\pi\tau_o}\,, \tag{27}$$

the determination of A_{oo} directly yields K.

As was the case with the conductivity problem described earlier in §2, using certain properties of the spherical harmonics we were able to derive a set of linear equations for the unknown coefficients in (25). This set was then solved either by successive approximations or by direct substitution. Thus, using the method of successive approximations we have obtained expressions for K to $O(c^{10})$ for the three cubic arrays. Again, in this case we found that for $\chi < 0.9$ both methods converged to the same answer. For $\chi > 0.9$, however, the direct substitution method converged rapidly whereas the successive approximations method did not converge at all. Of particular interest are the values of K when the particles are touching each other, i.e. when $\chi = 1$. For simple and body-centered cubic arrays the respective values of K are 42.1 and 170. Except for the case of nearly touching face-centered cubic arrays, all of our results agree well with those computed recently by Zick and Homsy [12].

ACKNOWLEDGEMENT

This work was supported in part by a grant by the National Science Foundation ENG-78-09241.

REFERENCES

1. Batchelor, G.K.; O'Brien, R.W.: Thermal or electrical conduction through a granular material. Proc. Roy. Soc. Lond. A355, 313-333 (1977).
2. Hasimoto, H.: On the periodic fundamental solutions of the Stokes equations and their application to viscous flow past a cubic array of spheres. J. Fluid Mech. 5, 317-328 (1959).
3. Jeffrey, D.J.: Conduction through a random suspension of spheres. Proc. Roy. Soc. Lond. A335, 355-367 (1973).
4. Levine, H.: The effective conductivity of a regular composite medium. J. Inst. Maths Applics. 2, 12-28 (1966)
5. McKenzie, D.R.; McPhedran, R.C.; Derrick, G.H.: The conductivity of lattices of spheres II. The body centered and faced centered lattices. Proc. Roy. Soc. Lond. A362, 211-232 (1978).
6. McPhedran, R.C.; McKenzie, D.R.: The conductivity of lattices of spheres I. The simple cubic lattice. Proc. Roy. Soc. Lond. A359, 45-62 (1978).
7. Meredith, R.E.; Tobias, C.W.: Resistance to potential flow through a cubical array of spheres. J. Appl. Phys. 31, 1270-1273 (1960).
8. O'Brien, R.W.: A method for the calculation of the effective transport properties of suspensions of interacting particles. J. Fluid Mech. 91, 17-39 (1979).
9. Runge, I.: On the electrical conductivity of metallic aggregates. Z. Tech. Physik. 6, 61-68 (1925).

REFERENCES (continued)

10. Sangani, A.S.; Acrivos, A.: The effective conductivity of a periodic array of spheres. To be submitted (1981).
11. Sangani, A.S.; Acrivos, A.: Slow flow through a periodic array of spheres. To be submitted (1981).
12. Zick. A.A.; Homsy, G.M.: Stokes flow through periodic arrays of spheres. (to appear) J. Fluid Mech. (1981).
13. Zuzovski, M.; Brenner, H.: Effective conductivities of composite materials composed of cubic arrangements of spherical particles embedded in an isotropic matrix. J. Appl. Math. Phys. (ZAMP) 28, 979-992 (1977).

DIELECTRIC AND ACOUSTIC RESPONSE OF ROCKS

P. N. Sen
Schlumberger-Doll Research
Ridgefield, CT 06877

ABSTRACT

A self-similar model is developed to explain the conductivity, dielectric, and acoustic response of rocks. The model explicitly takes into account the correlations that lead to a zero percolation threshold, and gives the d.c. conductivity of rock $\sigma(0) = \sigma_w(0)\phi^m$, with $\sigma_w(0)$ the d.c. conductivity of the brine, ϕ the porosity. The exponent m depends on the grain shape. Using well-known results, it is shown that a small concentration, η, of plate-like grains of aspect ratio $\delta, \delta << 1$, can give rise to a large ($\sim 10^4$) low-frequency dielectric constant ϵ_s, and explain its salinity and frequency dependences. The case $\delta < \eta$, $\eta \to 0$, corresponds to the well-known Maxwell-Wagner effect where a divergent ϵ_s is accompanied with a conductivity threshold $\sigma(0) \to 0$. The case $\delta \to 0, \eta \to 0, \delta > \eta$ gives rise to a new result, previously overlooked, with divergent ϵ_s and non-zero $\sigma(0)$. The velocity v of Biot second wave at high frequency and the fourth sound wave in ^{4}He saturated plug are given by $v = v_F/n$, v_F being the velocity in the free fluid, $n^2 = \phi\sigma_w(0)/\sigma(0)$ (exact) $= \phi^{1-m}$ (self-similar model).

Introduction

A class of inhomogeneous materials - sedimentary rocks - show a variety of interesting behavior. In this paper, a short description of a few of the properties of sedimentary rocks and explanations of those properties in terms of geometrical models are given. Most of the results presented here either have already been published, or are in the process of being published elsewhere.[1-4]

The real part of the dielectric constant, $\epsilon(\omega)$, of brine saturated rocks at low frequencies (kilohertz range) can be as great as 10^4. These gigantic values[5-6] are remarkable considering that ϵ of the individual constituents do not exceed 80: The ϵ of dry rock is about 10 and of brine about 80. Although ϵ of brine and rock separately are frequency independent up to 1 GHz, the brine saturated rock shows a great deal of dispersion. The d.c. conductivity of brine saturated sedimentary rocks obey a simple power law - known as Archie's law

$$\sigma(0) = \sigma_w(0)\phi^m \tag{1}$$

In Eq. (1), ϕ is the porosity and m depends on the grain aspect ratio, and $\sigma_w(0)$ and $\sigma(0)$ are the d.c. conductivities of brine and the rock. A first principles derivation of Archie's law has been given by Sen, Scala and Cohen[1] in terms of a self-similar model. A summary of the model is given below. The exponent m in Eq. (1) depends on grain shape. The value $m = 3/2$ for spheres agrees extremely well with experiments[1] on brine saturated fused glass beads.

We then show that the presence of high aspect ratio (thin plate-like) particles can explain[2-3] the experimentally observed frequency and salinity dependences and also the great values of dielectric constant at low frequencies. The shape of the particle is extremely important in general. This shows up in another context: Far infrared absorption by ultrafine metal particles is generally found much greater than that predicted for spheres. Needle-like structures formed by these spheres can explain these large absorptions. Surface effects become important at frequencies below the MHz range and also contribute to these large values of dielectric constant.

Several "potential" problems for porous media - dielectric response, thermal conductivity, etc. - are closely related. The conductivity can be used to predict the velocity of a sound mode in fluid saturated porous media.[4] This mode is known as the fourth sound in porous plugs saturated with ^4He and Biot slow wave in fluid saturated rocks.[7]

Surface Effects

The electrical properties of a rock-water mixture depends on the bulk electrical properties of the constituents, and then on...

a. Electrochemical/Interfacial Effects

b. Geometrical arrangement of the constituents

The interfacial effect is particularly prominent in clayey samples. The counter-ions on the clay surface become mobile in the presence of water and increase σ_w. These counter ions are polarizable, and contribute to $\epsilon(\omega)$. In this paper we denote the complex dielectric constant by $\epsilon = \epsilon + 4\pi\sigma/i\omega$. In the model of Schwarz,[8] which explains these large values, the basic mechanism is as follows: When an external field is applied, the interfacial charges move tangentially across the solid/liquid interface creating an excess of charge on one side, and a deficiency on the other. Since the distance, over which charges move, is macroscopic, the dipole moment induced (charge x distance) is great, hence contributing a large value of dielectric constant ($\epsilon \sim 10^3$ below 10^4 Hz). It can be shown[9] that the Schwarz model is correct only if the double layer is impenetrable to charges moving normally to the surface. Whether this is true remains to be shown. The hindered rotation of the water molecules adsorbed in the interfacial zone gives rise to a dielectric response in the MHz range, according to Schwan.[10] This is a mechanism not covered by the Schwarz model. Thus there are at least three types of contribution from the interfacial ions...

i. Free-ion-like contribution which gives the finite d.c. conductivity

ii. Polarization of the counter-ion layer

iii. Polarization of hindered water molecules

The detailed understanding of the interfacial effects needs more theoretical and experimental work, but the geometrical effects are better understood. The rest of the paper deals with geometrical effects.

Geometric Effects

In general, the macroscopic properties of a composite material containing phases with very different physical properties, depend not only on the volume fractions of the constituents, but are extremely sensitive to the geometry and topology of the boundary surfaces between the phases. The parameters of typical boundary surfaces in a mixture are not related, in any simple way, to the lowest order point distribution functions for the constituents, so that the mathematical problem of calculating the overall properties of the system is very ill—posed.[11]

The importance of the geometrical effects can be simply illustrated by the following *exact* results. The dielectric constant of a layered medium (with thicknesses smaller than the wavelength) or a medium consisting of tubes, when a field is appied parallel to the interface, is given by,

$$\epsilon^* = \sum_i f_i \epsilon_i^* \tag{2}$$

where f_i denotes the volume fraction of the i-th phase of dielectric constant ϵ_i^*. But when the field is applied perpendicular to the interface of the layered medium, ϵ^* is given by

$$\epsilon^* = \left[\sum_i f_i / \epsilon_i^*\right]^{-1} \tag{3}$$

If the layers are made up to conducting and insulating materials, in the first case we obtain a non-zero $\sigma(0)$, whereas $\sigma(0)$ is zero for the latter. Similarly, the dielectric constants are drastically different in two cases. The importance of the interface-shape can be further illustrated. The potential ϕ at a point $\mathbf{r}$, is given by[12]

$$\phi(\mathbf{r}) = \phi_0(\mathbf{r}) + \int \mathbf{P}\cdot\nabla' \frac{1}{|\mathbf{r}-\mathbf{r}'|} d^3r' \tag{4}$$

The term ϕ_o is from the fixed external sources (e.g., the charges on condenser plates) and $\mathbf{P}$ is the dipole moment per unit volume. Integrating by parts, (4) gives

$$\phi(\mathbf{r}) = \phi_o(\mathbf{r}) - \int \frac{\nabla'\cdot\mathbf{P}(\mathbf{r}')}{|\mathbf{r}-\mathbf{r}'|} d^3r' + \sum_\alpha \frac{\mathbf{P}\cdot d\mathbf{s}_\alpha'}{|\mathbf{r}-\mathbf{r}'|} \tag{5}$$

The sum over α denotes a sum over all the interfaces. Here we envision the material to be composed of non-overlapping cells, each cell having a constant polarizability. If $\mathbf{P}$ is uniform in each cell, which is the case in the single site approximations (both coherent potential approximation or average t-matrix approximation discussed below), *only the* interfacial terms contribute. In that case of $\nabla\cdot\mathbf{P} = 0$ in each cell, the field is given from (5) to be

$$E_i = -\partial\phi/\partial x_i = E_{oi} - 4\pi \sum_\alpha L_{\alpha ij} P_{\alpha j} \tag{6}$$

and $L_{\alpha ij}(r)$ is the generalized depolarization factor of the αth grain

$$L_{\alpha ij} = -\frac{1}{4\pi} \int \frac{r_i - r_i'}{|\mathbf{r} - \mathbf{r}'|^3} ds'_{\alpha j} \tag{7}$$

which depends on the grain shape. Thus, the shape of the interface is of paramount importance in determining the dielectric constant of an inhomogeneous material.

Equation (4) forms the basis of systematic iteration schemes for evaluating the dielectric constant of the composite material.[14] Since $\mathbf{P}$ is related to $\mathbf{E}$ via a constitutive relation $\mathbf{P} = (\epsilon^* - 1)\,\mathbf{E}/4\pi = \alpha^*\mathbf{E}$, (4) gives an integral equation for $\mathbf{E}$. This also is the starting equation for derivation of the Clausius-Mossotti formula following Ewald and Oseen.[15] The field at a given point is given by the external field plus the sum of fields due to each cell. In order to compute the field produced by each cell, we need to obtain the field which the cell itself is subjected to. In a single site approximation, each cell is assumed to be surrounded by an average medium of dielectric constant ϵ_o^*. In the average t-matrix approximation (ATA), this is taken to be the host material. The effective dielectric constant ϵ^* is computed by requiring that this material of dielectric constant ϵ^*, when embedded in ϵ_o^*, will give the same scattered field on the average. The other common single site approximation, coherent potential (CPA) or effective medium approximation (EMA) obtained by requiring that ϵ^* is such that the scattering vanishes on the average.

The formal iteration of (4) has been developed by many authors. These techniques have been summarized by M. Hori.[14] I give here a simple derivation due to Maxwell.[16] This derivation clarifies all the points which form the basis of the formal perturbation analysis and truncation schemes mentioned above. Suppose, spheres of radii a_i having dielectric constant ϵ_i^* are embedded in an assumed host ϵ_o^*. The total potential due to these spheres subjected to a field $E_o\hat{z}$ at infinity is

$$\phi = \sum_i \frac{\epsilon_i^* - \epsilon_o^*}{\epsilon_i^* + 2\epsilon_o^*} a_i^3 E_o \frac{\mathbf{r}\cdot\hat{z}}{r^3} - E_o(\mathbf{r}\cdot\hat{z}) \tag{8}$$

this is equated to the potential due a large sphere of dielectric constant ϵ^* and radius R that encloses all these small spheres:

$$\phi = \frac{\epsilon^* - \epsilon_o^*}{\epsilon^* + 2\epsilon_o^*} R^3 E_o \frac{\mathbf{r}\cdot\hat{z}}{r^3} - E_o(\mathbf{r}\cdot\hat{z}) \tag{9}$$

In (9), ϵ^* is the effective dielectric constant of the mixture. Equations (8) and (9) give the ATA results

$$\frac{\epsilon^* - \epsilon_o^*}{\epsilon^* + 2\epsilon_o^*} = \sum_i f_i \frac{\epsilon_i - \epsilon_o^*}{\epsilon_i^* + 2\epsilon_o^*} \tag{10}$$

Here f_i is the volume function of the i-th phase a_i^3 / R^3. If we require self-consistency i.e., $\epsilon_o^* = \epsilon^*$, then (8) gives, instead of (10)

$$\sum_i f_i \frac{\epsilon_i^* - \epsilon^*}{\epsilon_i^* + 2\epsilon^*} = 0 \tag{11}$$

In ATA, (10), ϵ_o^* is generally taken as the majority substance. Equation (11) implies, via Eq. (8), that the total potential produced by the spheres add up to zero. The key point in the derivation above is that the local field seen by each sphere is the same and is not affected by the *proximity* of the other spheres. This is the basic deficiency of the single-site theory. We describe below the self-similar model[1] which goes beyond the dilute limit. There are, of course, competing theories but we do not focus on them in this article.

The First Principles Derivation of Archie's Law: Self-Similar Model

In reference 1 we developed a theory for dielectric response of water- saturated rocks based on a realistic model of the pore space. The absence of a percolation threshold manifest in Archie's law, porecasts, electron-micrographs, and general theories of formation of detrital sedimentary rocks indicate that the pore spaces within such rocks remain interconnected to very low values of the porosity ϕ. In the simplest geometric model for which the conducting paths remain interconnected, each grain is envisioned to be coated with water. The dielectric constant of the assembly of water-coated grains is obtained by a self-consistent effective medium theory.

Using equation (11), the self-consistent dielectric constant of the rock made up of coated spheres is given by

$$0 = \sum_i f_i \frac{\epsilon^* - \epsilon_{cs}^{(i)}}{2\epsilon^* + \epsilon_{cs}^{(i)}} \tag{12}$$

Here f_i is the volume fraction of the i-th coated sphere in the assembly, i.e., volume of the i-th sphere divided by the total volume of the rock. The dielectric constant of a sphere i of material ϵ_m^* coated with water ϵ_w^* is[17]

$$\epsilon_{cs}^{(i)} = \epsilon_w \frac{\epsilon_m^* + 2\epsilon_w^* + 2\eta_i(\epsilon_m^* - \epsilon_w^*)}{\epsilon_m^* + 2\epsilon_w^* - \eta_i(\epsilon_m^* - \epsilon_w^*)} \tag{13}$$

Here η_i is the volume fraction of the inner sphere. In general, f_i and η_i are unrelated. The simplest case arises if we assume that $\eta_i = 1 - \phi$ is the same for all the spheres. Then separating real and imaginary parts of (12) gives in the d.c. limit

$$\sigma(0) = \sigma_w(0) \frac{2\phi}{3-\phi} \tag{14}$$

Equation (14) gives $\sigma \propto \phi^m$ with $m = 1$ for small ϕ. The empirical evidence for an exponent m greater than unity is overwhelming, both from laboratory and field data. Accordingly, we must regard the model of a sphere coated with a spherical shell of water of fixed proportions as overly simple.

The single-site or mean-field approximation such as CPA fails to take into account a variety of local environments of each type of rock grain. The systematic way around this problem is to consider bigger and bigger clusters, where a grain is not only surrounded by water but also surrounded by other rock grains. In practice this problem is extremely difficult and has been solved numerically only for a few special cases. We proceed here by a very simple intuitive method of incorporating the clustering effects in a single-site effective medium theory. This comes from the observation that each grain is not only surrounded by water but also by other grains which in turn are surrounded by others in a nested manner. This led to a self-similar model where the continuity of the conducting phase is guaranteed.

First we add a few grains of rock of any size to water. As long as the sizes are smaller than the smallest wavelength involved, the sizes are irrelevant. Then use this mixture to coat some new grains, and so on. At each step we add a small amount of grain, and we determine the dielectric constant self-consistently by generalizations of eqs. (13) and (12). This gives for an infinitesimal increase of grain volume,

$$\frac{d\epsilon^*}{\epsilon^*} = -\frac{d\phi}{\phi}\,\frac{\epsilon_m^*-\epsilon^*}{\epsilon_m^*+2\epsilon^*} \tag{15}$$

Integrating (15) with the boundary condition that $\epsilon^* = \epsilon^*_w$ at $\phi = 1$ gives

$$\left[\frac{\epsilon_m^*-\epsilon^*}{\epsilon_m^*-\epsilon_w^*}\right]\left[\frac{\epsilon_w^*}{\epsilon^*}\right]^{1/3} = \phi \tag{16}$$

The limiting case (as the frequency, ω, tends to zero) of equation (16) gives Archie's law. Separating the imaginary part gives

$$\sigma(0) = \sigma_w(0)\phi^{3/2} \tag{17}$$

For anisotropic rock with grains of fixed shape and orientation, equation (17) is easily generalized. For ellipsoidal grains with a depolarization ratio L (L is a function of grain shape[1-4,12,13] we find, under fairly general conditions,[1,3,4,18,19]

$$\left[\frac{\epsilon_m^*-\epsilon^*}{\epsilon_m^*-\epsilon_w^*}\right]\left[\frac{\epsilon_w^*}{\epsilon^*}\right]^{L} = \phi \tag{18}$$

This, in the d.c. limit, gives

$$\sigma(0) = \sigma_w(0)\phi^{\frac{1}{1-L}} \tag{19}$$

For isotropic rock, in which the spheroidal grains are randomly oriented, the cementation exponent is given by

$$m=\int \frac{5-3L_s}{3(1-L_s^2)} P(L_s)dL_s \tag{20}$$

Here $P(L_S)$ is the probability distribution for the depolarization factor L_S along the symmetry axis. Equation (16) is identical to that obtained by Bruggeman.[20] Bruggeman's procedure was generalized by Veinberg[18] for randomly oriented spheroids (and was applied to rocks by Mendelson and Cohen[19]) to obtain (18) and (20).

The procedure used by Bruggeman[20] and Veinberg[18] did not employ the coated spheres technique. They introduce grains, by an infinitesimal amount at each step, into the previous mixture and use the effective medium approximation (EMA) at each step. Since only an infintesimal amount of grain is introduced at each step (far below the percolation threshold), the medium remains conducting at each step. Furthermore, for coated spheres, EMA is equivalent to uncoated spheres with the Clausius-Mossotti (ATA). But, ATA in turn is the same as EMA in the dilute concentration limit. Thus the procedures of Sen, Scala and Cohen[1] and of Bruggeman[20]-Veinberg[18] are similar provided that we use the correct boundary condition, e.g., start with water at step zero.

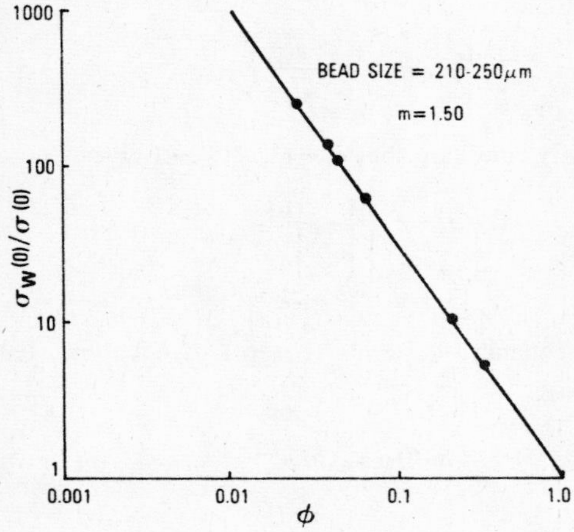

Fig. 1 - D.C. Conductivity of Glass Beads; Dots: Expt; Line: Theory (17)

The conductivity of the fused glass beads at low frequencies agree very well with equation (17) with an exponent of $m = 1.5$ (Fig. 1). Theory also predicts $\mathrm{Re}\,\epsilon^*$ for many cases extremely well.[1]

Frequency/Salinity Dependence and Large Values of Dielectric Constants: Thin Plate Effects

Despite the successes of the models discussed above, we have failed to predict the most spectacular property - the dielectric constant ϵ reaches values as high as 10^4 at low frequencies. In this section we show that the presence of high aspect ratio particles can explain both the observed frequency and salinity dependence and the high values of dielectric constants.

It is well-known that the low-frequency dielectric constant of a material made up of a layer of insulating material covered with a layer of conducting material can be extremely large when the concentration of the insulating region becomes small. This is known as the Maxwell-Wagner effect.[16,21] In this case, the material as a whole has a zero d.c. conductivity - because the layer of insulating material blocks the current path. We show that under certain circumstances $\mathrm{Re}\epsilon^*$ can be enormously large even when the sample remains conducting. This model is more appropriate to sedimentary rocks which remain conducting to very low values of porosity. We show that the controlling factor in determining the magnitude of $\mathrm{Re}\epsilon^*$ is the relative magnitude of the particle "aspect ratio" of the inclusion compared to the concentration of the inclusions. In order to compare our theoretical calculations with experimental values for $\mathrm{Re}\epsilon^*$, more reliable experimental values are required together with information on particle aspect ratios. The latter information may be obtained from electron microscopy and other techniques, or through use of carefully controlled techniques of artificial sample preparation.

Consider a rock with a few platey grains distributed in it, all oriented in a given direction as shown in Fig. 2.

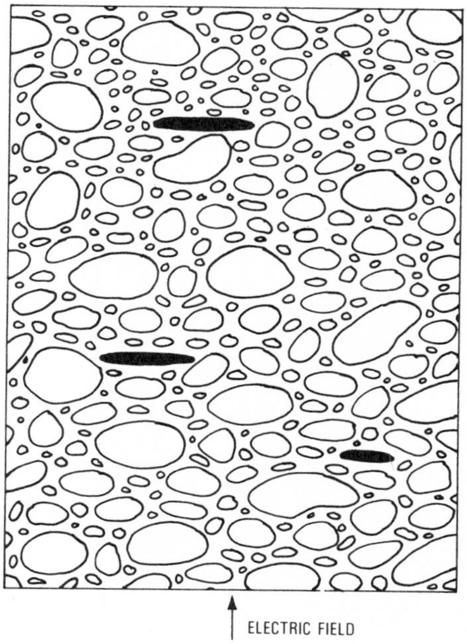

Fig. 2 - A Schematic of Rock With Platey Grains

The dielectric constant ϵ^* of a rock that contains concentration η of these plate-like particles, each of depolarization ratio L_S is given, via the self-consistent model (generalization of (11) as

$$(1-\eta)\frac{\epsilon_R^* - \epsilon^*}{L_s\epsilon_R^* + (1-L_s)\epsilon^*} + \eta\frac{\epsilon_m^* - \epsilon^*}{L_s\epsilon_m^* + (1-L_s)\epsilon^*} = 0 \tag{21}$$

Here ϵ_R^* denotes the dielectric constant of the rock if there were no plate-like objects present ($\eta = 0$). Solving (21) for ϵ^* in the low frequency limit, we find for $L_S = 1 - \delta$ ($\delta = \pi$ a/2b, a being the minor and b the major axes),

$$\sigma(0) = \lim_{\omega \to 0} \omega\epsilon_o \mathrm{Im}\epsilon^* = \frac{\delta - \eta}{\delta}\sigma_R(0) \tag{22}$$

$$\epsilon_s = \lim_{\omega \to 0} \mathrm{Re}\,\epsilon^* = \frac{\epsilon_m}{\delta - \eta} \tag{23}$$

Thus, if $\delta - \eta$ small and positive, we find $\sigma(0)$ is not zero and ϵ_s diverges. For example, if $\delta \sim 10^{-3}$, $\eta \sim 10^{-4}$, $\sigma(0)$ and $\epsilon_S \sim 10^4$, for $\epsilon_m = 10$.

In Fig. 3 we show ϵ [obtained from equation (21)] versus frequency for two values of water salinity. The relaxation frequency increases with salinity, which explains the salinity dependence of the dielectric constant observed in the MHz range. For example, if the measurements were made at 10 MHz, the dielectric would go up from 51 to 780 as the conductivity was increased from 1 mho/m to 10 mho/m.

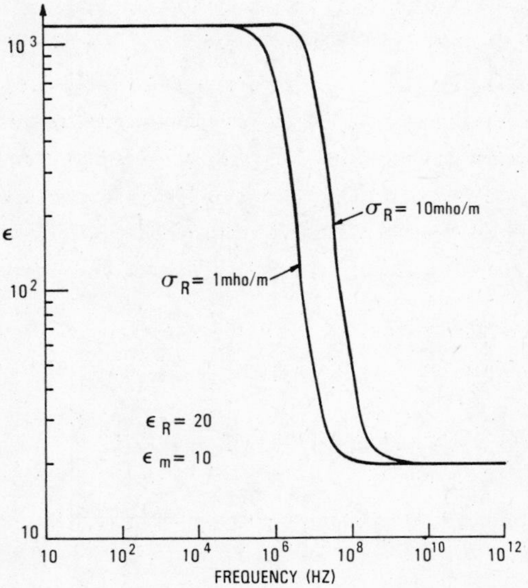

Fig. 3 - The Dielectric Constant of a Rock with Platey Grains

In actual rocks there will be a distribution of aspect ratios leading to a distribution of relaxation frequencies and a much smoother ϵ vs frequency response. An example is shown in reference 3.

The above calculation can be further improved by building up the concentration in infinitesimal steps as in the self-similar model. We incorporate, in this manner, a concentration η of plate-like grains distributed *isotropically* in a conducting rock. It can be shown that when the probability distribution is sharply peaked around $L_s = 1 - \delta$, for $\eta << 1$, $\delta << 1$,

$$\sigma(0) = \sigma_R e^{-\eta/3\delta} \tag{24}$$

$$\epsilon_S = \frac{\epsilon_m}{\delta}\,(1 - e^{-\eta/3\delta}) \tag{25}$$

There are two cases that are of particular interest:

Case (a) $\delta < \eta$, and, Case (b) $\delta > \eta$.

For case (a) when $\delta \to 0$, ϵ_s diverges, but since $\delta < \eta$, $\sigma(0) \to 0$. This is the well-known Maxwell-Wagner effect described above. However, for case (b) $\delta > \eta$, $\mathrm{Re}\,\epsilon^*$ can be very large while the system remains conducting. When $\delta \to 0$ and $\eta \to 0$, ϵ_s diverges as long as η/δ^s remains finite and $0 < s < 2$. For $0 < s < 1$, the divergence in ϵ_s is concomittant with a conductivity threshold (Maxwell-Wagner), but for $1 < s < 2$, we obtain the new result, completely overlooked before, i.e. a divergent ϵ_s with a non-zero $\sigma(0)$.

There are geometrical shapes, other than thin plates, that can give rise to extremely high values of ϵ_s. These are discussed in reference 3. For example, for a periodic structure of conducting spheres embedded in an insulating host, ϵ_s diverges as the spheres begin to touch each other. Here, again, thin insulating regions are trapped between conducting regions. The above calculation can be given a simple physical explanation: The thin plates, at low enough frequency, can be considered to be capacitors hung parallel to the rest of the rock, and hence give a large over all capacitance, i.e. a large dielectric constant. The R-C time constant depends on the salinity, which explains Fig. 3.

We can give a simple thought experiment to show how these capacitors are charged up. When external charges are brought into the system, Maxwell's equation gives the charge density as

$$\rho_{ex} = \nabla\cdot(\epsilon \mathbf{E}) = \frac{\epsilon}{\sigma}\nabla\cdot(\sigma \mathbf{E}) + \sigma \mathbf{E}\cdot\nabla\left\{\frac{\epsilon}{\sigma}\right\}. \tag{26}$$

The first term in the right side of (26) can be replaced by $-\partial\rho_{ex}/\partial t$, using the continuity equation,

$$\nabla\cdot J + \partial\rho_{ex}/\partial t = 0 \;. \tag{27}$$

Thus, the first term in (26) vanishes in the steady state. This shows a charging up effect when there is a discontinuity of ϵ/σ at a direction parallel to **E**. This is maximized for thin plates perpendicular to an external field.

Since one cannot unambiguously distinguish between conduction current and displacement current (see Purcell[22]) it is preferable to have another point of view. Another way of looking at this problem is as follows: At low frequencies, the in-phase part of the total current $\mathbf{J}^* = (\sigma + i\omega\epsilon / 4\pi)\mathbf{E}$ domonate. Note that Maxwell's equation for $\nabla \times B$ gives $\nabla \cdot (\nabla \times B) = 0$, i.e. the continuity equation for the total current $\nabla \cdot J^* = 0$. The scattering by an obstacle reduces the in-phase current (reducing the conductivity of the system), but gives a small out of phase component that enormously enhances $\mathrm{Re}\,\epsilon^*$ of the mixture.

It is easy to show, following Hori,[14] that the t-matrix for a thin plate-like object is given by

$$t = \epsilon_R^* \frac{\epsilon_m^* - \epsilon_R^*}{L\epsilon_m^* + (1-L)\epsilon_R^*} = \epsilon_R^*(4\pi\alpha^*) \tag{28}$$

where α^* is the polarizability of an ellipsoid subjected to an uniform field. In ATA, for example, we have,[14] when $\eta|\alpha^*| << 1$

$$\epsilon^* = \epsilon_R^* + \eta t = \epsilon_R^*[1 + 4\pi\eta\alpha^*] \tag{29}$$

At low frequency, $\omega \to 0$, Eq. (28)-(29) give

$$\epsilon_S = \epsilon_R(1 - \eta/\delta) + \eta\epsilon_R/\delta^2 \tag{30}$$

The in-phase term of α^* gives the blocking effect which reduces ϵ_R to $\epsilon_R(1-\eta/\delta)$, but the out of phase term gives $(\eta/\delta^2)\epsilon_m$ which increases enormously as $\delta \to 0$. The in-phase term of α^* describes the effect that a highly polarizable, i.e. conducting material has been replaced by an insulating material (plates). But the imaginary part of α^* when multiplied by the large $[4\pi\sigma_R/i\omega]$ term gives a large positive contribution.

Far Infra-red (FIR) Absorption by Metal Particles[23]

The FIR absorption by small metal particles ($\sim 1\ \mu m$) distributed in an insulating host is found to be much greater than that predicted for spheres. For larger particles, the eddy-current loss terms dominate the absorption. However, the eddy term is proportional to the particle size. Experimentally, it is found that the absorption is particle size independent.[24]

A picture of these ultrafine particles shows that they stick together to form needle-like structures. The electric dipole absorption by needle- like structures can explain the large absorption observed experimentally. The ω^2-frequency dependence of absorption for finite length needles is in agreement with the observed data.

The metal excludes the electric field, but the continuity of the tangential component of the electric field makes it possible for E-field to penetrate a needle parallel to it. In other words, the depolarization effects are minimum for fields parallel to the axis of needles.

Fourth Sound in ^{4}He and Biot Slow Wave [4]

The equivalence of the conduction problem and the sound propagation was known to Rayleigh[25] and Maxwell[16,25] in the last century. In a paper[25] on the conductivity of a periodic assembly of conducting spheres of a given conductivity packed in a conducting host of a different conductivity, Rayleigh mentioned this equivalence. (This semenal paper of Rayleigh[25] remains, to date, the basis of going beyond the dilute limit by taking the proximity effects in account.)

A class of the problems which are equivalent to each other have been listed byBatchelor.[26] All these problems entail solving the Laplace's equation. These problems are scale invariant, i.e. if the entire system is dialated or shrunk uniformly, the effective dielectric constant, etc. do not change. The permeability of a visicid fluid, on the other hand, depends on the actual size of the pores, and hence do not belong to the same class. However, when a sound wave is launched at a sufficiently high frequency such that the viscous skin depth is extremely small, and almost all of the fluid may be treated as an ideal fluid, the equivalence between dielectric properties and the sound velocity can be exploited. The so-called Biot slow wave,[7] at a sufficiently high frequency, can be treated in this manner. In this mode the solid frame and the fluid move out of phase with respect to each other. The problem simplifies enormously when the coupling between the solid and fluid components are minimal. This happens when there is a large mismatch of the acoustic impedance of solid and fluid parts. When the matrix is completely rigid in one extreme, or completely unconsolidated in the other, there is little coupling between solid and liquid motions. In the weak coupling case, the motion of fluid has one to one correspondence to the flow of electric current past insulating grains and the velocity of the Biot slow wave is given by

$$v = v_F/n \tag{31}$$

Here v_F is the velocity of sound in the free fluid and n is a refractive index which is related to the ratio conductivity of the brine saturated rock $\sigma(0)$ and that of the brine $\sigma_w(0)$(exact),

$$n^2 = \phi\sigma_w(0)/\sigma(0) \tag{32}$$

Similarly, the velocity of the fourth sound in ^{4}He saturated porous superleak below T_λ is given by (31). In this case the superfluid component has no viscosity, and the normal component is locked on to the substrate. In the fourth sound mode, the superfluid motion around the solid obstacle is exactly similar to the current flow skirting around the obstacle. Combining (1) and (30) gives

$$n^2 = \phi^{1-m} \tag{33}$$

Eq. (31) with m given by (20) agrees well with the experiments on fourth sound. More experiments, where conductivity, Biot slow wave/fourth sound are simultaneously measured, are needed.

It is a pleasure to thank my colleagues at Schlumberger-Doll Research, with whose collaboration some of the above works were done.

References

1. P. N. Sen, C. Scala and M. H. Cohen, Geophysics 46, 781 (1981).
2. P. N. Sen, Soc. Petrol. Eng., Talk No. 9379 (Fall Meeting, 1980).
3. P. N. Sen, Geophysics (to appear in Dec. 1981 issue).
4. D. L. Johnson and P. N. Sen (to appear in Phys. Rev. B).
5. J. H. Scott, R. D. Carroll and D. R. Cunningham, J. Geo. Res. 72, 5101, (1967).
6. J. Ph. Poley., J. J. Nooteboom and P. J. DeWaal, Log Analyst, 8 (May, 1978).
7. D. L. Johnson, Appl. Phys. Lett., 37, 1065 (1980).
8. G. Schwarz, J. Phys. Chem., 66, 2636 (1962).
9. P. N. Sen, unpublished.
10. H. P. Schwan, N.Y. Acad. of Sci., 303 (1977).
11. J. Ziman, *Models of Disorder*, Cambridge U. Press (N.Y.), 1979.
12. J. A. Stratton, *Electromagnetic Theory*: McGraw Hill (N.Y.) 1941.
13. E. Schlomann, J. Appl. Phys. 33, 2825 (1962).
14. M. Hori, J. Math. Phys. 18, 487 (1977) and references therein.
15. P. P. Ewald, Ann. der Physik, 49, 1 (1916); C. W. Oseen, Ibid 48, 1 (1915).
16. J. C. Maxwell, A Treatise on Electricity and Magnetism (1873), Dover Edition (1954) N.Y.
17. H. C. Van de Hulst, Light Scattering by Small Particles, Wiley, N.Y. (1957).
18. A. K. Veinberg, Soviet Phys. Doklady 11, 593 (1967).
19. K. Mendelson and M. H. Cohen (unpublished).
20. D. A. G. Bruggeman, Ann. Phy. Lpz. 24, 636 (1935).
21. C. Kittel, Introduction to Solid State Physics, 5th Ed., John Wiley, N.Y. (1976), p. 430.
22. E. M. Purcell, Electricity and Magnetism, McGraw Hill, N.Y. (1963).
23. P. N. Sen and D. B. Tanner: Unpublished.
24. D. B. Tanner: Private communication; N. E. Russel, J. C. Garland and D. B. Tanner, Phys. Rev. B 23, 632 (1981).
25. Lord J. W. S. Rayleigh, Phil. Mag. 34, 481 (1892).
26. G. K. Batchelor, Ann. Rev. Fluid Mech. 6, 227 (1974).

EFFECTIVE DIELECTRIC FUNCTION OF COMPOSITE MEDIA

Ping Sheng

Theoretical Sciences Group
Corporate Research Science Laboratories
Exxon Research and Engineering Co.
P. O. Box 45
Linden, N.J. 07036

Abstract

This article reviews the relationship between the microstructure of a composite, i.e. geometric shapes as well as topological arrangements of the different components, and two unique characteristics of metal-insulation composites -- the percolation threshold and the optical dielectric anomaly. It is demonstrated that the effective medium approach to the calculation of effective dielectric constant $\bar{\varepsilon}$ can yield realistic results provided that the microstructural information is properly taken into account.

It is well known that the electromagnetic response of a single-component, homogeneous system can be completely characterized by a complex dielectric function[1] $\varepsilon(\omega) = \varepsilon_R(\omega) + i(4\pi\,\sigma(\omega)/\omega)$, where ε_R is the dielectric constant, σ is the conductivity, and ω is the angular frequency of the electromagnetic wave. In the case of a random inhomogeneous composite, however, the task of characterizing the electromagnetic respons is generally much more involved due to the random scatterings of the probing wave by the inhomogeneities. Yet in the limit of $\lambda \gg \xi$, where λ is the wavelength and ξ the typical scale of inhomogeneities, a great conceptual simplification occurs because the waves cannot resolve the individual scattering centers. Therefore, the medium would appear uniform, characterized by an effective dielectric function $\bar{\varepsilon}(\omega)$.

In the literature there are two prevalent approaches for the calculation of the effective dielectric constant $\bar{\varepsilon}$. One is Bruggeman's effective medium theory[2]. The main idea of this theory can be described as follows. Consider a random composite consisti of two components, with dielectric constants ε_1 and ε_2, as schematically illustrated in Fig. 1(a).

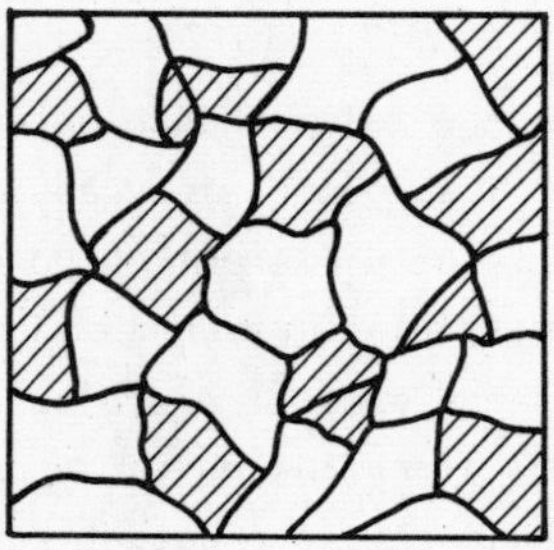

(a)

Fig. 1 (a) Schematic picture of a random composite.

Since an exact calculation of the electric field distribution for the random, infinite system is impossible, the Bruggeman approach is to focus attention on one of the grain (say a grain of component 1) and regard the rest of the composite as a homogeneous medium characterized by a yet undetermined effective dielectric constant $\bar{\varepsilon}$. This is shown in Fig. 1(b).

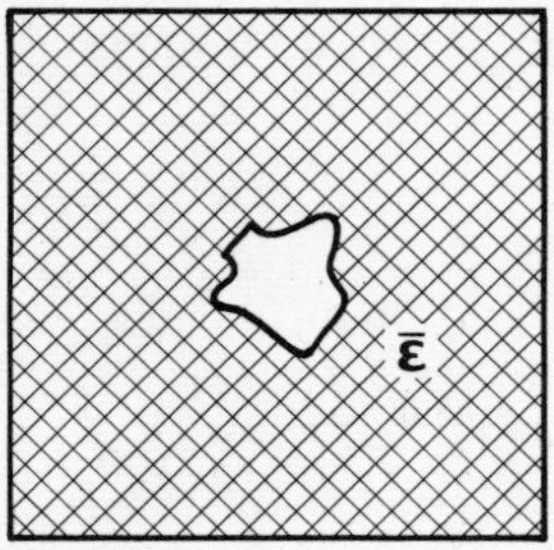

(b)

Fig. 1 (b) The rest of the medium is treated as homogeneous for the calculation of depolarization field of a single grain.

In the presence of an applied electric field, the single inclusion in the uniform medium $\bar{\varepsilon}$ will give rise to a dipole depolarization field which, in the spherical grain approximation, is proportional to $(\bar{\varepsilon} - \varepsilon_1)/(\bar{\varepsilon} + 2\varepsilon_1)$. Repeating the same problems with a grain of component 2 results in another dipole moment proportional to $(\bar{\varepsilon} - \varepsilon_2)/(\bar{\varepsilon} + 2\varepsilon_2)$. In order to be consistent with the initial assumption that the medium should appear homogeneous to the probing wave, the average depolarization field must vanish. That is,

$$p \frac{\bar{\varepsilon} - \varepsilon_1}{\varepsilon_1 + 2\bar{\varepsilon}} + (1-p) \frac{\bar{\varepsilon} - \varepsilon_2}{\varepsilon_2 + 2\bar{\varepsilon}} = 0, \tag{1}$$

where p is the volume fraction of component 1. It should be mentioned that Eq. (1) has been shown to be equivalent to the coherent potential approximation in the multiple-scattering formulism[3].

If the two components of the composite are metal and insulator, that is, $\varepsilon_1 = 1$ and $\varepsilon_2 = 0$, then it is easy to verify that the Bruggeman theory predicts an effective conductivity that vanishes at $p_c = 1/3$. This behavior, known as the percolation threshold, is physically related to the fact that below the threshold, there is no possibility for metal grains to form a connected network of infinite extent. In granular films[4-6], the percolation threshold behavior is reflected in the transmission electronmicroscope pictures of the microstructure as shown in Fig. 2.

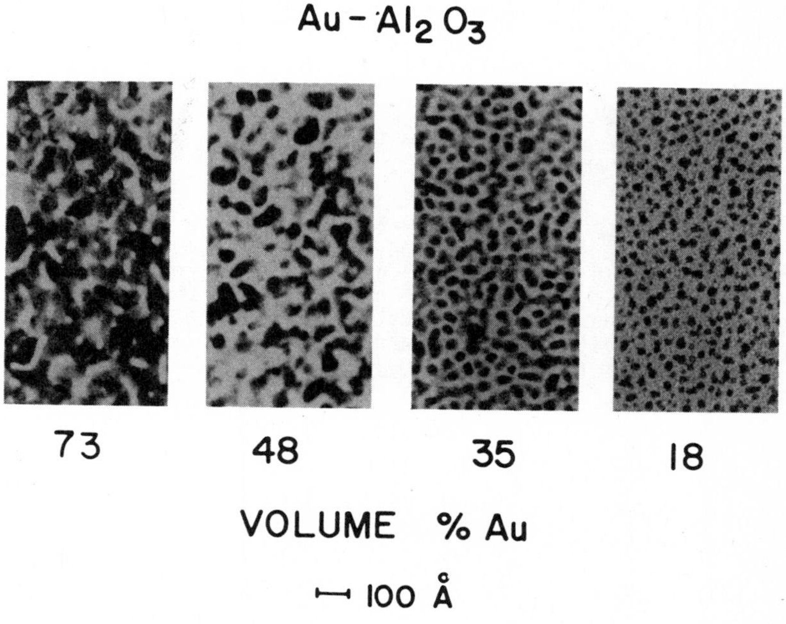

Fig. 2 Transmission electron micrographs of the granular cermets $Au\text{-}Al_2O_3$ for four different compositions.

It is seen that at p = 0.73, the Al_2O_3 (white) are disconnected inclusions in the metallic Au matrix (black). As p is decreased, a matrix inversion occurs until at p = 0.35 the metal particles are the inclusions in the insulator matrix. Between p = 0.48 and p = 0.35 we have a labyrinth structure, and somewhere between these two composition values it is clear that the last infinite metallic network must be broken and the dc conductivity vanishes. Therefore, Bruggeman's theory is qualitatively correct in predicting a percolation threshold. However, quantitatively the agreement is poor as shown in Fig. 3.

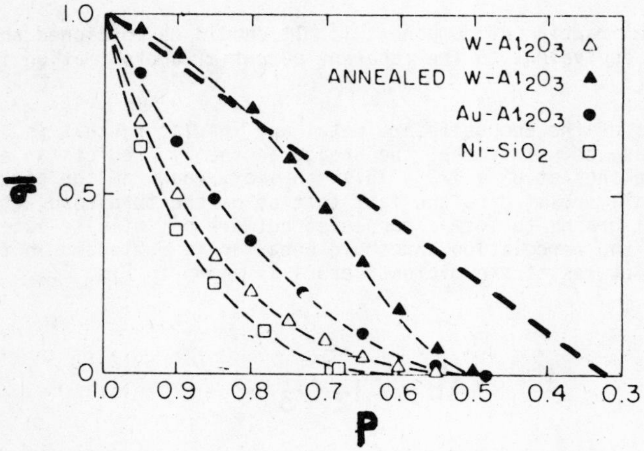

Fig. 3 Normalized dc conductivity of granular metals. The straight dashed line is the Bruggeman theory. After Abeles [1].

Another widely-used approach to the calculation of $\bar{\varepsilon}$ was the Maxwell-Garnett theory [7]. Its tranditional derivation relies on the analogy of metal-insulator composite as a polarizable medium in which the metal grains play the role of "atoms".

By using the familiar Clausius-Mosotti equation[8], one obtains a relationship between the dielectric constant of the composite and the polarizability of the metal particles:

$$\frac{\bar{\varepsilon} - \bar{\varepsilon}_2}{\varepsilon + 2\varepsilon_2} \quad \frac{4\pi n\alpha_1}{3\varepsilon_2} \quad . \tag{2}$$

Here n is the volume density of the metal particles, and α_1 is the polarizability of a metal grain. By substituting for α_1 the expression for polarizability of an isolated metal sphere immersed in a uniform medium of dielectric constant ε_2, we get the Maxwell-Garnett equation

$$\frac{\bar{\varepsilon} - \varepsilon_2}{\varepsilon_2 + 2\bar{\varepsilon}} = p\,\frac{\varepsilon_1 - \varepsilon_2}{\varepsilon_1 + 2\varepsilon_2} \quad . \tag{3}$$

Equation (3) is equivalent to the averaged T-matrix approximation in the multiple scattering formalism[3].

The Maxwell-Garnett theory can be easily shown not to yield a percolation threshold at any finite values of p. However, it does predict another peculiar property of the composite, the dielectric anomaly[9], which is absent in Bruggeman's theory.

To describe this effect, let us consider the frequency dependence for the real and imaginary parts of the metal dielectric function shown schematically in Fig. 4(a), where ω_p denotes the plasma frequency. If now we have a metal-insulator composite with insulator being the matrix component, the real and imaginary parts of the effective dielectric constant would look like Fig. 4(b) according to the Maxwell-Garnett theory.

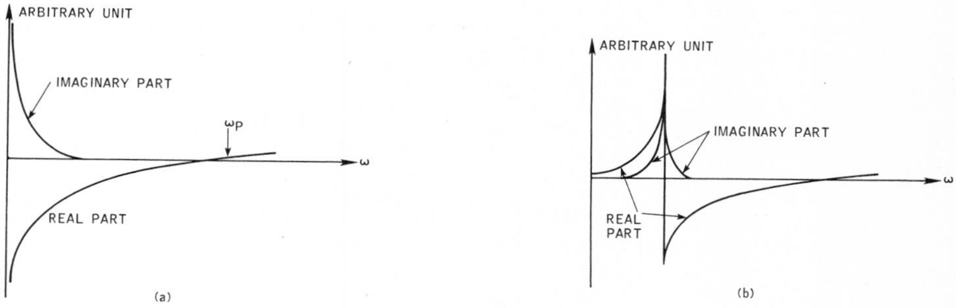

Fig. 4 Schematic illustration of the frequency dependence of the dielectric constant for (a) an ideal metal and (b) a metal-insulator composite according to the Maxwell-Garnett theory.

It is to be noted that there is a frequency ω_R above which the effective dielectric function behaves exactly like a metal. However, at $\omega \ll \omega_R$ the composite is more like an insulator. Therefore, ω_R is essentially a frequency threshold for the metal-insulator transition. At ω_R, there is an absorption peak arising from increased penetration of electromagnetic field into the metal grains.

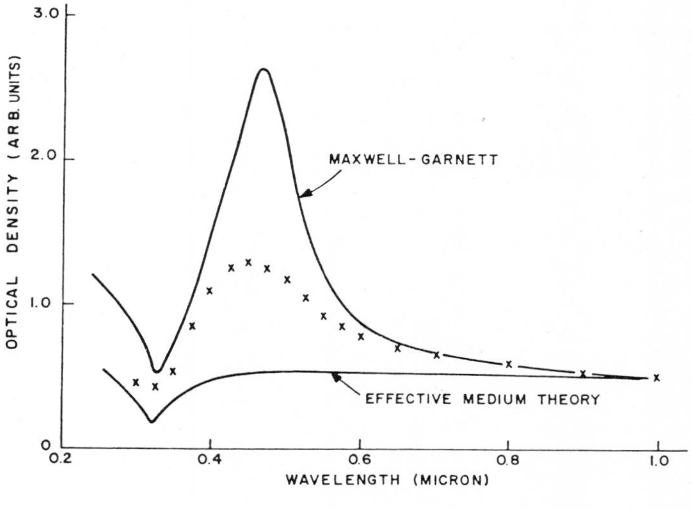

Fig. 5 Comparison of experimental results on the dielectric anomaly with both the Bruggeman and the Maxwell-Garnett theories. The sample is Ag-SiO_2 film containing 0.39 volume fraction Ag. After Ref. [10].

Evidence for the dielectric anomaly has been observed in granular Ag - SiO_2 and Au - SiO_2 films [9]. Figure 5 [10] compares the experimental result with both the Maxwell Garnett and the Bruggeman theories. It is seen that whereas the Maxwell-Garnett theory does produce the absorption peak at about the right frequency, the Bruggeman theory exhibits no peak at all.

Such comparisons raise the obvious question: what is the underlying physics responsi for this difference in behavior? The answer to this question is facilitated by the recognition in the last few years that the Maxwell-Garnett theory can be alternatively derived by using the effective - medium approach[11]. That is, if we consider the embedding of a coated sphere in an effective medium, then the condition of vanishing depolarization field would yield Eq. (3) provided that the sphere is of component 1 and the coating is of component 2 with a thickness determined directly by the relative volume fraction of the two components. This derivation of the Maxwell-Garnett formul immediately tells us that the difference between the two theories lies in the consider ation of two types of microstructures. Whereas the Bruggeman theory treats the two components in an equivalent manner, the basic structural unit of a coated sphere in the Maxwell-Garnett case implies an asymmetrical consideration of the two components. It is clear that if we make up a composite by the random placement of coated spheres, then the coating component would always remain the matrix constituent regardless of the composition. Therefore, it is not surprising that there is no percolation thresh in the Maxwell-Garnett theory. A schematic illustration of the microstructures implic treated by the Bruggeman and the Maxwell-Garnett theories are shown in Fig. 6.

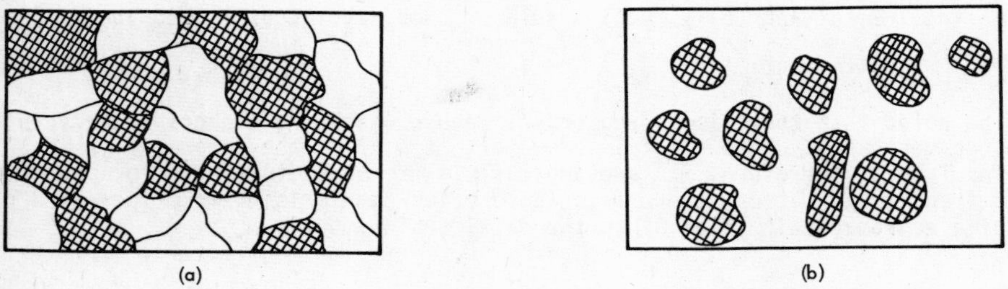

Fig. 6 Schematic illustration of the two types of microstructure treated by the Bruggeman theory (a) and the Maxwell-Garnett theory (b).

The elucidation of the difference between the two theories shows that there is only one underlying approach--the effective medium theory--to the approximate calculation of the effective dielectric constant. However, it also confronts the users of the effective medium theory with the following unresolved problems:

(1) Is the absence of the dielectric anomaly in the Bruggeman theory the physical consequence of its implicit microstructure or the result of the approximation used in its derivation?

(2) Does the fact that both theories cannot even qualitatively describe all the experimental results of granular metals signify (a) the need for carrying the effective medium approach to higher orders of approximation, or (b) the need for better modelling of the granular metal microstructure?

Consideration of the two problems shows that in both cases the question revolves arour the role of microstructure in the calculation of effective dielectric constant. In this article, we wish to use the concept of "structural units" as a means for incorporating the microstructural information, i.e. geometric shapes as well as topological arrangement of the grains, in a statistical manner. For example, the structural unit in the Maxwell-Garnett case is a coated sphere, but the basic unit in the Bruggeman theory is a grain of either component. An alternative choice of structural unit for (the microstructure of) the Bruggeman case is a two-grain combination[12] in which each grain can be either component 1 or component 2, denoted here as a pair-cluster. The possible advantages of using a two-grain pair-cluster unit rather than the usual one-grain unit lies in the explicit presence of the two-component interface in the basic structural units. As we may recall, in the derivation of the Bruggeman theory the only interfaces explicitly considered are those between the individual grain and the effective medium. Therefore, any phenomenon intrinsically associated with the interfaces between the two components, such as the dielectric anomaly, are not expected to be adequately accounted for. The use of the pair-cluster units thus holds promise to answer the first problem posed above.

To calculate the effective dielectric constant in the pair-cluster theory, we will approximate the geometry of the two-grain pair by a sphere in which each half can be either one of the two components. By embedding this sphere in an effective medium and calculating the resulting depolarization field, we obtain an equation for the effective dielectric constant $\bar{\varepsilon}$ after averaging over all two-hemisphere combinations and all orientations of the applied field relative to the plane separating the two hemispheres:

$$p^2 \frac{\bar{\varepsilon} - \varepsilon_1}{2\bar{\varepsilon} + \varepsilon_1} + (1-p)^2 \frac{\bar{\varepsilon} - \varepsilon_1}{2\bar{\varepsilon} + \varepsilon_2} \quad \frac{4}{9} p(1-p) \frac{2\bar{\varepsilon} - \varepsilon_1 - \varepsilon_2}{K\bar{\varepsilon}(\varepsilon_1 - \varepsilon_2)^2 + (\varepsilon_1 + \varepsilon_2 + 4\bar{\varepsilon})/3}$$
$$+ \frac{2}{9} p(1-p) \frac{\bar{\varepsilon}(\varepsilon_1 + \varepsilon_2) - 2\varepsilon_1\varepsilon_2}{H\bar{\varepsilon}(\varepsilon_1 - \varepsilon_2)^2 + 2(\varepsilon_1\varepsilon_2 + \varepsilon_1\bar{\varepsilon} + \varepsilon_2\bar{\varepsilon})/3} = 0, \qquad (4)$$

where

$$H = \frac{1}{4} \sum_{m=1}^{\infty} \frac{I_m}{[\, m(\varepsilon_1 + \varepsilon_2) + (2m + 1)\, \bar{\varepsilon}]}, \qquad (5)$$

$$K = \frac{1}{4} \sum_{m=1}^{\infty} \frac{I_m}{[\, 2m\, \varepsilon_1\varepsilon_2 + (n + 1/2)\,(\varepsilon_1 + \varepsilon_2)\bar{\varepsilon}]}, \qquad (6)$$

$$I_m = \frac{m(4m + 1)\,[(2m + 1)!]^2}{4^{2m}\,(m!)^4 (2m - 1)^2 (m + 1)^2\,(2m + 1)} \qquad (7)$$

Details of the derivation can be found in Ref. (12). To compare the results of the pair-cluster theory with the Bruggeman theory, we show in Fig. 7 the calculated optical transmission spectrum for a series of 500-A-thick Ag-SiO_2 films using realistic Ag and SiO_2 dielectric constant values. The optical dielectric constant of the composite is then evaluated by using both the pair-cluster theory (solid line) and the Bruggeman theory (dashed line). It is easily seen that the pair-cluster theory displays an extra absorption peak (or transmission dip) near $\lambda \simeq 0.37$ μm, the dielectric anomaly, which the Bruggeman theory does not have. The peak disappears, however, in the composition regime of $0.4 < p < 0.7$ where a matrix inversion occurs. Since in this particular composition range the two components is expected to exhibit a labyrinth structure, the disappearance of the absorption peak indicates that the dielectric anomaly may be associated with the particular microstructure of isolated inclusions embedded in a continuous matrix.

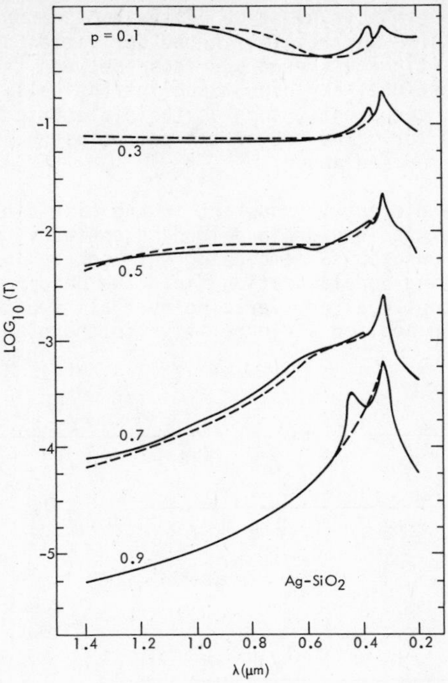

Fig. 7 Calculated optical transmission as a function of wavelength for a series of 500-A $Ag\text{-}SiO_2$ films. (-) pair-cluster theory, (---) Bruggeman theory. For clarity, the curves are displaced vertically with respect to each other.

To further accentuate the effect of microstructure, we show in Fig. 8 the transmission spectrum for the same $Ag\text{-}SiO_2$ films calculated in the Maxwell-Garnett theory. By comparing Figs. 7 and 8, it becomes clear that the position and the magnitude of the dielectric anomaly are drastically different. Therefore, although the complete absence of the dielectric anomaly in the Bruggeman theory can be ascribed to the neglect of two-component interfaces in its derivation, the microstructure, nevertheless, does govern the manifestation of the phenonemon.

Let us now consider the problem of constructing a realistic theory for the effective dielectric function of granular metals[6]. We will hypothesize at this stage that the discrepancies between the experimental results and the predictions of both the Bruggeman and the Maxwell-Garnett theories stem from the inadequacy of their structural units in modelling the cermet microstructure. To search for new structural units appropriate for these granular composites, we observe that the cermet is formed by the two-step process of surface diffusion and coalescence. That is, the molecules that land on the substrate (of the film) usually have excess energy and, therefore, tend to move about before they stick with other molecules of the same component and form the grain. The average distance of this surface motion is usually denoted as the surface diffusion length, which is the basic scale of inhomogeneity. Suppose now let us consider a spherical region with the dimension of a diffusion length. Inside such a region there can be a large number of molecules of either component. Therefore, statistically the relative volume fraction occupied by the two components should be close to the macroscopic average value. If a grain is formed inside this region through surface diffusion and coalescence, there are two possible outcomes as shown in Fig. 9. That is, component 1 may form the grain and component 2 the coating which we denote as a type 1 structural unit, or component 2 may form the grain and component 1 the coating, which we denote as a type 2 structural unit. At a given composition p, the relative probability of occurrence for the two types of structural unit can be estimated by counting the number of equally probable final configurations corresponding to different positions of the grain inside the region.

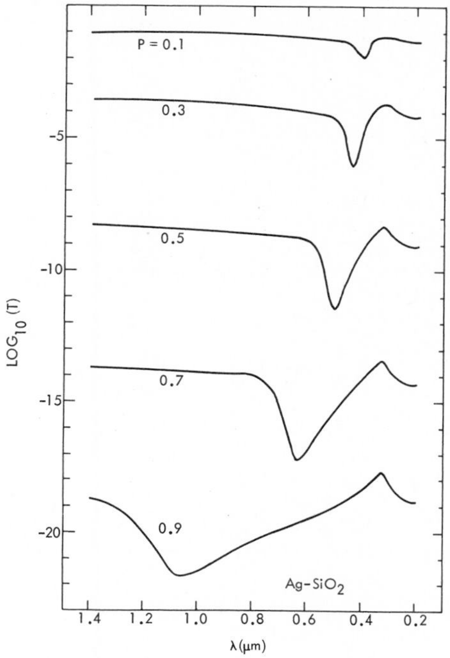

Fig. 8 Optical transmission as a function of wavelength for a series of 500-Å $Ag-SiO_2$ films, calculated by the Maxwell-Garnett theory. For clarity, the curves are vertically displaced with respect to each other.

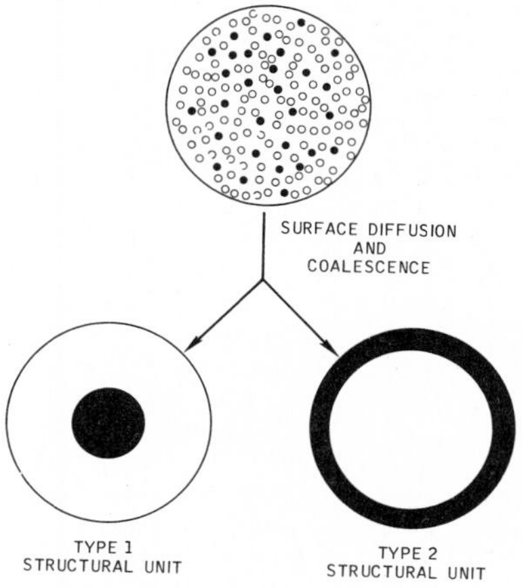

Fig. 9 Schematic illustration of the grain formation process in granular metal films.

By assuming the grain to be spherical, it is clear, that, in case of type 1, the number of configurations is proportional to $u_1 = (1-p^{1/3})^3$. By the same reasoning, the number of configurations for the type 2 unit is proportioned to $u_2 = [1 - (1-p)^{1/3}]^3$. It follows that the relative probability of occurrence for the type 1 unit is $f = u_1/(u_1 + u_2)$, and that for the type 2 unit is 1-f.

If we now build up a random composite from these two structural units according to their assigned statistical weights, the system will be dominated by type 1 structural units in the range $0.35 > p > 0$, since f only varies between 1 and 0.92. That is, the structural would essentially look like isolated grains of component 1 embedded in the continuous matrix of component 2. On the other hand, in the range of $1 > p > 0.65$ we expect the reverse structure in which component 2 becomes isolated and component 1 constitutes the continuous phase. Therefore, there is a matrix inversion occuring in the range $0.65 > p > 0.35$, where we can expect a labyrinth structure. Comparing the above structural description with Fig. 2, we see that the model composite is in reasonable accord with the reality. Now the construction of a theory for the effective dielectric function proceeds along the following three steps: (1) embedding the structural units in a uniform effective medium, (2) calculating the dipole moments of the structural units when they are polarized by a uniform applied field, and (3) requiring the average dipole moments to vanish. By approximating the dipole moment of a structural unit (in arbitrary configuration) by the dipole moment of the concentric configuration, $D_{1,2}$, we get the effective medium condition as

$$fD_1 + (1 - f)\, D_2 = 0 \qquad (8)$$

Since the arguments leading to the value of f remains unchanged if one relaxes the condition of spherical geometry and considers a speroidal particle enclosed in a similar-shaped region, Eq. (8) remains valid for spheroidal structural units. In that case $D_{1,2}$ stands for the orientationally averaged dipole moment of confocal spheroidal particles embedded in an effective medium $\bar{\varepsilon}$:

$$D_1 = \frac{2}{3} D[\bar{\varepsilon}, \varepsilon_1, \varepsilon_2, p, A(\alpha, u), B(\alpha)] + \frac{1}{3} D[\bar{\varepsilon}, \varepsilon_1, \varepsilon_2, p, 3\text{-}2A(\alpha,u), 3\text{-}2B(\alpha)] \qquad 9(a)$$

$$D_2 = \frac{2}{3} D[\bar{\varepsilon}, \varepsilon_2, \varepsilon_1, 1\text{-}p, A(\beta, v), B(\beta)] + \frac{1}{3} D[\bar{\varepsilon}, \varepsilon_2, \varepsilon_1, 1\text{-}p, B\text{-}2A(\beta,v), 3\text{-}2B(\beta)] \qquad 9(b)$$

Where α is the ratio between the minor (major) and major (minor) axes of the elliptic cross section for the type - 1 oblated (prolate) spheroidal unit, β is the similar quantity for the type - 2 unit, $u = (p/\alpha)^{1/2}$ and $v = ((1-p)/\beta)^{1/2}$. The functions D, A, and B have the following forms:

$$D(\bar{\varepsilon},x,y,\mu,A,B) = \frac{[A\bar{\varepsilon} + (3-A)y]\,[y-x]\mu + [Bx+(3-B)y]\,[\bar{\varepsilon}-y]}{A\,(3-A)\,(\bar{\varepsilon}-y)\,(y-x)\mu + [Bx + (3-B)y]\,[Ay + (3-A)\,\bar{\varepsilon}]} \qquad (10)$$

$$A(\gamma, w) = \frac{3}{2} \frac{1}{w^3(1-\gamma^2)} \left[\frac{1}{\sqrt{1-\gamma^2}} \tan^{-1} \frac{w\sqrt{1-\gamma^2}}{\sqrt{s^2+\gamma^2 w^2}} - w(s^2+\gamma^2 w^2)\right], \qquad (11)$$

$B(\gamma) = A(\gamma, 1/\gamma^{1/3})$ evaluated at s=0, where s is the solution of the equation $(s^2 + w^2)^2\,(s^2 + \gamma^2 w^2) = 1$. A and B assumes the special value of 1 for spherical geometry. It is noted that if f is set equal to 1, Eq. (8) becomes the generalized Maxwell-Garnett equation which specializes to Eq. (3) in the spherical case.

A comparison of the calculated optical transmissions spectrum for a series of thin Au-SiO_2 films with the experimental results are shown in Fig. 10. The theoretical inputs are the realistic dielectric constants of Au and SiO_2, $\alpha = 1$, $\beta = 2$, the p values marked to the right of each curve, and the film thickness marked above each cur It is seen that the new theory reproduces all the characteristics features of the data In particular, the position and magnitude of the dielectric anomaly, and its eventual disappearance for $p \geq 0.8$ are all in good agreement. For effective dc conductivity

$\bar{\sigma}$ and its variation with p, we show in Fig. 11 two sets of experimental data corresponding to the same sample before and after the annealing treatment and their best theoretical fits. The theory curves are obtained by using $\varepsilon_1 = 1$, $\varepsilon_2 = 0$, $\alpha = 1$, and the β values marked in the figure. The strikingly good agreement, especially the reproduction of the opposite curvatures in the two sets of experimental data, shows that the insulator inclusions before annealing are mostly in the form of oblate platelets, which becomes rounded upon annealing. The increase in the effective conductivity can be intuitively understood by recognizing that the platelets are more effective than spheres in impeding the current flow.

The success of the new theory in explaining both the optical and dc response of granular films brings us to the essential point of this article. Namely, the effective medium approach to the calculation of effective material parameters can indeed yield realistic results provided that the microstructure of the random composite is properly taken into account. This conclusion, which may seem natural in hindsight, nevertheless does emphasize the need for future research in quantifying the different types of microstructure in random inhomogeneous systems.

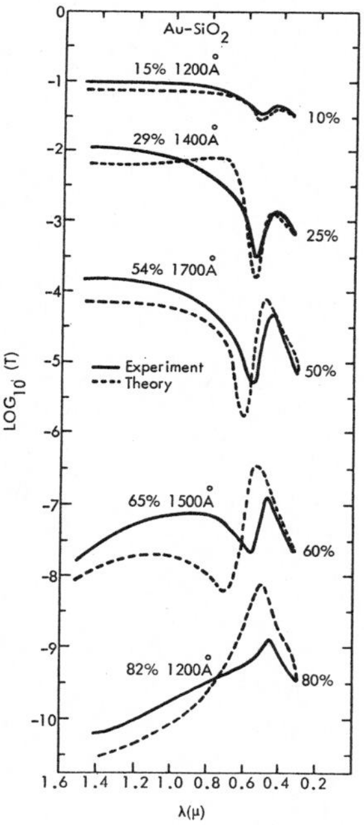

Fig. 10 Optical transmission as a function of light wavelength for a series of Au-SiO_2 samples. The metal volume fraction and the film thickness are labeled above each curve. Theoretical curves are normalized to the experimental values at 0.3 μm. For clarity, the curves are displaced with respect to one another.

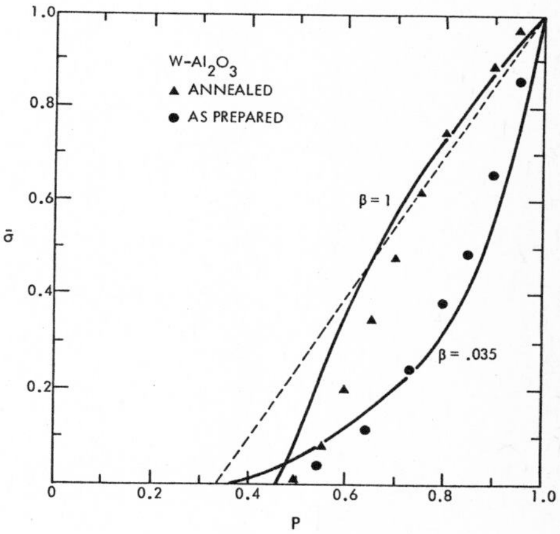

Fig. 11 Effective conductivity as a function of metal volume fraction p for samples of W-Al_2O_3 films. The solid lines are calculated from the theory. Dashed lines denote the Bruggeman result. The data are from Ref. [4].

References

[1] See, for example, Born and Wolf, Principles of Optics (Pergamon Press, New York, 1964).

[2] D. A. G. Bruggeman, Ann. Phys. (Leipzig) 24, 636 (1935).

[3] J. E. Gubernatis, AIP Conference Proc. No. 40, 84 (1978).

[4] B. Abeles, P. Sheng, M.D. Coutts, and Y. Arie, Adv. Phys. 24 407 (1975).

[5] P. Sheng, B. Abeles, and Y. Arie, Phys. Rev. Lett. 31, 44 (1973).

[6] P. Sheng, Phys. Rev. Lett. 45, 60 (1980).

[7] J. C. Maxwell-Garnett, Philos. Trans. Roy. Soc. London 203, 385 (1904).

[8] See, for example, J. D. Jackson, Classical Electrodynamics (John Wiley & Sons, New York, 1967).

[9] R. W. Cohen, G. D. Cody, M. D. Coutts, and B. Abeles, Phys. Rev. B8, 3689 (1973).

[10] J. I. Gittleman and B. Abeles, Phys. Rev. B15, 3273 (1977).

[11] D. Stroud, Phys. Rev. B12, 3368 (1975); see also W. Lamb, D. M. Wood, and N. W. Ashcroft, AIP Conf. Proc. No. 40, 240 (1978).

[12] P. Sheng, Phys. Rev. B22, 6364 (1980).

MACROSCOPIC AND MICROSCOPIC FIELDS IN ELECTRON AND ATOM TRANSPORT

R. S. Sorbello
Department of Physics
University of Wisconsin-Milwaukee
Milwaukee, WI 53201/USA

Introduction

Application of an electric field to a metal gives rise to transport of electrons and atoms. If the metal is homogeneous, one usually regards the electric field and the particle currents to be macroscopic quantities satisfying some generalized Ohm's law, where the conductivities are spatially constant parameters in the theory. On the microscopic- or atomic-scale, however, the fields and currents are not homogeneous. The nature of these microscopic inhomogeneities was first described by Landauer[1] in an important paper in 1957. Here we base our discussion on some recent work which delves more deeply into the structure of these microscopic inhomogeneities from a quantum-mechanical viewpoint.[2]

The important quantity which connects the microscopic and macroscopic levels is the residual resistivity dipole (RRD). According to Landauer,[1,3] the RRD is set-up around each scattering center in the presence of current flow. The RRD charge is analogous to the dipolar surface-polarization charge which is set-up on the surface of poorly conducting inclusions in an otherwise homogeneous conductor. When these individual RRD fields are added, their space-average gives the macroscopic electric field that is needed to drive the current past these scatterers. The picture is thus self-consistent.

Before studying the RRD, we present the phenomenological equations (generalized Ohm's Law) for homogeneous media. This allows us to define a useful parameter Z^* which is a measure of the driving field in electromigration.[4] Various questions and controversies are discussed, and the structure of the RRD is examined. We then show how the RRD can be used to calculate Z^*. Finally, we apply the RRD to Landauer's method of calculating the conductivity of one-dimensional disordered media.[5]

Phenomenological Equations

The linear phenomenological equations of transport theory relate fluxes and forces. The fluxes in our problem are the electronic particle current $\vec{J}_e$ and the ionic particle current $\vec{J}_i$. (For simplicity we assume only one ionic species). The forces are the gradients of the electrochemical potentials. The equations for a system at constant temperature are

$$\vec{J}_i = -L_{ii}\nabla(\mu_i+q_i\phi) - L_{ie}\nabla(\mu_e+q_e\phi) \tag{1}$$

$$\vec{J}_e = -L_{ei}\nabla(\mu_i+q_i\phi) - L_{ee}\nabla(\mu_e+q_e\phi) \tag{2}$$

where the L's are the generalized conductivities, μ_i and μ_e are the chemical potentials of ions and electrons, respectively, and ϕ is the macroscopic electrical potential. The charge of the ion and of the electron are q_i and q_e, respectively. The macroscopic electric field is given by $\vec{E} = -\nabla\phi$.

In the usual conductivity experiment, there are no concentration gradients ($\nabla\mu_i = \nabla\mu_e = 0$), and the electrons and ions undergo simple drift. One can write

$$\vec{J}_i = L_{ii}\, q_i^* \vec{E} \tag{3}$$

where we have defined the effective charge

$$q_i^* = q_i + \frac{L_{ie}}{L_{ii}} q_e \tag{4}$$

One can interpret $\vec{J}_i$ to be the current which would arise from an effective local field $\vec{E}_L = (q_i^*/q_i)\vec{E}$ which acts directly on the ions without any additional cross-contribution due to electron-ion coupling.

In a diffusion experiment for which $\vec{E} = 0$ and $\nabla\mu_e = 0$, we have $\vec{J}_i = -L_{ii}\nabla\mu_i$ and $\vec{J}_e = -L_{ei}\nabla\mu_i$. The charge transported per ion is

$$\frac{q_i\vec{J}_i + q_e\vec{J}_e}{J_i} = q_i^* \tag{5}$$

where the equality follows upon use of the Onsager relation[4] $L_{ie} = L_{ei}$.

Experimental results for q_i^* are traditionally expressed in terms of the parameter $Z^* = -q_i^*/q_e$. Determination of this "effective valence" is the prime concern of workers in electromigration.[4]

Questions and Controversies

There has been considerable effort in electromigration theory to try to determine Z^*, or equivalently, the local field $\vec{E}_L$ seen by an ion. It was suggested by Landauer and Woo[6] that the RRD may be the dominant contribution. This would lead to an intensification of the local field at the ion, and hence we would expect $Z^* > 0$. Yet for simple metals where band structure can be ignored, measurements reveal that $Z^* < 0$ and moreover $|Z^*|$ is typically larger than unity.[4] What has happened to the RRD? Where are the negative electrical charges which would accompany a moving ion according to Eq. (5)?

Other questions and controversies include the role of external vs. internal charges in setting up the macroscopic field.[7] What kind of charge is set up at the boundaries or electrodes? When one performs a quantum calculation and introduces into the hamiltonian the perturbation due to the electric field, is this a microscopic, macroscopic, or external field?

Questions such as these are often ignored in traditional treatments of transport theory. Landauer's viewpoint puts these questions in a deservedly central position in the theory.

Calculation of the Local Field

We consider the local field $\vec{E}_L(\vec{r})$ at some position $\vec{r}$ measured from an origin taken to be the nucleus of the ion in question. The ion is visualized as the only scatterer within some small region of an electron gas. Since we ignore explicit scattering by other agents in this region, we should restrict our attention to $r < \ell$, where ℓ is the electron mean-free-path. The relaxation time τ and the Fermi velocity v_F are related to ℓ according to $\ell = v_F\tau$.

In general, we can write $\vec{E}_L(\vec{r}) = -\nabla\Phi(\vec{r})$, where we introduce the local electrical potential Φ. The quantum mechanical expression for Φ is of the same form as the classical one, namely,[2]

$$\Phi(\vec{r}) = \int n(\vec{r}\,')v_b(\vec{r}\,'-\vec{r})d^3r' \tag{6}$$

where $n(\vec{r}\,')$ is the electron density at position $\vec{r}\,'$ and where $v_b(\vec{r}\,'-\vec{r}) = -e/|\vec{r}\,'-\vec{r}|$ is the bare potential for point charges. $e = |q_e|$ is the magnitude of the electron charge. The integration is over all space. Here and in what follows we are only concerned with that part of the density which is linear in $\vec{E}$.

Quantum-mechanical calculation of the electron density in the presence of $\vec{E}$ is difficult. Schaich[8] has derived from first principles an expression which had been used[9] or suggested[10] by others. His result for the <u>non-interacting</u> electron gas is

$$n_o(\vec{r}) = \sum_k g_{\vec{k}}\, |\psi_{\vec{k}}(\vec{r})|^2 \tag{7}$$

where $g_{\vec{k}}$ is the standard solution of the Boltzmann equation for the electron distribution in an electric field and $\psi_{\vec{k}}(\vec{r})$ is the electron scattering state which gives the wavefunction of an incoming electron $\exp(i\vec{k}\cdot\vec{r})$ being scattered by the ion. The sum is over all states, and the system is assumed to have unit volume.

To convert the non-interacting density $n_o(\vec{r})$ to the correct self-consistent density $n(\vec{r})$ one needs to include the screening response. Using Fourier-transformed quantities, this implies $n(\vec{q}) = n_o(\vec{q})/\varepsilon(q)$ where $\varepsilon(q)$ is the dielectric function. Assuming small screening length, these results can be used in Eq. (6) to yield[2]

$$\Phi(\vec{r}) = -(4\pi e/q_{TF}^2)n_o(\vec{r}) \tag{8}$$

where q_{TF} is the Thomas-Fermi wavevector. Explicitly, $q_{TF}^2 = 4\pi e^2/(\partial\varepsilon_F/\partial\bar{n})$ where ε_F

is the Fermi energy and $\overline{n}$ is the average density of electrons in equilibrium. For the 3-d electron gas $q_{TF}^2 = 4mk_F e^2/\pi\hbar^2$ where k_F is the Fermi wavevector and m is the electron mass. Expression (8) is valid beyond the small screening length assumption ($q_{TF} >> k_F$) if one resorts to averaging fields and densities over distances of several electron wavelengths, as in Landauer's work.[1]

In the asymptotic limit $n_o(\vec{r})$ and $\Phi(\vec{r})$ are easily evaluated.[2,8] Using the asymptotic form $\psi_k(\vec{r}) \to \exp(i\vec{k}\cdot\vec{r}) + [f(\theta')/r]\exp(ikr)\cos\theta'$, where θ' is the angle between $\vec{k}$ and $\vec{r}$, and f is the scattering amplitude one finds

$$\Phi(\vec{r}) = -\frac{p_o\cos\theta}{r^2} + (\text{osc. terms}) \tag{9}$$

where

$$p_o = S\ell\, E/4\pi \tag{10}$$

and S is the (momentum-weighted) cross-section for scattering of electrons by the ion. θ is the angle between $\vec{r}$ and $\vec{E}$. The value of p_o is precisely the value of the RRD dipole moment calculated by Landauer. The leading oscillatory terms in (9) are of the form $r^{-2}\sin(2k_F r+\gamma)$ where γ is a phase angle.

We have obtained Landauer's form of the potential due to the RRD provided that we ignore the oscillatory terms in Eq. (9). The latter terms will wash away if one averages $\Phi(\vec{r})$ over regions whose dimensions are several wavelengths long. This coarse graining is just what one must do to calculate the macroscopic field. Landauer's conclusions thus apply to the macroscopic field. In particular, the macroscopic field generated by a density N of ions (in a slab-geometry) is given by $\Delta E = 4\pi N p_o$, and this is precisely the field needed to overcome the scattering by the ions, i.e., $\Delta E \doteq eJ_e\Delta\rho$ where $\Delta\rho$ is the extra resistance introduced by the ions.

While the oscillatory terms in $\Phi(\vec{r})$ are not important for the macroscopic field, they are important for the microscopic field. We have verified this in a calculation of $n_o(r)$ and $\Phi(\vec{r})$ close to the ion for the model of s-wave scattering.[2] We found that the dipole-moment p_o cannot really be said to arise from a well defined localized charge distribution near the ion. This lack of a pure dipole near the ion explains why the local field is not strongly intensified by the RRD as one would have expected were there pure dipole behavior close-in to the ion. Thus $Z^* > 0$ does not follow.

An implication of the existence of RRD's is that boundary effects might not be adequately considered in quantum treatments of transport. The "external" field one inserts into the hamiltonian finds its origins in the RRD's, and not in external charges. There are complicated distributions of surface charges which are not strictly external, but rather are vestiges of RRD's and the response to RRD's. Traditional treatments ignore RRD's except as they might implicitly set up a macroscopic "external" field. This leaves uncertain whether interactions between RRD's are correctly accounted for in the traditional treatments. It remains to be seen whether these quantum mechanical analyses correctly treat such effects as the Lorentz corrections obtained by Landauer and Woo.[11]

Relationship Between Z^* and RRD

We have seen that the local field is not entirely governed by the RRD. There are other contributions to $\vec{E}_L$ or Z^*. It is possible, however, to derive an expression for Z^* solely from RRD considerations. To this end, consider the diffusion of ions through a slab under open circuit conditions for electrons. Let the ions be diffusing at average velocity $\vec{v}_o$ through the slab. The macroscopic field which is set up is the average field of the RRD's in the equivalent problem where the ions are fixed but the electrons move at a drift velocity of $-\vec{v}_o$. The equivalence follows by Galilean transformation. Now the RRD field set-up from a density of N ions per unit volume is $E = 4\pi N p_o$, where p_o is the RRD moment appropriate to an electron drift velocity $-\vec{v}_o$. Thus p_o is given by Eq. (10) with E in that equation replaced by $mv_o/\tau e$ which is the usual expression for the electric field in terms of a drift velocity v_o. The sample is thus acting as an open circuit battery which generates a field $4\pi N p_o$ internally. If we now short-out the battery to force $E = 0$, we will induce an electron current $J_e = 4\pi N p_o/\rho e$, where $\rho = m/\bar{n}e^2\tau$ is the resistivity. If we now use this value of J_e in Eq. (4) for q_i^*, we find

$$q_i^* = q_i - e\bar{n}\ell S \tag{11}$$

which is precisely the Huntington-Fiks result[4] for q_i^*. This result has also been derived quantum mechanically by others.[12,13] Our line of reasoning shows how the result follows from RRD considerations.

Landauer's 1-d Conductivity Formula

The RRD is also relevant to a very powerful 1-d conductivity formula which was derived by Landauer[5] and which is currently being debated in the literature.[14] To obtain the conductivity for a sample of length L one calculates $\sigma = J/E$ where J is now the electronic charge current and E is the average field across the sample. An electron of velocity v incident on the sample from the left of the sample is described by the plane wave exp(ikx), where $v = \hbar k/m$. There is a reflected wave r exp(-ikx) to the left and a transmitted wave t exp(ikx) coming out of the sample on the right, where $|r|^2 + |t|^2 = 1$. The current is thus $J = -ev|t|^2$. Landauer derived his formula by calculating the diffusion coefficient appropriate to this current and relating it to σ via the Einstein equation. Here we follow the more direct route[3] and evaluate E from our expression (8) for $\Phi(x)$.

The density to the left of the sample is $n_o(x) = |\exp(ikx) + r\exp(-ikx)|^2$. To the right, $n_o(x) = |t\exp(ikx)|^2$. Since the density, and thus the potential, are constant on the right we conveniently choose $\Phi(x) = 0$ there. Then everywhere to the left we find from Eq. (8) and our $n_o(x)$ expressions the result

$$\Phi(x) = -\left(\frac{4\pi e}{q_{TF}^2}\right)[2|r|^2 + r\,e^{-2ikx} + r^*\,e^{2ikx}] .$$

Upon averaging over a few wavelengths, we can ignore the oscillatory terms. The average potential drop across the sample is thus $8\pi e|r|^2/q_{TF}^2$. Equating this to -EL, we determine E. The conductivity $\sigma = J/E$ then becomes $\sigma = q_{TF}^2 v|t|^2 L/8\pi|r|^2$, which is precisely Landauer's formula when one makes use of the fact that $q_{TF}^2 = 4\pi e^2/(\partial\varepsilon_F/\partial\bar{n})$.

We see that the RRD is playing a central role in setting up the self-consistent potential $\Phi(x)$ in the sample. If one ignores the self-consistency aspects of the problem one obtains an incorrect expression for σ.[14,15]

Conclusion

We have seen that macroscopic and microscopic fields are naturally related via the RRD. Although the calculations based on expression (7) for $n_o(\vec{r})$ are only appropriate for distances within a distance ℓ from the ion, the results for the electric field coarse-grained over several wavelengths is the same as Landauer's. To extend the results for $n_o(\vec{r})$ and hence $\Phi(\vec{r})$ to distances $r > \ell$ is difficult within a strict quantum mechanical framework which treats background scattering as incoherent. One possible approach is to add the $\Phi(\vec{r})$ from each scatterer (ions and background) independently and to disregard incoherent scattering in damping the electron propagation and consequently $n_o(\vec{r})$ and $\Phi(\vec{r})$. The incoherence would instead enter as a result of the random placement of RRD's. This would effectively wash-out the oscillatory terms in Eq. (9) and restore Landauer's result for the macroscopic field automatically. Such an approach is quite different from the traditional one employed in quantum mechanical analyses of electron transport. A correct quantum calculation in the $r \geq \ell$ region requires careful consideration of RRD-RRD interactions. As we pointed out earlier it is not clear that the traditional quantum approach can handle these interactions correctly. In terms of conductivity corrections, or deviations from Matthiessen's rule, we are here dealing with the notoriously difficult corrections of order $1/k_F\ell$ or of second-order in impurity ion density.

Although we have not obtained new answers here by appealing to a picture based on microscopic inhomogeneities, we have gained better insight into the response of electrons and atoms to electric fields. More important, these considerations have led us to identify some possible inconsistencies in traditional quantum mechanical approaches to transport theory.

Acknowledgments

I am grateful to Rolf Landauer for valuable discussions and correspondence. This work was supported by the Graduate School of the University of Wisconsin-Milwaukee.

References

1. R. Landauer, IBM J. Res. Dev. 1, 223 (1957).
2. R. S. Sorbello, Phys. Rev. B23, 5119 (1981). Note the following corrections: The right-hand side of Eqs. (12) and (16) should be divided by $\hbar$. The right-hand side of Eq. (26) should be multiplied by $e/\hbar$. In Eq. (20), p_0 should read $2p_0$. In Eq. (21), 2m should read $\hbar/m$. The values of p/p_0 on the vertical axis of Fig. 3 should be doubled.
3. R. Landauer, Z. Physik B21, 247 (1975).
4. See, for example, H. B. Huntington in Diffusion in Solids: Recent Developments, edited by A. S. Nowick and J. J. Burton (Academic, New York, 1974).
5. R. Landauer, Philos. Mag. 21, 863 (1970).
6. R. Landauer and J.W.F. Woo, Phys. Rev. B10, 1266 (1974).
7. R. Landauer, Phys. Rev. B16, 4698 (1977).
8. W. L. Schaich, Phys. Rev. B13, 3350 (1976).
9. C. Bosvieux and J. Friedel, J. Phys. Chem. Solids 23, 123 (1962).
10. R. S. Sorbello, Comments Solid State Phys. 6, 117 (1975).
11. R. Landauer and J.W.F. Woo, Phys. Rev. B5, 1189 (1972).
12. L. Sham, Phys. Rev. B12, 3142 (1975).
13. See P. R. Rimbey and R. S. Sorbello, Phys. Rev. B21, 2150 (1980) and references cited therein.
14. See D. S. Fisher and P. A. Lee, Phys. Rev. B23, 6851 (1981) and references cited therein.
15. R. Landauer, pre-print.

Propagation and Attenuation in Composite Media*

Victor Twersky**

Mathematics Department
University of Illinois
Chicago, Illinois 60680

1. INTRODUCTION

In recent papers[1,2] we analyzed coherent scattering of waves by randomly distributed pair-correlated particles. We worked with the ensemble average of the functional equation for the field of a scatterer within the distribution in terms of its field when isolated, and derived representations and deterministic approximations for the coherent wave (the ensemble average of the solution over the realizable configurations of particles). The associated bulk parameters and corresponding indices of refraction specify a composite medium whose physical properties depend on the particles' properties and their distribution, as well as on the properties of the embedding medium. In the present paper we summarize some of the earlier results and consider later developments.

We start with results for the long wavelength (λ) limit which include refraction and absorption effects but no scattering losses. Then we introduce the leading term of the scattering losses to obtain results for absorbing particles that suffice if the net attenuation is small, and also consider generalizations. To facilitate applications, we first emphasize the different bases for anisotropy of the bulk parameters and indices of refraction of general composites and the physical import of key approximations, and then consider the analytical aspects on which the development is based.

For an ensemble of configurations of identical aligned scatterers, we specify the one-particle statistics by the average number (ρ) of scatterers in unit volume, and the two-particle statistics by $\rho f(\underset{\sim}{R})$ with $f(\underset{\sim}{R})$ as the distribution function for the separation $(\underset{\sim}{R})$ of pairs. The minimum separation of centers as a function of $\hat{R}$ specifies the exclusion surface $\underset{\sim}{R} = \underset{\sim}{b}(\hat{R})$; we require $f(\underset{\sim}{R}) = 0$ for $R < |\underset{\sim}{b}(\hat{R})|$, and $f(\underset{\sim}{R}) \sim 1$ for $R \sim \infty$. If $\underset{\sim}{b} = b\hat{R}$ is a sphere with radius $b \geq 2a$, then $f(R)$ is the usual radial distribution function. We use $f(R)$ not only for spherical obstacles of radius a but also for more general shapes $\underset{\sim}{R} = \underset{\sim}{a}(\hat{R})$ (identical and aligned, or averaged over shape, alignment, etc.) regarded as if enclosed in transparent coatings whose outer surfaces are spheres of diameter b. The transparent shell has no direct influence on an isolated obstacle's scattering properties but preserves the radial symmetry of the distribution, and specifies a minimum separation between centers as an additional parameter. For cases where we average over alignment we assume the distribution of alignments to be uniform and uncorrelated with position or separation; similarly for averages over particle shapes, sizes, or parameters. The general case we consider corresponds to differently aligned nonsimilar scatterer $(\underset{\sim}{a})$ and exclusion $(\underset{\sim}{b})$ surfaces, such that $\underset{\sim}{R} = \underset{\sim}{b}/2$ (the outer surface of the transparent coating) represents impenetrable statistical-mechanics particles whose shape and volume fraction govern the configurational aspects. Exclusive of the properties of the embedding space, the shape and parameters (tensors, in general) of aligned particles and the shape of the exclusion surface provide three distinct bases for anisotropy.

For ellipsoidal particles specified by major semidiameters $a_j (j = 1,2,3)$ along the directions $\hat{\xi}_j$ with associated depolarization integrals $q_j = q_j(a_j)$ normalized to satisfy $\Sigma q_j = 1$, we write the particle's shape parameter as a dyadic $\tilde{q} = \Sigma q_j \hat{x}_j \hat{x}_j$. Similarly we specify the particle's material properties by a dyadic parameter $\tilde{p} = \Sigma p_j \hat{\xi}_j \hat{\xi}_j$, or by two such dyadics (for the general electromagnetic case), or by one dyadic and one scalar parameter (for general small-amplitude acoustics). The exclusion-correlation surface b_i with depolarization factors $Q_j = q_j(b_j) = q_j(b_j/2)$ is specified by the dyadic $\tilde{Q} = \Sigma Q_j \hat{X}_j \hat{X}_j$. In general, the orthogonal sets of unit vectors $\hat{x}_j$, $\hat{\xi}_j$, and $\hat{X}_j$ do not coincide, and the cocentered ellipsoids q and Q (the surfaces of the particle and of its transparent coating) are not necessarily similar or conformal.

2. LOW-FREQUENCY LIMITS

For monopole particles each of volume v specified by a scalar parameter c in an embedding medium c_0, the bulk parameter C is given by

$$C = c_0 + w(c-c_0) = wc + (1-w)c_0, \quad w = \rho v \tag{1}$$

where w is the volume fraction occupied by particles (parameter c), and 1-w is the void (c_0) fraction of the composite (C). The second equality in (1), the representation of the bulk parameter C as the volume-weighted mean of the parameters of the components goes back to Archimedes. In terms of shifted normalized values, we rewrite (1) as

$$\Gamma = w\gamma; \quad \gamma = (c/c_0)-1, \quad \Gamma = (C/c_0)-1 \tag{2}$$

In general the parameters are complex, $c = c_r + ic_i$ such that c_r and $c_i = |c_i|$ are real; c_i accounts for absorption. In small amplitude acoustics, in the simplest case, c is the compressibility.

For aligned dipole particles with parameter $\tilde{p}$ in an embedding medium $\tilde{p}_0$, for particle and exclusion surfaces $\tilde{q}$ and $\tilde{Q}$, the analog of (2) is

$$\tilde{\Delta} = w\tilde{\delta}\cdot(\tilde{I}+\tilde{D}\cdot\tilde{\delta})^{-1}; \; \tilde{\delta} = \tilde{p}\cdot\tilde{p}_0^{-1} - \tilde{I}, \; \tilde{\Delta} = \tilde{P}\cdot\tilde{p}_0^{-1} - \tilde{I}, \; \tilde{D} = \tilde{q} - w\tilde{Q}. \tag{3}$$

Here $\tilde{I}$ is the identity, and $\tilde{D}$ is the compound depolarization factor. In small amplitude acoustics, in the simplest case p^{-1} is the particle's mass density; more generally, $p = p_r - ip_i$ with p_r and $p_i = |p_i|$ real. In electromagnetics, $\tilde{p}$ represents either the particle's dielectric parameter $\tilde{\epsilon}'$ or its magnetic parameter $\tilde{\mu}'$, $\tilde{p}_0$ represents $\tilde{\epsilon}_0$ or $\tilde{\mu}_0$, and $\tilde{P}$ the corresponding bulk value $\tilde{\epsilon}$ or $\tilde{\mu}$; we emphasize the dielectric case $p = \epsilon' = \epsilon_r' + i\epsilon_i'$, $\epsilon_i' = |\epsilon_i'|$.

To first order in $\tilde{\delta}$, (3) reduces to $\tilde{\Delta} \approx w\tilde{\delta}$, i.e., the tensor version of (2). The factor $(\tilde{I}+\tilde{D}\cdot\tilde{\delta})^{-1}$ in (3) arises from the dipole character of the particle. To make the physical content more accessible we decompose $\tilde{D}$ and rewrite (3) as

$$\tilde{\Delta}\cdot(\tilde{I}+\tilde{Q}\cdot\tilde{\Delta})^{-1} = w\tilde{\delta}\cdot(\tilde{I}+\tilde{q}\cdot\tilde{\delta})^{-1}. \tag{4}$$

If the principal axes of all dyadics coincide, then for each component,

$$\frac{\Delta_j}{1 + Q_j\Delta_j} = w\,\frac{\delta_j}{1 + q_j\Delta_j} \tag{5}$$

In particular for spherical particles $(q_j = 1/3)$ and spherical coats $(Q_j = 1/3)$ and isotropic parameters $(\delta_j = p/p_0-1)$, we have

$$\frac{P - p_0}{P + 2p_0} = w\,\frac{p - p_0}{p + 2p_0} \tag{6}$$

i.e., the classical form obtained originally by Maxwell[3] (and attributed to Clausius, Mossotti, Lorenz, and Lorentz). Maxwell constructed (6) by equating the potential of a sphere of volume V and equivalent parameter P to the uncoupled sum of the potentials of N spheres of volume v and parameter p within V. Writing $w = \rho v = Nv/V$ in

(6) and multiplying through by V gives Maxwell's initial form; although useful as a mnemonic device, we stress that the left side of (6) corresponds to the radially symmetric shape of the region excluding the centers of all neighbors of one particle $(Q_j = 1/3)$, and not to the volume of the distribution, and that there is no equivalent sphere. The same construction suffices for (5) and (4); these serve to relate the sum of the distant potentials of N ellipsoids $(v, \tilde{p}, \tilde{q})$ to an equivalent ellipsoid $(V, \tilde{P}, \tilde{Q})$, but actually $\tilde{Q}$ corresponds to the shape of the exclusion region and not to that of the distribution as a whole, and there is no equivalent ellipsoid.

For scalar waves and composites specified by bulk parameters C and $\tilde{P}$ the index of refraction (η) corresponding to a direction of propagation $\hat{K}$ is determined by

$$\eta^2 = C/\hat{K}\cdot\tilde{P}\cdot\hat{K} \tag{7}$$

See References 1 and 4 for illustrations and applications to a plane wave incident at an arbitrary angle on a slab region of composite corresponding to (7). For electromagnetic waves and composites with bulk parameters $\tilde{\epsilon}$ and $\tilde{\mu}$, the index is specified by the determinantal equation

$$|(\tilde{I}-\hat{K}\hat{K})\eta^{-2} + (\hat{K}\times\tilde{\mu}^{-1}\times\hat{K})\cdot\tilde{\epsilon}^{-1}| = 0. \tag{8}$$

If $\tilde{\mu}' = \tilde{\mu}_0$, then $\tilde{\mu} = \tilde{I}$, and (8) simplifies to

$$\hat{K}\cdot(\tilde{I}\eta^{-2} - \tilde{\epsilon}^{-1})^{-1}\cdot\hat{K} = 0. \tag{9}$$

See References 2 and 5-7 for illustrations and applications.

3. SCATTERING LOSSES

To include the leading term in $k = 2\pi/\lambda$ corresponding to scattering losses we replace (2) by

$$\Gamma = w\gamma + i\alpha\gamma^2 w\mathfrak{b} \tag{10}$$

where α depends on k, and $\mathfrak{b}$ depends on the volume fraction W of the coated particles. For scatterers in the form of slabs, cylinders, or ellipsoids $(n = 1,2,3)$ we have

$$\alpha = k^n v_n/d_n,\ v_n = \{2a_1, \pi a_1 a_2, 4\pi a_1 a_2 a_3/3\},\ d_n = \{2,4,4\pi\} \tag{11}$$

corresponding to incidence perpendicular to a generator for $n = 1,2$. The function

$$\mathfrak{w} = 1 + \rho \int [f(\underset{\sim}{R})-1] d\underset{\sim}{R} \tag{12}$$

is a statistical mechanics packing factor proportional to the variance (fluctuations) in the number (N_c) of particles in a central region (V_c), i.e., $\mathfrak{w} = [\langle N_c^2 \rangle - \langle N_c \rangle^2]/\langle N_c \rangle$ such that $\langle N_c \rangle / V_c = N/V = \rho$. The dependence of $\mathfrak{w}$ on W is illustrated by results based on the scaled-particle equations[8] of state $(\mathcal{E})$ for impenetrable particles of volume fraction W governed by radially symmetric pair statistics, or by lattice gas statistics. From $\mathfrak{w} = (\partial \mathcal{E}/\partial \rho)^{-1}$ we construct

$$\mathfrak{w}_n = (1-W)^{n+1}/[1+(n-1)W]^{n-1}, \quad \mathfrak{w}_0 = 1-W \tag{13}$$

The cases $n = 3,2,1$ correspond to correlated spheres, cylinders, slabs, and 0 to uncorrelated space-occupying particles (random lattice gas). More generally, we could determine $\mathfrak{w}$ were $f(\underset{\sim}{R})$ known, e.g., by numerical computations based on the Percus-Yevick equation.

The corresponding form for the dyadic parameter (3) is

$$\tilde{\Delta} = w\underline{\tilde{\delta}} + i\beta\underline{\tilde{\delta}}\cdot\underline{\tilde{\delta}} w\mathfrak{w}, \quad \underline{\tilde{\delta}} = \tilde{\delta}\cdot(\tilde{I}+\tilde{D}\cdot\tilde{\delta})^{-1} = (\tilde{I}+\tilde{\delta}\cdot\tilde{D})^{-1}\cdot\tilde{\delta} \tag{14}$$

For acoustic dipoles with index as in (7), we have

$$\beta = -\alpha/n = -k^n v_n/d_n^a, \quad d_n^a = \{2,8,12\pi\} \tag{15}$$

with α as in (11). For dielectric poles $\underset{\sim}{P} = \tilde{\epsilon}$ with index as in (9),

$$\beta = k^n v_n/d_n^e, \quad d_n^e = \{2,4 \text{ or } 8,6\pi\}. \tag{16}$$

The two values of $n = 2$ correspond to incident polarization $(\underset{\sim}{E})$ parallel or perpendicular to a generator; for $\underset{\sim}{E}$ parallel, the cylinder is a monopole (as is the slab for normal incidence), i.e., $\tilde{D} = 0$.

4. COMPLEX DIELECTRICS

Initially we apply (3) and (4) for the components of $\tilde{P} = \tilde{\epsilon}$ for complex $\tilde{p} = \tilde{\epsilon}'$, and then we include the scattering loss terms (14). For coincident dielectric and geometric axes, and polarization parallel to a principal axis, we write

$$\Delta = \frac{w\delta}{1+\delta D}, \quad D = q - wQ, \quad \delta = \delta_r + i\delta_i, \quad \delta_i \geq 0 \tag{17}$$

where the common subscript j has been suppressed. For uncoated infinite slabs or circular cylinders and longitudinal polarization, $q = Q = 0$, and (17) reduces to $\Delta = w\delta$, the form in (2); for transverse polarization, $q = Q = 1$ for the slab, and $1/2$ for the cylinder. The results for the slab are due to Maxwell[3], and those for the cylinder to Rayleigh.[9] Wiener[10] applied the results for the corresponding indices to investigate form effects in birefringence and dichroism studies.

From (17) we have

$$\Delta = \Delta_r + i\Delta_i, \quad \Delta_r = w\,\frac{(\delta_r + |\delta|^2 D)}{|1+\delta D|^2}, \quad \Delta_i = \frac{w\delta_i}{|1+\delta D|^2}. \tag{18}$$

For the cases at hand the indices of refraction are equal to the square roots of the parameters, i.e., $\eta^2 = P$, $\eta'^2 = p$, $\eta_0^2 = p_0$. We take η_0 as real and construct the corresponding bulk indices:

$$\eta = \eta_r + i\eta_i = \eta_0(1+\Delta)^{1/2}, \quad \begin{Bmatrix}\eta_r\\ \eta_i\end{Bmatrix} = \eta_0\left[\frac{|1+\Delta| \pm (1+\Delta_r)|}{2}\right]^{1/2}. \tag{19}$$

To express (19) in terms of η', we introduce

$$\nu = \frac{\eta'}{\eta_0} - 1 = \left(\frac{\eta'_r}{\eta_0} - 1\right) + i\,\frac{\eta'_i}{\eta_0} = \nu_r + i\nu_i \tag{20}$$

such that $\delta = (\eta'/\eta_0)^2 - 1 = \nu(2+\nu)$, and work with

$$\delta_r = 2\nu_r + \nu_r^2 - \nu_i^2, \quad \delta_i = 2\nu_i(1+\nu_r). \tag{21}$$

To include the scattering terms of (14), we replace (17) by

$$\Delta \approx w(\underline{\delta} + i\underline{\delta}^2 S), \quad \underline{\delta} = \delta/(1+\delta D), \quad S = \beta\mathfrak{w} \tag{22}$$

with $\mathfrak{w}$ as in (12) and $\beta = \beta(k^m)$ as in (16). Because δ of (22) does not include terms of order k^2 (higher order terms of the dipole at hand, contributions of other multipoles, and from k-dependent distribution integrals), we retain at most only the first order term of $S = S(k^m)$ in the corresponding η. To delineate conditions on the retention of S, we first consider non-absorbing particles, $\delta_i = 0$. For such cases

$$\Delta_r = w\delta_r/(1+\delta_r D), \quad \Delta_i = wS\delta_r^2/(1+\delta_r D)^2 \tag{23}$$

where the next terms in k of Δ_r and Δ_i are of order k^2 and k^{m+2} respectively. If $|\Delta_i| \ll |1+\Delta_r|$, then from (19),

$$\frac{\eta_r}{\eta_0} \approx (1+\Delta_r)^{1/2} = \left[1 + \frac{w\delta_r}{1+\delta_r D}\right]^{1/2}, \quad \frac{\eta_i}{\eta_0} = \frac{\Delta_i\eta_0}{2\eta_r} = \frac{\Delta_i}{2(1+\Delta_r)^{1/2}} \tag{24}$$

where the corrections in k to η_r, η_i are of order k^2, k^{m+2} respectively.

To third order in δ_r, we construct

$$\eta_r/\eta_0 = 1 + \frac{1}{2}w\delta_r - \frac{1}{8}w\delta_r^2(w+4D) + \frac{1}{16}w\delta_r^2(w^2+4wD+8D^2) \tag{25}$$

and

$$\eta_i/\eta_0 = \frac{1}{2}wS\delta_r^2\left[1 - \frac{1}{2}\delta_r(w+4D)\right] . \tag{26}$$

Substituting for δ_r from (21), we see that $\delta_i = 0$ implies either $\nu_i = 0$ or $1 + \nu_r = 0$; for the first, $\delta_r = 2\nu_r + \nu_r^2$ as before[6], but for the second, $\delta_r = -\nu_i^2 - 1$.

If the particles are absorbing, then from (22) for $\delta = \delta_r + i\delta_i$,

$$\Delta_r = w\underline{\delta}_r(1-2\underline{\delta}_i S), \quad \Delta_i = w[\underline{\delta}_i + (\underline{\delta}_r^2-\underline{\delta}_i^2)S]$$

$$\underline{\delta}_r = (\delta_r+|\delta|^2D)/\mathcal{D}, \ \underline{\delta}_i = \delta_i/\mathcal{D}, \ \mathcal{D} = |1+\delta D| = (1+\delta_r D)^2 + (\delta_i D)^2. \tag{27}$$

Although we may use these heuristically in (19), they are inconsistent in k except for $S = 0$ or $\delta_i = 0$. For small δ_i, we use (27) to generalize (24). Thus, if $\delta_i \ll \delta_r$, then to first order in δ_i and S

$$\Delta_r = w\delta_r/(1+\delta_r D), \quad \Delta_i = w(\delta_i+\delta_r^2 S)/(1+\delta D)^2. \tag{28}$$

Substituting into (19), we obtain (24) in terms of the present Δ_i. Explicitely, we have

$$\frac{\eta_r}{\eta_0} = \left[\frac{1 + \delta_r(w+D)}{1 + \delta_r D}\right]^{1/2}, \quad \frac{\eta_i}{\eta_0} = \frac{w(\delta_i+\delta_r^2 S)}{2(1+\delta_r D)^{3/2}[1+\delta_r(w+D)]^{1/2}} \tag{29}$$

which suffice for situations where the net attenuation is small. The real index η_r is the same as in (24)-(25). For η_i, to order $\delta_i\delta_r^2$ we have

$$\frac{2\eta_i}{w\eta_0} = S\delta_r^2[1 - \frac{1}{2}\delta_r(w+4D)] + \delta_i[1 - \frac{1}{2}\delta_r(w+4D) + \frac{3}{8}\delta_r^2(w^2+4wD+8D^2)]. \tag{30}$$

The net attenuation coefficient $2k\eta_i$ consists of two sets of terms, one corresponding to scattering losses and one to absorption losses.

To display the different bases of the anistropic effects explicitly, and to obtain simple forms for practical applications, we work with the leading terms of the power series expansions of P and η. To avoid repetition we keep terms to third order in δ_i and first order in S with the understanding that if $S = 0$ we work with terms through δ_i^3, but if $S \neq 0$ we keep only the leading terms in δ_i and drop the coupling terms in $\delta_i S$ because of the neglected k^2 terms. The coupling terms are not significant for numerical computations but we display them for physical import. Thus, from (22) to third order in δ we have

$$(P-p_0)/p_0 w = \Delta/w = \delta - \delta^2(D-iS) + \delta^3 D(D-2iS) \quad (31)$$

Introducing $\delta = \delta_r + i\delta_i$, the real part of (31) is the real dielectric contrast

$$\Delta_r/w = \delta_r(1-\delta_r D+\delta_r^2 D^2) + \delta_i^2 D(1-3\delta_r D) - 2S\delta_r\delta_i + 2S\delta_i(3\delta_r^2-\delta_i^2)D \quad (32)$$

and the imaginary part is the normalized net loss

$$\Delta_i/w = \delta_i[1-2\delta_r D+(3\delta_r^2-\delta_i^2)D^2] + S(\delta_r^2-\delta_i^2) - 2S\delta_r(\delta_r^2-3\delta_i^2)D. \quad (33)$$

In Δ_r, the terms that involve δ_i^2 and not S arise from absorption and those that involve $\delta_i S$ arise from the interplay of absorption and scattering losses. To order S, there are no pure scattering loss corrections in Δ_r; if $\delta_i = 0$, all loss corrections vanish to third order in δ and first order in S. In Δ_i, there are essentially three kinds of attenuation processes: pure absorption, pure scattering, and the interaction terms that depend on $\delta_i^2 S$.

From (31), we construct the corresponding representations of η in terms of $\nu = (\eta'-\eta_0)/\eta_0$ by using $\delta = \nu(2+\nu)$ and expanding $\eta/\eta_0 = (1+\Delta)^{1/2}$ to third order in ν and first order in S. Thus,

$$\frac{\eta-\eta_0}{\eta_0 w} = \frac{n}{w} = \nu + \frac{1}{2}\nu^2(A+i4S) + \frac{1}{2}\nu^3(B+i4AS). \quad (34)$$

$$A = 1 - (w+4D), \qquad B = -(1-w)(w+4D) + 8D^2.$$

Writing $\nu = \nu_r + i\nu_i$, we separate real and imaginary parts to obtain the refractive contrast $n_r = (\eta_r-\eta_0)/\eta_0$,

$$n_r/w = \nu_r\left(1+\frac{1}{2}\nu_r A+\frac{1}{2}\nu_r^2 B\right) - \frac{1}{2}\nu_i^2(A+3\nu_r B) - 4S\nu_i\nu_r - 2S\nu_i(3\nu_r^2-\nu_i^2)A \quad (35)$$

and the normalized attenuation $n_i = \eta_i/\eta_0$,

$$n_i/w = \nu_i\left[1 + \nu_r A + \frac{1}{2}(3\nu_r^2-\nu_i^2)B\right] + 2S(\nu_r^2-\nu_i^2) + 2S\nu_r(\nu_r^2-3\nu_i^2)A. \quad (36)$$

The grouping of terms in (35) and (36) is the same as in (32) and (33), and similar considerations apply. As before for Δ, if $S = 0$ we work with terms through ν_i^3, but if $S \neq 0$ we keep only the leading terms in ν_i and drop the coupling terms in $\nu_i S$ because of the neglected k^2 terms.

To stress physical import, we write (32) to second order terms as

$$\Delta_r = w\delta_r - w|\delta|^2 D + 2w\delta_i(\delta_i D - \delta_r S). \tag{37}$$

The leading term is a volumetric (intrinsic) contribution, and the second term $-w|\delta|^2 D = -w|\delta|^2 q + w^2|\delta|^2 Q$ consists of a decrease determined by particle shape (q) and an increase arising from pair (w^2) interactions governed by the correlation-congifurational exclusion surface (Q). The third term shows the effects of absorption and the interaction of absorption and scattering: the pure absorption term shows an increase arising from shape and a decrease arising from correlations; the mixed term for positive δ_r represents a decrease, with S determined by fluctuations. Similarly for

$$\Delta_i = w\delta_i + wS|\delta|^2 - 2w\delta_i(\delta_r D + \delta_i S), \tag{38}$$

the leading term is the volumetric (intrinsic) absorption, and the second term is the scattering loss arising from fluctuations. The term in $\delta_i\delta_r D$ for $\delta_r > 0$ involves a decrease due to particle shape and an increase because of correlations, and the mixed term is a decrease arising from interaction of the two loss mechanisms.

For the corresponding bulk refractive index contrast, we have

$$h_r = w\nu_r + w|\nu|^2 A/2 - w\nu_i(\nu_i A + 4S\nu_r) \tag{39}$$

$$A = 1-w - 4D = (1-4q) - w(1-4Q)$$

where if $Q = q$, then $A = (1-w)(1-4Q)$ is positive for $Q < 1/4$, negative for $Q > 1/4$. Splitting A into 1-w and -4D, we see that the D and S contributions show the same trends as in (37), but now there are two additional effects symmetrical in volume and void fractions, i.e., in $w(1-w)\left(\frac{1}{2}|\nu|^2 - \nu_i^2\right)$. Similarly for

$$h_i = w\nu_i + 2wS|\nu|^2 + w\nu_i(\nu_r A - 4\nu_i S) \tag{40}$$

where we decompose wA in w(1-w) and -w4D for comparison with (38).

We apply the above results elsewhere in terms of η_0 as the variable to analyze birerfingence ($\eta_{r1}-\eta_{r2}$) and dichroism ($\eta_{i1}-\eta_{i2}$) measurements in which the index of the embedding medium is varied to help determine the properties and shapes of the particles.

5. ANALYTICAL ASPECTS

In previous sections we considered the key forms (10) and (14) for the bulk parameters and (7)-(9) for the associated indices of refraction with emphasis on their physical content and implications for propagation and attenuation of the coherent field. Now we indicate essential steps of their analytical derivation. For brevity we restrict initial considerations to the three-dimensional case of bounded obstacles.

For the scalar problem of Helmholtz's equation, we take the field incident on an arbitrary obstacle as the plane wave

$$\phi = e^{i\underset{\sim}{k}\cdot\underset{\sim}{r}}, \quad \underset{\sim}{r} = r\hat{r}(\theta,\varphi), \quad \underset{\sim}{k} = k\hat{k} \tag{41}$$

We write the corresponding scattered wave as a radiative function in the form of an integral over the obstacle's surface $\mathfrak{S}(\underset{\sim}{r}')$,

$$u(\underset{\sim}{r}) = \{h(k|\underset{\sim}{r}-\underset{\sim}{r}'|),u(\underset{\sim}{r}')\},$$
$$h(x) = h_0^{(1)}(x) = \frac{e^{ix}}{ix}, \quad \{h,u\} = \frac{k}{i4\pi}\int(h\nabla u-u\nabla h)\cdot d\underset{\sim}{\mathfrak{S}}. \tag{42}$$

We take the origin of $\underset{\sim}{r}$ as the center of the smallest sphere circumscribing the obstacle. For $r \sim \infty$,

$$u \sim h(kr)g(\hat{r},\hat{k}), \quad g(\hat{r},\hat{k}) = \{e^{-i\underset{\sim}{k}_r\cdot\underset{\sim}{r}'},u\} \quad \underset{\sim}{k}_r = k\hat{r} \tag{43}$$

with g as the normalized scattering amplitude. An inverse to $g[u]$ of (43), at least for r greater than the scatterer's projection on $\hat{r}$, is the complex spectral representation, say $u[g]$:

$$u(\underset{\sim}{r}) = \int_c e^{i\underset{\sim}{k}_c\cdot\underset{\sim}{r}}g(\hat{r}_c,\hat{k}), \quad \underset{\sim}{k}_c = k\hat{r}_c, \quad \int_c = \frac{1}{2\pi}\int d\Omega(\theta_c,\varphi_c) \tag{44}$$

with contours as for $h_0^{(1)}$.

We use analogous forms for the corresponding multiple scattered wave of one element (say s) of a configuration of N obstacles, i.e., $U_s[G_s]$ and $G_s[U_s]$. Applying Green's operator $\{u,v\}$ as in (42) to the solution of Helmholtz's equation for one obstacle and N-obstacles we obtain the key representation

$$G_t(\hat{r}) = g_t(\hat{r},\hat{k})e^{i\underset{\sim}{k}\cdot\underset{\sim}{r}_t} = \sum_{s\neq t}\int g_t(\hat{r},\hat{r}_c)G_s(\hat{r}_c)e^{i\underset{\sim}{k}_c\cdot\underset{\sim}{R}_{ts}}, \; \underset{\sim}{R}_{ts} = \underset{\sim}{r}_t - \underset{\sim}{r}_s \tag{45}$$

where g_t and G_t are the isolated and multiple scattering amplitudes of obstacle t. Equation (45) determines G in terms of g (the direct problem), or g in terms of G (the inverse), at least if the scatterer's projections on $\hat{R}_{ts}$ do not overlap.

The ensemble average of $\Psi = \phi + \Sigma U_s$ yields a system of hierarchy integrals. The first expresses the ensemble average $\langle\Psi\rangle$, the coherent wave (a set of plane waves with propagation coefficients $\underset{\sim}{K}_i = k\eta_i\hat{K}_i$), in terms of an integral over $\langle G_t\rangle_t$, the ensemble average of G_t with obstacle-t fixed. The second relates $\langle G_t\rangle_t$ to an integral over

$\langle G_s \rangle_{st}$, the ensemble average of G_s with obstacles s and t fixed, etc. For a distribution with one-particle and two-particle statistics specified by $\rho = N/V$ and $f(\underset{\sim}{R}_{ts}) = f(\underset{\sim}{R})$, we truncate the hierarchy system by $\langle G_s \rangle_{st} \approx \langle G_s \rangle_s$ to obtain a deterministic approximation; see References 11-13. In terms of the radiative function

$$u = \int_c g(\hat{r}, \hat{r}_c) \mathfrak{G}(\underset{\sim}{k}_c | \underset{\sim}{K}) e^{i\underset{\sim}{k}_c \cdot \underset{\sim}{R}}, \quad \underset{\sim}{K} = k\eta\hat{K} \tag{46}$$

we determine η by

$$\mathfrak{G}(\underset{\sim}{k}_r | \underset{\sim}{K}) = -\frac{i4\pi\rho}{k^3(\eta^2-1)} \left\{ e^{-i\underset{\sim}{K}\cdot\underset{\sim}{R}}, u \right\}_s + \rho \int_{V-\bar{v}} [f(\underset{\sim}{R})-1] e^{-i\underset{\sim}{K}\cdot\underset{\sim}{R}} u \, d\underset{\sim}{R} \tag{47}$$

where $s(k)$ and $\bar{v}(k)$ are the exclusion surface and volume, and V is taken as the volume of all space.

See Reference 1 for detailed derivation of (47), and for volume integral representations of the bulk parameters C and $\tilde{P}$. These representations show that if the obstacles are specified solely by the parameter c, then $\tilde{P} = \tilde{I}$ and $\eta^2 = C$; similarly, if $c = 1$, then $C = 1$ and $\eta^{-2} = \hat{K}\cdot\tilde{P}\cdot\hat{K}$. Here and in the following we have set c_0, p_0 and η_0 equal to unity so that all quantities C, $\tilde{P}$, and η are normalized values relative to those of the embedding medium. See also Reference 1 for reduction of (47) to obtain the results for η^2 derived originally by Rayleigh, Reiche, Foldy, and by Lax, and for the relations to the procedures of Lax[11] and of Keller.[12]

To derive (10) and (14), we start with

$$g(\hat{r}, \hat{k}) \approx a_0 + \hat{r}\cdot\tilde{a}_1\cdot\hat{k} \tag{48}$$

$$a_0 \approx a_0'(1+a_0') \qquad a_0' = i\alpha\gamma$$

$$\tilde{a}_1 \approx \tilde{a}_1'\cdot(\tilde{I}+\tilde{a}_1'/n), \quad \tilde{a}_1' = -i\alpha(\tilde{I}+\tilde{\delta}\cdot\tilde{q})^{-1}\cdot\tilde{\delta}$$

where γ, $\tilde{\delta}$, and $\alpha = \alpha(k^n)$ are given in (2), (3) and (11). [Note for $c_0 = 1$ and $\tilde{p}_0 = \tilde{I}$, we have $\gamma = c-1$ and $\tilde{\delta} = (\tilde{p}-\tilde{I})$ in terms of relative particle parameters.] This approximation is correct to lowest order in k for the real and imaginary parts of g. The imaginary part of g, i.e., the k^n terms for the isotropic case $\tilde{\delta} = \delta\tilde{I} = (p-1)\tilde{I}$ were obtained by Rayleigh[14] from the corresponding potential theory problems.[3]

If $\tilde{p} = \tilde{I}$, then $g = a_0$, and similarly for the multiple scattering amplitude $\mathfrak{G} = A_0$. For this case of pure monopoles, $u = a_0 h_0(kR) A_0$, and from (47) to lowest order in k for the real and imaginary parts,

$$\eta_m^2 = C \approx 1 + w\gamma + i\alpha\gamma^2 w\mathfrak{w}, \; w = \rho v, \; \mathfrak{w} = 1 + \rho\int_V [f(\underset{\sim}{R})-1] d\underset{\sim}{R} \tag{49}$$

where $\eta_m^2 = C$ follows from (7) because $\tilde{p} = \tilde{I}$ yields $\tilde{P} = \tilde{I}$. Here $\mathfrak{w}$ involves $f(\underset{\sim}{R})$ for elliptically symmetric statistics determined by the shape dyadic $\tilde{Q}$ of the exclusion surface; the integration over all space (V) corresponds to the volume integral over $V - \bar{v}$ in (47) minus $\rho \int_{\bar{v}} d\underset{\sim}{R}$ as generated by evaluating the surface integral form.

If $c = 1$, then from $g(\hat{r},\hat{R}) = \hat{r}\cdot\tilde{a}_1\cdot\hat{R}$ and $\mathfrak{G}(k\hat{R}|\underset{\sim}{K}) = \hat{R}\cdot\tilde{A}\cdot\hat{K} = \hat{R}\cdot\underset{\sim}{A}$, we have $u = \hat{r}\cdot\tilde{a}_1\cdot\tilde{h}_a(kR)\cdot\underset{\sim}{A}$ with $\tilde{h}_a = [\tilde{I}h_0 + (\tilde{I}-n\hat{R}\hat{R})h_2]/n$ as the acoustic dipole propagator. From (47), and using $c = 1$ implies $C = 1$, we obtain

$$\eta_d^{-2} = \hat{K}\cdot\tilde{P}\cdot\hat{K},\ \tilde{P} \approx \tilde{I} + w\underset{\sim}{\tilde{\delta}} - i\alpha\underset{\sim}{\tilde{\delta}}\cdot\tilde{\delta}w\mathfrak{w}/n,\ \underset{\sim}{\tilde{\delta}} = [\tilde{I} + \tilde{\delta}\cdot(q-w\tilde{Q})]^{-1}\cdot\tilde{\delta}\ . \quad (50)$$

To the present order in k, the index of refraction for distributions of scatterers specified by two-parameters c and $\tilde{p}$ is the product of the indices for pure monopoles and pure dipoles. Writing the results in (49) and (50) in terms of real quantities as $C = C_1 + iC_2$, $\hat{K}\cdot\tilde{P}\cdot\hat{K} = P_1 - iP_2$ with $C_2 > 0$ and $P_2 > 0$, we have

$$\eta^2 = \frac{C}{\hat{K}\cdot\tilde{P}\cdot\hat{K}} \approx \eta_m^2\eta_d^2 \approx \frac{C_1}{P_1}\left[1 + i\left(\frac{C_2}{C_1} + \frac{P_2}{P_1}\right)\right]. \quad (51)$$

See Reference 1, Equations (126)-(128) for values of η corresponding to an arbitrary direction of incidence on a slab distribution of finite thickness.

For the corresponding electromagnetic problems, we use the vector and dyadic analogs of (41)-(47) developed in Reference 2. In particular, we work with the present (47) in terms of vector functions $\underset{\sim}{\mathfrak{G}}$ and $\underset{\sim}{u}$. The resulting equation, say (47)′, determines η via

$$\underset{\sim}{u} = \int_c \tilde{g}(\hat{r},\hat{r}_c)\cdot\underset{\sim}{\mathfrak{G}}(\underset{\sim}{k}_c|\underset{\sim}{K})e^{i\underset{\sim}{k}_c\cdot\underset{\sim}{R}} \quad (52)$$

as in Reference 2, Equation (62) [henceforth (2:62) for brevity] in terms of the isolated dyadic scattering amplitude $\tilde{g}$. The representations for the bulk dyadics $\tilde{\epsilon}$ and $\tilde{\mu}$ and their connections with η^2, as developed in (2:36)-(2:52), show that if the particle is specified by one parameter either $\underset{\sim}{\epsilon}'$ or $\underset{\sim}{\mu}'$, then so is the distribution (i.e., the composite).

To derive (14) for $\tilde{\epsilon}$ or $\tilde{\mu}$, we consider the isolated scattering amplitude as a sum of an electric and a magnetic dipole. Correct to lowest order in k for the real and imaginary parts, we use[15]

$$\tilde{g}(\hat{r},\hat{k}) = \tilde{g}_e + \tilde{g}_m, \; \tilde{g}_e(\hat{r},\hat{k}) = (\tilde{I}-\hat{r}\hat{r})\cdot\tilde{a}_e\cdot(\tilde{I}-\hat{k}\hat{k}) \equiv \tilde{T}_r\cdot\tilde{a}_e\cdot\tilde{T}_k$$
$$\tilde{g}_m(\hat{r},\hat{k}) = -(\tilde{I}\times\hat{r})\cdot\tilde{a}_m\cdot(\tilde{I}\times\hat{k}) \equiv -\tilde{J}_r\cdot\tilde{a}_m\cdot\tilde{J}_k$$
$$\tilde{a} \approx -\tilde{a}'_1\cdot[\tilde{I}-\tilde{a}'_1(n-1)/n)] \qquad (53)$$

where $\tilde{a}'_1$ is the form in (48) with $\tilde{\delta} = \tilde{\epsilon}' - \tilde{I}$ for $\tilde{a}'_e$ and $\tilde{\mu}' - I$ for $\tilde{a}'_m$. Both operators $\tilde{T}_r = (I-\hat{r}\hat{r})$ and $\tilde{J}_r = \tilde{I}\times\hat{r} = \hat{r}\times\tilde{I}$ are planar dyadics transverse to $\hat{r}$. If $\tilde{\mu}' = \tilde{I}$, then from $\tilde{g} = \tilde{g}_e(\hat{r},\hat{R}) = \tilde{T}_r\cdot\tilde{a}_e\cdot\tilde{T}_R$ and $\underset{\sim}{\mathfrak{g}}(k\hat{R}) = \tilde{T}_R\cdot\underset{\sim}{A}$ we have $\underset{\sim}{u}_e = \tilde{T}_r\cdot\tilde{a}_e\cdot\tilde{h}_e\cdot\underset{\sim}{A}$ with $\tilde{h}_e = \tilde{I}h_0-\tilde{h}_a$ as the electric dipole propagator. Substituting into (47)$'$, we obtain the form $\tilde{M}\cdot A = 0$ with $|\tilde{M}| = |\tilde{T}_K[\eta^{-2}\tilde{I} - (\tilde{I}+\tilde{\Delta})^{-1}]| =$ with $\tilde{\Delta}$ as in (14) for two cases of β of (16) corresponding to $d_3^e = 6\pi$ and $d_2^e = 8$ (the dipole cases). Comparing with (8) for $\tilde{\mu} = \tilde{I}$, and using $\hat{K}\times\tilde{I}\times\hat{K} = \tilde{J}_K\cdot\tilde{J}_K = -\tilde{I}_K$ we identify $\tilde{\epsilon}$ as $\tilde{I} + \tilde{\Delta}(\tilde{\epsilon}')$. Similarl if $\tilde{\epsilon}' = \tilde{I}$, then from $\tilde{g} = \tilde{g}_m(\hat{r},\hat{R}) = -\tilde{J}_r\cdot\tilde{a}_m\cdot\tilde{J}_R$ and $\underset{\sim}{\mathfrak{g}}(k\hat{R}) = -\tilde{J}_R\cdot\underset{\sim}{A}'$, we have $\underset{\sim}{u}_m = -\tilde{J}_R\cdot\tilde{a}_m\cdot\tilde{h}_e\cdot\underset{\sim}{A}'$. Substituting into (47)$'$ we obtain $|\tilde{J}_K\cdot[-\tilde{I}\eta^{-2} + (\tilde{I}+\tilde{\Delta})^{-1}]\cdot\tilde{J}_K| = 0$, or equivalently $|\tilde{T}_K\eta^{-2} + \hat{K}\times(\tilde{I}+\tilde{\Delta})^{-1}\times\hat{K}| =$ by using $(\tilde{I}\times\hat{K})\cdot\underset{\sim}{V} = \hat{K}\times\underset{\sim}{V}$. Comparing with (8) for $\tilde{\epsilon} = \tilde{I}$ we identify $\tilde{\mu}$ as $\tilde{I} + \tilde{\Delta}(\tilde{\mu}')$. To the present order in k, we use (8) in terms of $\tilde{\mu} = I + \tilde{\Delta}(\tilde{\mu}')$ and $\tilde{\epsilon} = I + \tilde{\Delta}(\tilde{\epsilon}')$ to determine η^2. See Reference 2 for detailed applications and special cases.

ACKNOWLEDGEMENTS

Part of this work was done in the Mathematics Departments of The Weizmann Institute of Science in Israel and of Stanford University during sabbatical leave from the University of Illinois. I am grateful to the Departments for their hospitality, and to the John Simon Guggenheim Foundation and to the National Science Foundation for their support.

FOOTNOTES.

* Work supported in part by National Science Foundation Grant MCS-79-01718.

** Fellow of the John Simon Guggenheim Foundation, 1979-1980.

REFERENCES.

[1] V. Twersky, "Coherent Scalar Field in Pair-Correlated Random Distributions of Aligned Scatterers", J. Math. Phys. 18, 2468-2486 (1977).

[2] V. Twersky, "Coherent Electromagnetic Waves in Pair-Correlated Random Distributions of Aligned Scatterers", J. Math. Phys. 19, 215-230 (1978).

[3] J.C. Maxwell, A Treatise on Electricity and Magnetism, (Cambridge, 1873; Dover, N.Y., 1954); spheres are considered in Section 314, and slabs in Section 321.

[4] V. Twersky, "Acoustic Bulk Parameters in Distributions of Pair-Correlated Scatterers", J. Acoust. Soc. Am. 64, 1710-1719 (1978).

[5] V. Twersky, "Form and Intrinsic Birefringence", J. Opt. Soc. Am. 65, 239-245 (1975).

[6] V. Twersky, "Intrinsic, Shape, and Configurational Birefringence", J. Opt. Soc. Am. 69, 1199-1205 (1979).

[7] V. Twersky, "Propagation in Pair-Correlated Distributions of Small-Spaces Lossy Scatterers", J. Opt. Soc. Am. 69, 1567-1572, (1979).

[8] H. Reiss, H.L. Frisch, and J.L. Lebowitz, "Statistical mechanics of rigid spheres", J. Chem. Phys. 31, 379-380 (1959); E. Helfand, H.L. Frisch, and J.L. Lebowitz, "Theory of the two- and one-dimensional rigid sphere fluids", J. Chem Phys. 34, 1037-1042 (1961).

[9] Lord Rayleigh, "On the Influence of Obstacles Arranged in Rectangular Order upon the Properties of a Medium", Philos. Mag. 34, 481-501 (1892).

[10] O. Wiener, "Formdoppelbrechung bei Absorption", Kolloidchemische Beihefte 23, 189-198 (1926); "Die Theorie des Mischkorpers fur das Feld der stationaren Stromung", Sachische Akad. Wiss. 32, 507-604 (1912).

[11] M. Lax, "Multiple Scattering of Waves", Rev. Mod. Phys. 23, 287-310 (1951); "The Effective Field in Dense Systems", Phys. Rev. (2) 88, 621-629 (1952).

[12] J.B. Keller, "Wave Propagation in Random Media", Proc. Sympos. Appl. Math. Vol. 13, 227-246, Amer. Math. Soc., Providence, R.I. (1962); "Stochastic Equations and Wave Propagation in Random Media", Vol. 16, 145-170 (1964).

[13] V. Twersky, "On Propagation in Random Media of Discrete Scatterers", Proc. Sympos. Appl. Math. Vol. 16, 84-116, Amer. Math. Soc., Providence, R.I. (1964).

[14] Lord Rayleigh, "On the Incidence of Aerial and Electric Waves upon small Obstacles", Phil. Mag. 44, 28-52 (1897).

[15] V. Twersky, "Multiple Scattering of Electromagnetic Waves by Arbitrary Configurations", J. Math. Phys. 8, 589-610 (1967).

FREQUENCY DEPENDENT DIELECTRIC CONSTANTS OF DISCRETE RANDOM MEDIA

V.V. Varadan†, V.N. Bringi* and V.K. Varadan†
Wave Propagation Group
Boyd Laboratory
The Ohio State University, Columbus, Ohio 43210

Abstract

Numerical computations of the effective dielectric constant of discrete random media are presented as a function of frequency. Such media have a complex dielectric constant giving rise to absorption of a propagating wave both due to geometric dispersion or multiple scattering as well as absorption, if any, due to the viscosity of the particles and the matrix medium. We are concerned with the absorption due to multiple scattering. The scattering characteristics of the individual particles are described by a transition or T-matrix. The effects of two models of the pair correlation function which arises in the multiple scattering analysis are considered. We conclude that the well stirred approximation (WSA) is good for sparse concentrations and/or high frequencies whereas the Percus-Yevick approximation (P-YA) is preferred for higher concentrations.

Introduction

The study of the frequency dependence of the effective dielectric constant of statistically inhomogeneous media is important for practical applications such as geophysical exploration, artificial dielectrics etc. In such dielectrics a propagating electromagnetic wave undergoes dispersion and absorption. Some materials are naturally absorptive due to viscosity whereas inhomogeneous media exhibit absorption due to geometric dispersion or multiple scattering.

In this paper the effective, complex frequency dependent dielectric constant of a discrete random medium containing a distribution of aligned spheroidal dielectric scatterers in free space is calculated for different concentrations of the scatterers as well as for different material properties of the scatterers. We use a multiple scattering formalism analogous to that used by Twersky[1] but use the concept of a transition matrix or T-matrix to characterize the scattering from a single obstacle. All details of the geometry and material properties of the scatterer are contained in the T-matrix leaving the general formalism independent of the type of scatterer. Spherical statistics are used even though the scatterers may be non-spherical. Lax's[2] quasi-crystalline approximation (QCA) is used to truncate the heirarchy of equations that result when an ensemble average is performed on the multiply scattered field.

The resulting equations for the average field require a knowledge of the pair correlation function of the dielectric scatterers. In previous work[3,4,5], we assumed that the particles did not penetrate each other but were otherwise uncorrelated. Willis[6] has called this the well stirred approximation (WSA). However, the WSA lead to unphysical results for the absorption coefficient of the average medium for scatterer concentrations $c > 0.125$. In many artificial dielectrics, the scatterer concentration is often greater than 0.125. In this paper, we have also considered the Percus-Yevick[7] approximation (P-YA) to the pair correlation function. Wertheim[8] has provided a semi-analytical solution of the resulting integral equation for a system of hard spheres. Throop and Bearman[9] have provided tabulated values of the pair correlation function for different values of the concentration as a function of the inter particle distance. We have used these tabulated values in the numerical computations.

Calculations are presented for a system of polyethylene spheres and spheroids as well as ice particles for $0 < c < 0.26$ for several values of the non-dimensional wavenumber $ka = \frac{\omega a}{c}$ ranging from 0 to 5.0. ('a' is a characteristic dimension of the scatterer). Two types of results are presented. In the first instance the validity of the WSA and P-YA and their effect on the absorption coefficient is studied as a function of concentration and frequency. Secondly, the complex plane locus of the effective dielectric constant is plotted for the systems considered. For artificial dielectrics the locus deviates dramatically from the circular arc locus commonly noticed for ordinary solids and liquids that exhibit absorption due to viscosity.

Wave propagation in a discrete random medium

Consider N identical rotationally symmetric dielectric scatterers that are aligned but distributed randomly in free space (see Fig. 1). Let 0 be the origin of a coordinate system located outside the scatterers whose centers are denoted by $0_1, 0_2, 0_3 \ldots 0_N$. Monochromatic plane electromagnetic waves of frequency ω propagate along the symmetry axis of the scatterers which is taken to be the z-axis. Since the medium is isotropic about the z-axis there are no depolarization effects. The time dependence of the incident and hence the fields scattered by the individual scatterers is all of the form $\exp(-i\omega t)$ and this is suppressed in the equations that follow.

Let $\vec{E}_o(\vec{r})$ be the electric field arising from the incident plane wave and $\vec{E}_i^s(\vec{r})$ the field scattered by the i-th scatterer. The total field at a point $\vec{r}$ outside all the N scatterers, denoted by $\vec{E}(\vec{r})$ is given by

$$\vec{E}(\vec{r}) = \vec{E}^o(\vec{r}) + \sum_{i=1}^{N} \vec{E}_i^s(\vec{r}) \tag{1}$$

The field incident on or exciting the i-th scatterer is given by

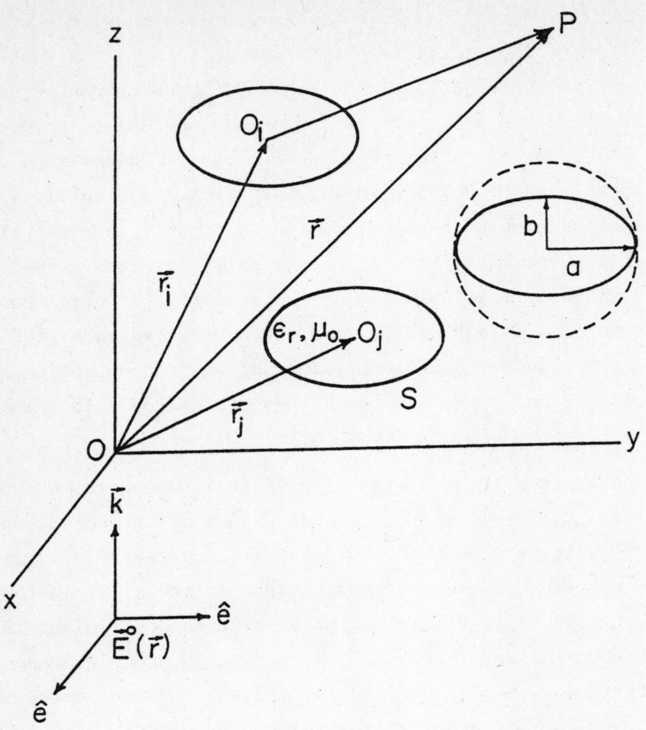

Figure 1. Scattering geometry

$$\vec{E}_i^e(\vec{r}) = \vec{E}^o(\vec{r}) + \sum_{j\neq i}^{N} \vec{E}_j^s(\vec{r}) \quad ; \ a \leq |\vec{r}-\vec{r}_i| < 2a \tag{2}$$

where 'a' is a typical dimension of the scatterer. From Eqs. (1) and (2) we note that

$$\vec{E}(\vec{r}) = \vec{E}_i^e(\vec{r}) + \vec{E}_i^s(\vec{r}) \tag{3}$$

We need an additional equation relating $\vec{E}_i^e$ and $\vec{E}_i^s$ in order to make the fields microscopically self-consistent.

Vector spherical functions are used to expand the exciting and scattered fields associated with each scatterer with respect to an origin at the center of that scatterer. Thus

$$\vec{E}_i^e(\vec{r}) = \sum_{\tau=1}^{2} \sum_{\ell=1}^{\infty} \sum_{m=0}^{\ell} \sum_{\sigma=1}^{2} A^i_{\tau\ell m\sigma} \mathrm{Re}\ \vec{\psi}_{\tau\ell m\sigma}(\vec{\rho}_i) \ ; \ a \leq |\vec{\rho}_i| < 2a \tag{4}$$

and

$$\vec{E}_i^s(\vec{r}) = \sum_{\tau=1}^{2} \sum_{\ell=1}^{m} \sum_{m=0}^{\ell} \sum_{\sigma=1}^{2} F^i_{\tau\ell m\sigma} \mathrm{Ou}\ \vec{\psi}_{\tau\ell m\sigma}(\vec{\rho}_i) \quad ; \ |\vec{\rho}_i| \geq 2a \tag{5}$$

where

$$\vec{\rho}_i = \vec{r}-\vec{r}_i \tag{6}$$

and the vector spherical functions are defined as

$$\left\{\begin{matrix} Ou \\ Re \end{matrix} \vec{\psi}_{1\ell m\sigma}(\vec{r})\right\} = \nabla x \left\{\vec{r} \begin{matrix} h_\ell(kr) \\ j_\ell(kr) \end{matrix}\right\} Y_{\ell m\sigma}(\theta\phi) \ ; \tag{7}$$

$$\left\{\begin{matrix} Ou \\ Re \end{matrix} \vec{\psi}_{2\ell m\sigma}(\vec{r})\right\} = \frac{1}{k} \nabla x \left\{\begin{matrix} Ou \\ Re \end{matrix} \vec{\psi}_{1\ell m\sigma}(\vec{r})\right\} \tag{8}$$

In Eqs. (7) and (8) j_ℓ and h_ℓ are the spherical Bessel and Hankel functions and the $Y_{\ell m\sigma}(\theta,\phi)$ are the normalized spherical harmonics defined with real angular functions. To make the notation more compact we introduce a super index 'n' to represent $\{\tau\ell m\sigma\}$ as follows

$$\begin{matrix} Ou \\ Re \end{matrix} \vec{\psi}_{\tau\ell m\sigma} = \begin{matrix} Ou \\ Re \end{matrix} \vec{\psi}_n$$

We observe that the coefficients of expansion A_n^i and F_n^i associated with the exciting and scattered fields depend on the position of all N scatterers. Further, since Eq. (3) is satisfied, we can relate the two sets of coefficients by means of the T-matrix as defined by Waterman[10]. We have

$$F_n^i = \sum_m T_{nm}^i A_n^i \tag{9}$$

The T-matrix depends on the frequency of the wave exciting the scatterer as well as its geometry and material properties.

If Eqs. (4), (5) and (9) are substituted in Eq. (2), we would need the translation addition theorems for the vector spherical functions in order that we may refer all expansions in Eq. (2) to a common origin. In compact form

$$Ou\ \vec{\psi}_n(\vec{\rho}_j) = \begin{cases} \sum_{n'} \sigma_{nn'}(\vec{r}_{ij})\ Re\ \vec{\psi}_{n'}(\vec{\rho}_i) \ ; & |\vec{r}_{ij}| > |\vec{\rho}_i| \\ \sum_{n} R_{nn'}(\vec{r}_{ij})\ Ou\ \vec{\psi}_{n'}(\vec{\rho}_i) \ ; & |\vec{\rho}_i| > |\vec{r}_{ij}| \end{cases} \tag{10}$$

where $\vec{r}_{ij} = \vec{r}_i - \vec{r}_j$ is the vector connecting 0_j to 0_i and $\sigma_{nn'}$ is the translation matrix for the vector functions and $R_{nn'}$ is the same as $\sigma_{nn'}$ with the spherical Hankel functions in $\sigma_{nn'}$ replaced by spherical Bessel functions. Detailed expressions for the matrices are given by Boström[11].

The incident electric field $\vec{E}^o$ can be expanded with respect to an origin at 0_i as

$$\vec{E}^o(\vec{r}) = e^{ikz} = e^{ik\hat{z}\cdot\vec{r}_i} \sum_n a_n\ Re\ \vec{\psi}_n(\rho_i) \tag{11}$$

where the coefficients a_n are known (see for example Morse and Feshbach[12]). We observe that for a plane wave propagating in the z-direction the only non-zero values of $a_n = a_{\tau\ell m\sigma}$ are $a_{1\ell 12}$ and $a_{2\ell 11}$; $\ell\varepsilon[1,\infty]$, all other coefficients being zero.

Using Eqs. (4),(5),(9)-(11) in Eq. (2), using the orthogonality of the vector spherical functions we obtain

$$A_n^i = e^{ik\hat{z}\cdot\vec{r}_i}\, a_n + \sum_{j\neq i}^{N} \sum_{n'} \sum_{n''} \sigma_{n'n}(\vec{r}_{ij})\, T^j_{n'n}\, A^j_{n''} \tag{12}$$

Equation (12) is a set of coupled algebraic equations for the exciting field coefficients associated with each scatterer. If the number of scatterers N is finite and the position of all the scatterers is known, then Eq. (12) can be solved in principle. But we wish to consider the case $N\to\infty$, $V\to\infty$ such that $N/V = n_o$ is a finite number density. Since N is large, we are only interested in the configurational average of Eq. (12) over the positions of all particles

The coherent field

The average of Eq. (12) over the position of all scatterers (the average exciting field) is the same as the ensemble average, where the ensemble is composed of different possible configurations of the scatterers. Equation (12) is averaged over the position of all particles except the i-th. But the right hand side of Eq. (12) explicitly depends on the position of the j-th particle. Hence we must specify the two particle joint probability density $P(\vec{r}_j|\vec{r}_i)$ Further, we assume that all scatterers are identical, so that

$$\left\langle A_n^i \right\rangle_i = e^{ik\hat{z}\cdot\vec{r}_i} a_n + (N-1) \sum_{n'} \sum_{n''} T_{n'n''} \int_V P(\vec{r}_j|\vec{r}_i)\, \sigma_{n'n}(r_{ij}) \left\langle A_n^j \right\rangle_{ij} d\vec{r}_j \tag{13}$$

We note that the average exciting field with one scatterer held fixed is given in terms of the average with two scatterers held fixed, leading to a heirarchy that requires knowledge of higher order probability densities. It has been customary to truncate the heirarchy by invoking the 'quasi crystalline approximation' (QCA) first introduced by Lax[2]. According to this approximation

$$\left\langle A^j_{n''} \right\rangle_{ij} \simeq \left\langle A^j_{n''} \right\rangle_j \tag{14}$$

Specifically the QCA neglects multiple scattering between pairs of scatterers. Improvements to the QCA have been suggested by Twersky[1] and in previous work by us[5].

The joint probability density is defined as

$$P(\vec{r}_j|\vec{r}_i) = \begin{cases} \frac{1}{V}\, g(|\vec{r}_j-\vec{r}_i|) & ; \ |\vec{r}_j-\vec{r}_i| \geq 2a \\ 0 & ; \ |\vec{r}_j-\vec{r}_i| < 2a \end{cases} \tag{15}$$

Equation (15) implies that the particles are hard (no-interpenetration) and the excluded volume is a sphere of radius 'a' although the particles themselves may be non-spherical. The function $g(|\vec{r}_i-\vec{r}_j|)$ is called the pair correlation function and depends only on $|\vec{r}_j-\vec{r}_i|$ due to translational invariance of the system under consideration.

We assume that the coherent field propagates in the same direction as the incident field with a new, effective wavenumber K that is complex and frequency dependent. Hence

$$\left\langle \vec{E}_i^{e}(\vec{r}) \right\rangle_i = A\, e^{iK\hat{z}\cdot\vec{r}} \tag{16}$$

where A is the amplitude of the coherent wave. Thus the average exciting field coefficient

$$\left\langle A_n^i \right\rangle_i = \left\langle A_{\tau\ell m\sigma}^i \right\rangle_i = e^{iK\hat{z}\cdot\vec{r}_i}\, X_{\tau\ell m\sigma}\, \delta_{m1}\, [\delta_{\tau 1}\delta_{\sigma 2} + \delta_{\tau 2}\delta_{\sigma 1}] \tag{17}$$

The Kronecker deltas in Eq. (17) indicate that only the azimuthal index m=1 contributes, since the coherent wave propagates in the z-direction and those in the square bracket indicate that there is no depolarization.

Equations (14)-(17) are substituted in Eq. (13). Since the T-matrix of a rotational symmetric scatterer is block diagonal in the azimuthal index (see Waterman[10]) and the coherent field propagates in the z-direction, the sums associated with the azimuthal indices of the super indices n' and n" in Eq. (13) are removed. Further, as shown in previous work by us[5] as well as Twersky[1], the extinction theorem can be used to cancel the incident wave term in Eq. (13) with the contribution of the integral at infinity. Finally Eq. (13) can be written in the form

$$X_{\tau\ell m\sigma}\, \delta_{m1}\, [\delta_{\tau 1}\delta_{\sigma 2} + \delta_{\tau 2}\delta_{\sigma 1}] = \frac{N-1}{V} \sum_{\tau'\ell'\sigma'} \sum_{\tau''\ell''\sigma''} T_{\tau'\ell' 1\sigma',\tau''\ell'' 1\sigma''} \cdot$$

$$\sum_{\lambda=|\ell-\ell'|}^{\ell+\ell'} I(K,k,c,\lambda)\, X_{\tau''\ell'' 1\sigma''}\, D^{(\lambda)}_{\tau'\ell' 1\sigma',\tau\ell 1\sigma} \tag{18}$$

where

$$I(K,k,c,\lambda) = \frac{6c}{(Ka)^2-(ka)^2}\,[2ka j_\lambda(2Ka) h'_\lambda(2ha) - 2Ka h_\lambda(2ka) j'_\lambda(2Ka)] +$$

$$24c \int_1^\infty x^2 [g(x)-1]\, h_\lambda(kx) j_\lambda(Kx) dx \tag{19}$$

and

$$D^{(\lambda)}_{\tau'\ell'1\sigma',\tau\ell 1\sigma} = i^{\ell-\ell'+\lambda} \left[\frac{(2\lambda+1)(2\ell+1)}{2\ell(\ell+1)}\right] \left[\frac{\ell'(\ell'+1)}{\ell(\ell+1)}\right]^{1/2} \begin{bmatrix} \ell & \ell' & \lambda \\ 1 & -1 & 0 \end{bmatrix} \cdot$$

$$\left\{ \begin{bmatrix} \ell & \ell' & \lambda \\ 0 & 0 & 0 \end{bmatrix} [\ell'(\ell'+1) + \ell(\ell+1) - \lambda(\lambda+1)]\, \delta_{\tau\tau'}\, \delta_{\sigma\sigma'} + i \begin{bmatrix} \ell' & \ell & \lambda-1 \\ 0 & 0 & 0 \end{bmatrix} \cdot \right.$$

$$\left. (\lambda^2-(\ell'-\ell)^2) \quad (\ell'+\ell+1)^2-\lambda^2)^{1/2} \quad (1-\delta_{\tau\tau'})\, (\delta_{\sigma'1}\delta_{\sigma 2} - \delta_{\sigma'2}\delta_{\sigma 1}) \right\} \qquad (20)$$

In Eq. (19) $c = \frac{4\pi}{3} n_o a^3$ is the effective spherical concentration of the particles and in Eq. (20) $\begin{bmatrix} j_1 & j_2 & j_3 \\ m_1 & m_2 & m_3 \end{bmatrix}$ is the Wigner 3-j symbol.

If the integral in Eq. (19) can be evaluated for suitable models of the pair correlation function, then Eq. (18) is a set of coupled, homogeneous, algebraic equations for the coherent field expansion coefficients. For a non-trivial solution, the determinant of the coefficient matrix must vanish. This yields the required dispersion equation for the effective or average medium. In general the system of equations can be solved only numerically to yield the effective wave number K as a function of frequency ($k=\omega/c$) which is complex ($K = K_1+iK_2$). The real part K_1 yields the phase velocity in the medium and the imaginary part K_2 leads to damping of a propagating wave due to geometric dispersion as well as real losses if any, associated with the discrete particles. We now proceed to consider the evaluation of the integral in Eq. (19).

The Percus-Yevick pair correlation function

The pair correlation function for an ensemble of particles depends on the nature and range of the interparticle forces. The average of several measurements of a statistical variable that characterizes an ensemble will depend on the pair correlation function. As we have seen, the coherent or average electric field in an ensemble of dielectric scatterers depends on the pair correlation function (Eqs. (18)-(19)). To obtain expressions for the pair correlation function, one needs a description of the interparticle forces. In our case we assume that the dielectric scatterers behave like effective hard spheres (where the radius 'a' is that of the sphere circumscribing the scatterer). Percus and Yevick[7] have obtained an approximate integral equation for the pair correlation function of a classical fluid in equilibrium. Wertheim[8] has obtained a series solution of the integral equation for an ensemble of hard spheres. The statistics of the fluid are then same as those of the ensemble of discrete hard particles that we are considering.

Although integral expressions for the correlation functions also result in a heirarchy, Percus and Yevick have truncated the heirarchy by making certain approximations that result in a self-consistent relation between the pair correlation function $g(x)$ and the direct correlation function $C(x)$. The direct correlation function may be interpreted as the correlation function resulting from an 'external potential' that produces a simultaneous density fluctuation at a point and the external potential is taken to be the potential seen by a particle given that there is a particle fixed at another site. Fisher[13] comments that the Percus-Yevick approximation is a strong statement of the extremely short range nature of the direct correlation function. The integral equation has the form

$$\tau(x) = 1 + n_o \int_{x<2a} \tau(x')dx' - n_o \int_{\substack{x'<2a \\ |x-x'|>2a}} \tau(x')\tau(x-x')dx' \tag{21}$$

where

$$\begin{aligned} \tau(x) &= g(x) \ ; \ x > 2a \\ g(x) &= 0 \ ; \ x < 2a \\ \tau(x) &= -C(x) \ ; \ x < 2a \\ C(x) &= 0 \ ; \ x > 2a \end{aligned} \tag{22}$$

Wertheim[8] has solved the integral equation by Laplace transformation that results in an analytic expression for C(n) in the form

$$C(x) = -(1-\eta)^{-4} \left[(1+2\eta)^2 - 6\eta(1+\tfrac{1}{2}\eta)^2 x + \eta(1+2\eta)^2 x^3/2\right] \ ; \ \eta = c/8 \tag{23}$$

where 'c' is the effective spherical concentration of the particles. The Percus-Yevick approximation fails as the concentration approaches the close packing factor for spheres and is expected to be good for $c < 0.3$ or 0.4.

Equation (23) can be substituted back into Eq. (21) to yield a series solution for g(x) in the form[8]

$$g(x) = \sum_{n=1}^{\infty} g_n(x) \tag{24}$$

where

$$g_n(x) = \frac{1}{24\eta x i} \int e^{t(x-n)} [L(t) \mid S(t)]^n \, tdt \tag{25}$$

where

$$S(t) = (1-\eta^2)t^3 + 6\eta(1-\eta)t^2 + 18\eta^2 t - 12\eta(1+2\eta) \tag{26}$$

and

$$L(t) = 12\eta \left[(1+\eta 2)t + (1+2\eta)\right]. \tag{27}$$

Throop and Bearman[9] have tabulated g(x) as a function of x for values of $\eta = c/8$. A few representative plots of the pair correlation function are shown in Fig. 2. These tabulated values were used in evaluating the integral in Eq. (19).

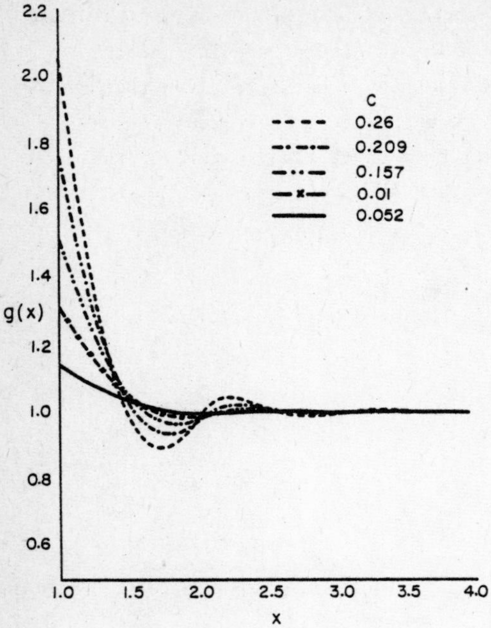

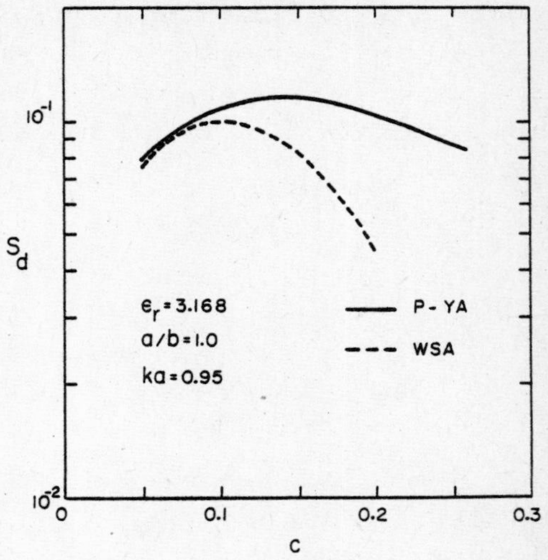

Figure 2. The Percus-Yevick pair correlation function g(x)

Figure 3. Coherent attenuation vs. concentration for spherical ice particles

Comparison of WSA and P-YA

The homogeneous system of algebraic equations for the effective exciting field were solved numerically for two different models of the correlation integral I appearing in Eq. (18). In eq. (19) if the second term is set equal to zero, we just have a system of uncorrelated hard particles. This is what we have referred to as the well stirred approximation (WSA)[6] earlier. Computations were also performed by numerically evaluating the integral in eq. (19) by using the tabulated values of the Percus-Yevick approximation to the pair correlation functions provided by Throop and Beerman[9].

In Fig. 3, the specific damping $S_d = 4\Pi K_2/K_1$ is plotted as a function of concentratio for a random distribution of numerical ice particles ($\epsilon_r = 3.168$) in free space at ka = 0.55. The WSA agrees with the P-YA solution only up to concentrations $C \backsim 0.075$ and then there is a marked difference and the WSA fails completely at $C > 0.125$ leading to unphysical results. In Fig. 4, the calculations are repeated

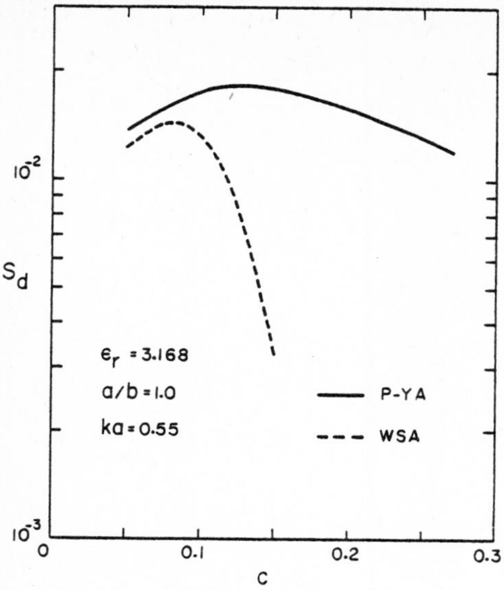

Figure 4. Coherent attenuation vs. concentration for spherical ice particles

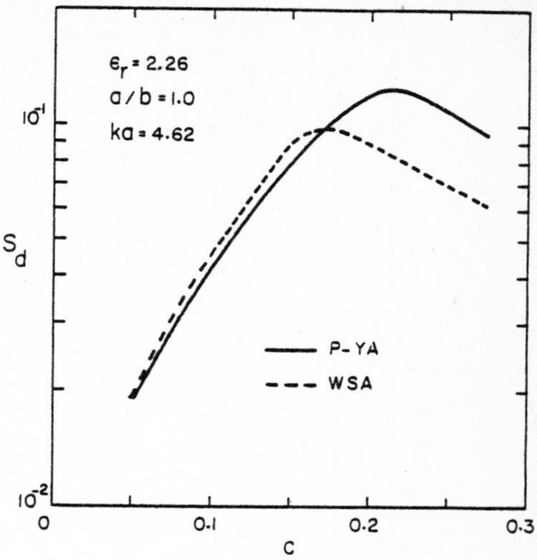

Figure 5. Coherent attenuation vs. concentration for polyethylene spheres

for the same system at a higher value of ka = 0.95. Here the WSA agrees with P-YA up to C = 0.1 and in Fig. 5 similar calculations were performed for polyethylene spheres ($\varepsilon_r = 2.26$) at ka = 4.62. For this case WSA and P-YA results agree up to C = 0.15. From these results it would appear that although the WSA is very poor at higher scatterer concentrations, the results improve dramatically at higher values of ka, yielding reasonably good results for higher concentrations. The natural explanation is that at higher values of ka, multiple scattering effects between pairs of particles become smaller and thus pair correlation effects are not significant and the QCA also becomes more exact. But for arbitrary concentration and frequency it is safer to use the Percus-Yevick approximation.

The effective dielectric constant

Once the effective complex wavenumber K has been computed by solving Eq. (18) numerically, we can proceed further and evaluate the effective dielectric constant of the medium which is also complex and frequency dependent. In the usual way, the dielectric constant $\varepsilon_r^*(\omega)$ of the random medium is defined as

$$\varepsilon_r^*(\omega) = \frac{K^2}{k^2} = \varepsilon_1(\omega) + i\varepsilon_2(\omega)$$

where ε_1 and ε_2 are the real and imaginary parts of the dielectric constant and the

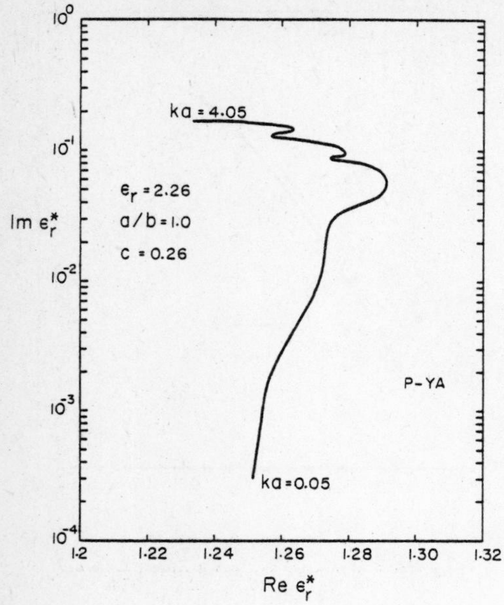

Figure 6. Complex plane locus of the effective dielectric constant for a system of polyethylene spheres

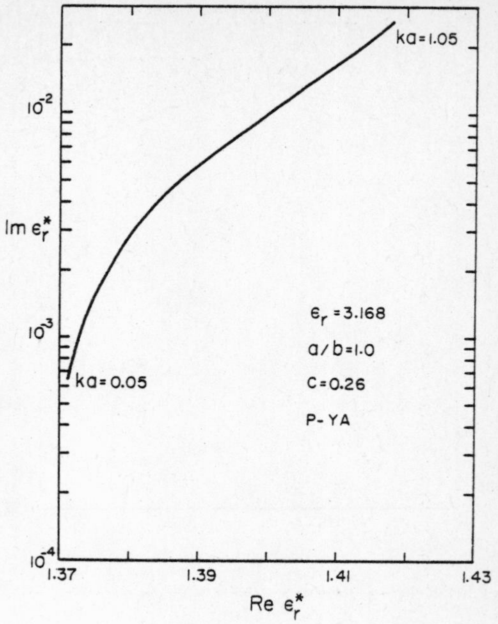

Figure 7. Complex plane locus of the effective dielectric constant for a system of spherical ice particles

subscript on ε_r^* denotes 'relative to the matrix medium'. The real part ε_1 is related to the refractive index and phase velocity in the artifical medium and the imaginary part ε_2 accounts for the damping in the medium. In real materials, the damping is intrinsic to the system and is due to macroscopic viscosity of the dielectric. For the artificial or effective medium under consideration, in addition to natural losses there is damping due to geometric dispersion or scattering.

Cole and Cole[14] have given a convenient representation of the dispersion and absorption in a dielectric by means of an Argand diagram or a plot in the complex ε-plane of ε_1 versus ε_2, each point of the plot being characteristic of a particular frequency. For many types of loss mechanisms, the locus of the points is a semi circle with its center on the real axis or a circular arc. In Ref. 14, the complex dielectric constant of several liquids and solids is plotted conforming to the circular arc.

In the present case the complex dielectric constant $\varepsilon(\omega)$ corresponding to the effective wavenumber K of the effective medium is studied for several values of the frequency. Overall results show a dramatic deviation from the circular arc locus. This is to be expected since the medium is artificial.

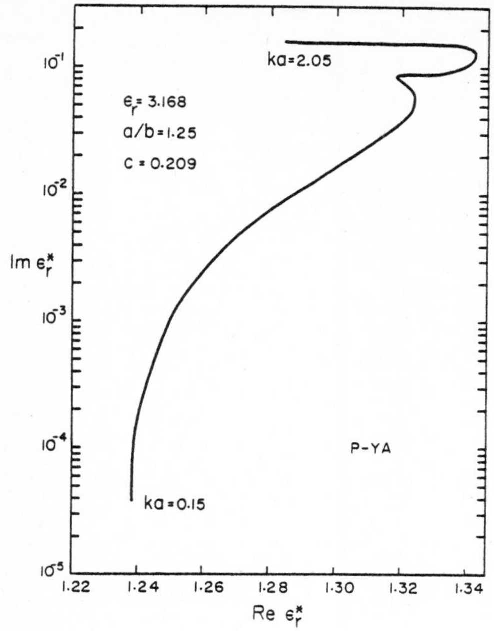

Figure 8. Complex plane locus of the effective dielectric constant for a system of oblate spheroidal ice particles

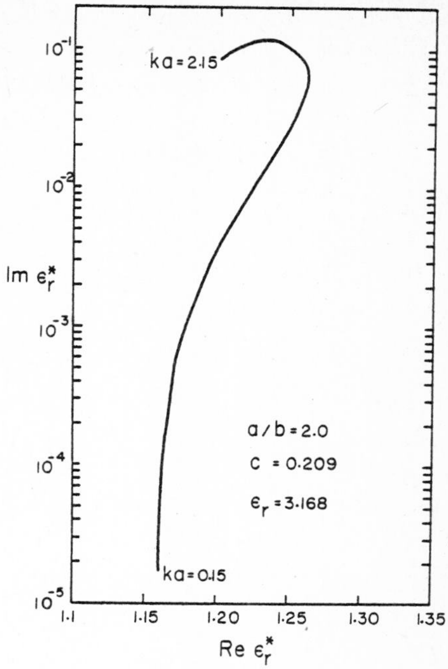

Figure 9. Complex plane locus of the effective dielectric constant for a system of oblate spheroidal ice particles

In Fig. 6 the complex plane locus of the relative dielectric constant of a random distribution of polyethylene spheres in free space is presented at a concentration of 26%. The calculations were done using the Percus-Yevick approximation (P-YA) for the pair correlation function from ka = 0.05 to 4.05. As can be seen, the figure bears no resemblance to a circular arc locus. By extrapolating the locus at the low value of ka, one can find the intercept on the Re ε_r^* axis which is equal to the static dielectric constant of the effective medium. Since the dielectric constant of the spherical particles is assumed to be real, the effective medium shows no absorption at low frequencies. The static dielectric constant thus obtained will correspond to the one that can be obtained from mixture theory. In real media displaying a circular arc locus the high frequency value of ε^* also intercepts the real axis and this yields the optical limit or ε_∞ for the material. In our case, it is not at all clear at what value of ka, if at all, the locus will intercept the real axes.

In Figs. 7,8 and 9 the complex plane locus of the effective dielectric constant of spherical and oblate spheroidal ice particles is presented where 'a' and 'b' are the semi major and semi minor axes respectively. They all show marked deviation from the circular arc locus and it is unclear what ε_∞ will be for these effective media.

At the present time there are no experimental results available to verify these calculations. The practical applications of these computations are many. Such calculations will provide reasonable estimates of the frequency dependence dielectric constant as a function of particle concentration, size and shape for inhomogeneous media.

Acknowledgements

This work was supported in part by NOAA under grant No.: 04-78-Bol-21, NRL(USRD) contract No: N00014-80-C-0483, and NRL (Washington) contract No: N00014-80-C-0835. The use of the Instructional and Research Computer Center at the Ohio State University is gratefully acknowledged.

† Department of Engineering Mechanics

* Department of Electrical Engineering and now at Colorado State University, Fort Collins, Colorado.

References

1. V. Twersky, J. Math Phys. 18, 2468 (1977), and 19, 215 (1978).

2. M. Lax, Phys. Rev. 85, 621 (1952).

3. V.K. Varadan, V.N. Bringi, and V.V. Varadan, Phys. Rev. D 19, 2480 (1979).

4. V.V. Varadan and V.K. Varadan, Phys. Rev. D 21, 388 (1980).

5. V.N. Bringi, T.A. Seliga, V.K. Varadan and V.V. Varadan, 'Bulk propagation characteristics of discrete random media' in Multiple Scattering of Waves in Random Media (Edited by P.L. Chow, W.E. Kohler and G. Papanicolaou), North-Holland Publishing Company, Amsterdam (1981).

6. D.R.S. Talbot and J.R.Willis, Proc. Roy. Soc. Lond. A 370, 351 (1980).

7. J.K. Percus and G.J. Yevick, Phys. Rev. 110, 1 (1958).

8. M.S. Wertheim, Phys. Rev. Letters 10, 321 (1963).

9. G.J. Throop and R.J. Bearman, J.Chem. Phys. 42, 2408 (1965).

10. P.C. Waterman, Phys. Rev. D. 3, 825 (1971).

11. A. Bostrom, J. Acoust. Soc. Am. 67, 399 (1980).

12. P.M. Morse and H. Reshbach, Methods of Theoretical Physics, Vol II., McGraw Hill Book Company Inc., New York (1953).

13. I.Z. Fisher, Statistical Theory of Liquids, The University of Chicago Press, Chicago, 309 (1965).

14. K.S. Cole and R.H. Cole, J. Chem. Phys. 9, 341 (1941).

A VARIATIONAL METHOD TO FIND EFFECTIVE COEFFICIENTS FOR PERIODIC MEDIA. A COMPARISON WITH STANDARD HOMOGENIZATION

Michael Vogelius*
Courant Institute of Mathematical Sciences
New York University
New York, N. Y. 10012

1. Introduction

Elliptic boundary value problems with rapidly varying coefficients (i.e. coefficients varying significantly on a small scale) naturally occur in dealing with problems involving composite materials. A problem of utmost practical importance is then to find an effective constant coefficient operator (or generally one with slowly varying coefficients) to replace the rapid variations. In the literature this process is often denoted homogenization and is traditionally performed by means of some sort of multiple scale asymptotic expansion.

As we shall point out later techniques of asymptotic expansion have certain deficiencies in connection with the determination of boundary layers and solutions containing singularities.

Accurate knowledge of the boundary layers or singularities is important for many practical applications, such as predicting whether a composite will delaminate or cracks occur.

In this paper we shall review an alternate method for homogenization which is much better suited to determine the boundary layers or the solution near a singularity. This method is based on a combination of an asymptotic expansion and a variational principle. We compare this method to the more standard both theoretically and computationally.

As our model problem we consider the following of diffusion in a periodic structure.

$$(1)\qquad -\nabla_x \cdot (\underline{\underline{A}}(x/\varepsilon)\ \nabla_x u^\varepsilon)(x) + b(x/\varepsilon)u^\varepsilon(x) = f(x) \text{ in } \Omega$$

$$u^\varepsilon(x) = 0 \quad \text{on } \partial\Omega.$$

Ω is a bounded domain in $\mathbb{R}^n$ with a Lipschitz boundary. The components of $\underline{\underline{A}}$ and the single function b are periodic elements of $L^\infty(\mathbb{R}^n)$ with a period = 1. $\underline{\underline{A}}$ is uniformly positive definite and b is nonnegative.

* This work was supported by the Army Research Office under Contract No. DAAG29-78-G-0177.

Since both $\underline{\underline{A}}$ and b for convenience are taken to be functions only of the "fast" variable x/ε and not explicitly of the slow variable x the result of homogenization shall be a constant coefficient operator.

2. The "standard" approach.

Introduce the new independent variable $y = x/\varepsilon$. The problem (1) then "transforms" into

(2) $$-(\nabla_x + 1/\varepsilon\,\nabla_y)\cdot(\underline{\underline{A}}(y)(\nabla_x + 1/\varepsilon\nabla_y)U^\varepsilon(x,y)) + b(y)U^\varepsilon(x,y) = f(x)$$

in $\Omega \times [0,1]^n$

U^ε is periodic in y with period $= 1$, and

$U^\varepsilon(x,y) = 0 \quad \forall (x,y) \in \partial\Omega \times [0,1]^n$.

To be more exact, if $U^\varepsilon(x,y)$ is a solution to (2) then $u^\varepsilon(x) = U^\varepsilon(x,x/\varepsilon)$ solves (1).

Formally one can now expand $U^\varepsilon(x,y)$ in powers of ε. By matching terms of same order of ε in the equation (2) one obtains

(3) $$u^\varepsilon(x) = U^\varepsilon(x,x/\varepsilon) = u(x) + \varepsilon\underline{\chi}(x/\varepsilon)\cdot\nabla_x u(x) + \dots ,$$

where $\underline{\chi}(y) = (\chi_1(y),\dots,\chi_n(y))$ is a periodic solution to

$$-\nabla_y\cdot(\underline{\underline{A}}(y)\ \nabla_y\underline{\chi}) = \nabla_y\cdot\underline{\underline{A}}(y) ,$$

and the function u solves

(4) $$-\nabla_x\cdot(\underline{\underline{\mathfrak{a}}}\nabla_x u(x)) + \bar{b}u(x) = f(x) \text{ in } \Omega$$
$$u(x) = 0 \text{ on } \partial\Omega ,$$

with $\underline{\underline{\mathfrak{a}}} = \overline{\underline{\underline{A}}(y) + \underline{\underline{A}}(y)\nabla_y\underline{\chi}(y)}$.

Here we have adopted the notation $\overline{\quad}$ for averages.

The reference [3] contains a very rigorous analysis of this approach particularly concerning convergence properties as $\varepsilon \to 0$.

Let $W^{s,p}(\Omega)$ denote the Sobolev space of order $s > 0$ based on $L^p(\Omega)$ (cf. [1]). For the particular case $p = 2$ we shall use the more standard notation $H^s(\Omega)$ instead of $W^{s,2}(\Omega)$. The norm on $H^s(\Omega)$ is denoted $\|\cdot\|_s$. By $\overset{\circ}{H}{}^1(\Omega)$ we understand the set of functions $u \in H^1(\Omega)$ such that $u = 0$ on $\partial\Omega$.

One can show that $u^\varepsilon(x) - u(x) \to 0$ (weakly) in $H^1(\Omega)$ and therefore strongly in $H^s(\Omega)$ for any $s < 1$. It is however easy to see that $u^\varepsilon(x) - u(x)$ does not in general converge strongly in $H^1(\Omega)$.

In order to obtain strong H^1-convergence (which is the equivalent of L^2-convergence of stresses or strains for problems of elasticity) we have to include the next term in the asymptotic expansion. For specific information concerning the boundary layers it is important to satisfy the boundary conditions exactly. This can within the present framework most naturally be accomplished by introducing a cut-off function $m^{\varepsilon}(x) = m(d(x)/\varepsilon)$, where $m \in W^{1,\infty}([0,\infty[)$ satisfies $m(0) = 0$ and $m(t) = 1$ for $t \geq 1$ and $d(x) = \mathrm{dist}(x,\partial\Omega)$. The exact formulation of a convergence result is as follows.

Theorem 2.1. Assume that $\chi_j \in W^{1,\infty}([0,1]^n)$, $1 \leq j \leq n$. If $u \in H^2(\Omega) \cap W^{1,p}(\Omega)$ then there exists C such that

(5) $$\| u^{\varepsilon}(x) - (u(x) + \varepsilon \sum_{j=1}^{n} \chi_j(x/\varepsilon) m^{\varepsilon}(x) \frac{\partial}{\partial x_j} u(x)) \|_1 \leq C\varepsilon^{\nu}$$

for any ε in $]0,1]$, where $\nu = \max\{(1-1/p)/2,\ 1/4\}$.

Some clear disadvantages of this approach and the result in this theorem are

(a) The cut-off function m^{ε} can be chosen rather arbitrarily. This leaves totally open the very difficult question of how to select it optimally in order to obtain a smallest possible error.

(b) If the right hand side f is not in $L^2(\Omega)$ so that $u \notin H^2(\Omega)$, then in general the expression

(6) $$u(x) + \varepsilon \sum_{j=1}^{n} \chi_j(x/\varepsilon)\ m^{\varepsilon}(x) \frac{\partial}{\partial x_j} u(x)$$

does not make sense in $H^1(\Omega)$, and standard homogenization thus fails to produce a family s^{ε} such that $\| \mathbf{u}^{\varepsilon} - \mathbf{s}^{\varepsilon} \|_1 \to 0$.

(c) If the boundary $\partial\Omega$ is not sufficiently smooth that $u \in H^2(\Omega)$ then this result does not apply. The expression (6) still makes sense in $H^1(\Omega)$ provided f is at least in $L^2(\Omega)$ (and for an appropriately chosen cut-off function m^{ε}), but a separate analysis is needed to see if we recover the convergence in $H^1(\Omega)$ and at what rate. An analysis of this sort is to the author's best knowledge not found in the literature.

(d) The estimate (5) is asymptotic in ε. As shall be clear from numerical evidence later on this reflects a deficiency in the very approach. The formula (6) does not in general perform well for large to moderate size ε .

(e) From a computational point of view (6) is not very natural either. If u is approximated by a Finite Element Method as u^{Δ} , then for the above expression to be in $H^1(\Omega)$ requires that $\frac{\partial}{\partial x_j} u^{\Delta} \in H^1(\Omega)$, $1 \leq j \leq n$.
This means we have to use C^1-elements or alternatively treat (4) by a so called mixed method.

In the following section we shall introduce the variationally based method and review some of its basic properties. In short: this alternate method has the same approximation properties as (6) for smooth data and small ε, and at the same time all the problems we mentioned above do no longer occur. The review will be very brief, for more details see [5].

3. The variationally based approach.

As before we introduce the new independent variable $\underline{y} = x/\varepsilon$. The corresponding equation (2) has the following natural variational formulation.

(7) $$B_\varepsilon(U^\varepsilon, V) = \iint_{\Omega \times [0,1]^n} f(x) V(x,\underline{y})\, dx\, d\underline{y}$$

$\forall\, V \in H^1(\Omega \times [0,1]^n)$, with the properties that

V is periodic (period = 1) in $\underline{y}$

$V = 0$ on $\partial\Omega \times [0,1]^n$.

$B_\varepsilon(U,V)$ here denotes the bilinear form

$$\iint_{\Omega \times [0,1]^n} (\underline{\underline{A}}(y)(\nabla_x + 1/\varepsilon\, \nabla_y) U(x,\underline{y}) (\nabla_x + 1/\varepsilon\, \nabla_y) V(x,\underline{y}) + b(y) U(x,\underline{y}) V(x,\underline{y}))\, dx\, d\underline{y}\,.$$

We intend to utilize part (but not all) of the information contained in the asymptotic expansion (3), namely the fact that

$$U^\varepsilon(x,y) = u_0^\varepsilon(x) + \varepsilon \sum_{j=1}^{n} \chi_j(y)\, u_j^\varepsilon(x) + \ldots$$

We do this by restricting the U^ε and V appearing in (7) to elements of the space

(8) $$\{v_0(x) + \varepsilon \sum_{j=1}^{n} \chi_j(y) v_j(x) \mid v_0, \{v_j\}_1^n \subseteq \overset{0}{H}{}^1(\Omega)\}$$

As a result we obtain the following system

(9) $$-\nabla_x \begin{Bmatrix} \overline{\underline{\underline{A}}} & \varepsilon\overline{\underline{\underline{A}}\chi_j} \\ \varepsilon\overline{\underline{\underline{A}}\chi_k} & \varepsilon^2\overline{\underline{\underline{A}}\chi_k\chi_j} \end{Bmatrix} \nabla_x \begin{Bmatrix} u_0^\varepsilon \\ \{u_j^\varepsilon\} \end{Bmatrix} + \begin{Bmatrix} 0 & -\nabla_x \cdot \overline{\underline{\underline{A}}\nabla_y\chi_j} \\ \overline{\underline{\underline{A}}\nabla_y\chi_k} \cdot \nabla_x & \overline{\underline{\underline{A}}\nabla_y\chi_k\nabla_y\chi_j} \end{Bmatrix} \begin{Bmatrix} u_0^\varepsilon \\ \{u_j^\varepsilon\} \end{Bmatrix}$$

$$+ \begin{Bmatrix} \overline{b} & \varepsilon\overline{b\chi_j} \\ \varepsilon\overline{b\chi_k} & \varepsilon^2\overline{b\chi_k\chi_j} + \varepsilon\overline{\underline{\underline{A}}(\nabla_y\chi_k\chi_j - \chi_k\nabla_y\chi_j)} \end{Bmatrix} \begin{Bmatrix} u_0^\varepsilon \\ \{u_j^\varepsilon\} \end{Bmatrix} = \begin{Bmatrix} f \\ \{\varepsilon\overline{\chi}_k f\} \end{Bmatrix}$$

with the boundary conditions that

$$u_0^\varepsilon \ , \ \{u_j^\varepsilon\} = 0 \text{ on } \partial\Omega \ .$$

Note. In order for the system (9) to be a positive definite elliptic system it is necessary and sufficient that the functions $1 \cup \{\chi_j\}_{j=1}^n$ are linearly independent.

In case they are not linearly independent the problem actually simplifies:

"Pick $1 \cup \{\chi_{j_k}\}_{k=1}^m (m \leq n)$ to be a maximally linearly independent subset of $1 \cup \{\chi_j\}_{j=1}^n$ and instead of (8) use the space

$$\{v_0(x) + \varepsilon \sum_{k=1}^m \chi_{j_k}(y) v_k(x) \ | \ v_0, \{v_k\} \subseteq \overset{\circ}{H}{}^1(\Omega)\}$$

as test and trial functions for the variational procedure."

Whether the set $1 \cup \{\chi_j\}_{j=1}^n$ consists of linearly independent functions can easily be checked once the χ_j's are computed. For the case of an isotropic material it can directly be seen already in the original equation (1). The matrix $\underline{\underline{A}}(y)$ is here given by $a(y)\underline{\underline{I}}$ ($\underline{\underline{I}}$ denoting the identity), and linear dependency in the set $1 \cup \{\chi_j\}_{j=1}^n$ is equivalent to the fact that there exists $\underline{\beta} \in \mathbb{R}^n \setminus \{0\}$ with $\underline{\beta} \cdot \nabla_y a(y) = 0$.

One can prove the following convergence result for the alternate method given by the system (9).

Theorem 3.1 Assume that $\partial\Omega$ is sufficiently smooth and that the functions $\{\chi_j\}_{j=1}^n$ are elements of $W^{1,\infty}([0,1]^n)$ and satisfy a certain nondegeneracy condition. If $f \in H^s(\Omega)$, $-1 < s$, then for any $t < s$ there exists C_t such that

$$\| u^\varepsilon(x) - (u_0^\varepsilon(x) + \varepsilon \sum_{j=1}^n \chi_j(x/\varepsilon) \ u_j^\varepsilon(x)) \|_1 \leq C_t \varepsilon^\nu \tag{10}$$

for any ε in $]0,1]$, where $\nu = \min\{(1+t)/N, 1/2\}$ and $N = \max\{n,2\}$.

Note. The nondegeneracy required is essentially that the set $1 \cup \{\chi_j\}_{j=1}^n$ be linearly independent, for more details see [5]. The existence of linear dependent exceptional cases to the theorem is quite unimportant, especially in the light of the previous note, which showed how linear dependency among $1 \cup \{\chi_j\}_{j=1}^n$ actually represented a simplification (for which the theorem is now valid). A question of much more interest is how the estimate, i.e., the constant C_t, depends on the distance to an exceptional set of χ_j's. This is also analyzed in [5], although only for the case $n = 1$, and it is shown that the constant C_t is totally independent of how close the $\underline{\underline{A}}$ gets to an element for which the χ degenerates. (In this case that happens if $\underline{\underline{A}}$ is a constant.) We conjecture that the same is true in more dimensions.

If f and $\partial\Omega$ are sufficiently smooth both Theorem 2.1 and Theorem 3.1 give a convergence rate of $\varepsilon^{1/2}$ (which incidentally in both cases is the optimal rate). This is the basis for our claim that the two procedures perform equally well for smooth data and small ε.

We shall now individually address the points (a)-(e) which were brought up in the previous section and see how these are resolved for the variationally based procedure.

(a) Let u be sufficiently regular and assume that none of the derivatives $(\partial/\partial x_j)u$ vanish identically on $\partial\Omega$. Let $\{m_j^\varepsilon\}_{j=0}^n$ be an "optimal cut-off" for $u(x) + \varepsilon \sum_{j=0}^n \chi_j(x/\varepsilon)(\partial/\partial x_j)u(x)$. By this we mean that $m_j^\varepsilon(x) = m_j(d(x)/\varepsilon)$ where $d(x) = \text{dist}(x,\partial\Omega)$, and where $m_j \varepsilon W^{1,\infty}([0,\infty[)$, satisfying $m_j(0) = 0$, $1 \leq j \leq n$, $m_j = 1$ on $[1,\infty[$, $0 \leq j \leq n$, have been selected such as to minimize the energy norm of

$$u^\varepsilon(x) - (m_0^\varepsilon(x)u(x) + \varepsilon \sum_{j=1}^n \chi_j(x/\varepsilon)\, m_j^\varepsilon(x) \frac{\partial}{\partial x_j} u(x))$$

asymptotically for small ε. We can show that the boundary layer behavior of

$$u_0^\varepsilon(x) + \varepsilon \sum_{j=1}^n \chi_j(x/\varepsilon) u_j^\varepsilon(x)$$

near $\partial\Omega$ is the same as that of

$$m_0^\varepsilon(x)u(x) + \varepsilon \sum_{j=1}^n \chi_j(x/\varepsilon) m_j^\varepsilon(x) \frac{\partial}{\partial x_j} u(x) \quad .$$

(b) As is clear from Theorem 3.1 the variational approach has no problem if f is not in $L^2(\Omega)$. As long as $f \varepsilon H^s(\Omega)$, $-1 < s$, it will always produce a sequence $s^\varepsilon \ \varepsilon \ \overset{\circ}{H}{}^1(\Omega)$ such that $\| u^\varepsilon - s^\varepsilon \|_1 \to 0$. Note that $f \varepsilon H^{-1}(\Omega)$ is necessary in order for our original problem (1) to have a solution in $\overset{\circ}{H}{}^1(\Omega)$.

(c) Theorem 3.1 only covers the case that $\partial\Omega$ has a certain degree of smoothness. For the practically very important case of a <u>piece-wise</u> smooth domain with a finite number of corners of arbitrarily large angles (here the dimension n = 2) we have been able to show that

$$\| u^\varepsilon(x) - (u_0^\varepsilon(x) + \varepsilon \sum_{j=1}^n \chi_j(x/\varepsilon) u_j^\varepsilon(x)) \|_1$$

still converges to 0 as $\varepsilon \to 0$. (The actual convergence rate of course depends on the maximal angle as well as the regularity of f). The proof of this fact is based on (i) the stability of the variationally based approach even for very rough boundaries (ii) a nontrivial interpolation result for a scale of Sobolev spaces on a domain with corners, and finally (iii) regularity properties of solutions to constant coefficient elliptic boundary value problems on nonsmooth domains.

(d) The numerical experiments that are reported in the next section clearly show that the expression

$$u_0^\varepsilon(x) + \varepsilon \sum_{j=1}^{n} \chi_j(x/\varepsilon) u_j^\varepsilon(x)$$

generally performs much better than

$$u(x) + \varepsilon \sum_{j=1}^{n} \chi_j(x/\varepsilon) m^\varepsilon(x) \frac{\partial}{\partial x_j} u(x)$$

for large to moderate size ε. This is not quite unexpected, since the first expression is almost a projection.

(e) From a computational point of view

$$u_0^\varepsilon(x) + \varepsilon \sum_{j=1}^{n} \chi_j(x/\varepsilon) u_j^\varepsilon(x)$$

is also very desirable. By a Finite Element approximation we can use C^0-elements for all the fields u_0^ε, $\{u_j^\varepsilon\}$. The variationally based approach has many similarities to so called mixed methods for the Finite Element Method (cf.[4]), which commonly treat a function and its derivatives as independent variables. We also refer to [2] for a discussion of a conceptually very related procedure used by some engineers.

4. Numerical experimentation

Let $E^\varepsilon(\cdot)$ denote the energy functional associated with our original problem (1), i.e.,

$$E^\varepsilon(\cdot) = 1/2 \iint_\Omega [\underline{\underline{A}}(x/\varepsilon)\nabla_x \cdot \nabla_x \cdot + b(x/\varepsilon)(\cdot)^2]\, dx - \iint_\Omega f \cdot dx$$

For any $v \in \overset{o}{H}{}^1(\Omega)$ it is clear that

$$\|u^\varepsilon - v\|_E^2$$

(11) $$= \iint_\Omega [\underline{\underline{A}}(x/\varepsilon)\nabla_x(u^\varepsilon - v)\nabla_x(u^\varepsilon - v) + b(x/\varepsilon)(u^\varepsilon - v)^2]dx$$

$$= 2[E^\varepsilon(v) - E^\varepsilon(u^\varepsilon)] .$$

(This illustrates the close connection between reliable upper bounds for the energy and good approximation of the fluxes.)

The relative error in (energy-norm)^2:

$$\|u^\varepsilon - v\|_E^2 \,/\, \|u^\varepsilon\|_E^2$$

and the relative error in energy

$$(E^\varepsilon(u^\varepsilon) - E^\varepsilon(v))/E^\varepsilon(u^\varepsilon)$$

are as consequence of (11) the same.

The computations performed here are for the simplest case n = 1. As our domain Ω we picked the interval [0,1]. $\underline{\underline{A}}(y) = a(y)$ and b(y) are two-phased, i.e., both are periodic with period 1,

$$a(y) = \begin{cases} a_0, & 0 \le y < D \\ a_1, & D < y \le 1 \end{cases}$$

and the same for b, with constants b_0 and b_1 respectively. Here D (the "volume" fraction of a_0 and b_0) is a given number in [0,1].

The cut-off function m that enters in the sum

$$u(x) + \varepsilon\chi(x/\varepsilon)\, m^\varepsilon(x)\frac{d}{dx}u(x)$$

is defined as

$$m^\varepsilon(x) = m(d(x)/\varepsilon)$$

where

$$d(x) = \min\{1-x, x\}$$

and

$$m(t) = \begin{cases} t & \text{for } 0 \le t \le 1 \\ 1 & \text{for } 1 \le t \end{cases}$$

Without performing any analysis to find the optimal m this seems a most natural and simple choice.

The function χ, being the periodic solution to

$$-\frac{d}{dy}a(y)\frac{d}{dy}\chi(y) = \frac{d}{dy}a(y)$$

with $\overline{a\chi} = 0$, is expressed explicitly as a function of y parametrized by a_0, a_1 and D.

Based on this we can compute values for the coefficients of the system (9) and the effective diffusivity entering the equation (4).

Actually the effective diffusivity can be derived directly from a by means of the formula $(a_0^{-1}D + a_1^{-1}(1-D))^{-1}$ (the harmonic average), but it should be noted that this is a strictly one-dimensional phenomenon.

The equations (4) and (9) respectively are now solved by a Finite Element Method. In the first case we use piecewise cubic C^1-elements as test and trial functions (since we have to compute a derivative), in the second case we only use piecewise linear C^0-elements as test and trial functions for the two fields. In both cases we take more than 1000 degrees of freedom. We now form the two expressions

$$\text{I:} \quad u(x) + \varepsilon\chi(x/\varepsilon)m^{\varepsilon}(x)\,\frac{d}{dx}\,u(x)$$

$$\text{II:} \quad u_0^{\varepsilon}(x) + \varepsilon\chi(x/\varepsilon)u_1^{\varepsilon}(x) \quad ,$$

where u denotes the solution to (4) and $(u_0^{\varepsilon}, u_1^{\varepsilon})$ the solution to (9).

In order to compare the performance of these two expressions we also compute the true solution u^{ε} to (1). This is done also by a Finite Element Method with piecewise linear C^0 test and trial functions and more than 1000 degrees of freedom. The solutions of all the involved algebraic equations are performed by Gauss elimination.

The numbers in the following tables are the obtained values for the relative error in energy

$$(E^{\varepsilon}(u^{\varepsilon}) - E^{\varepsilon}(v))/E^{\varepsilon}(u^{\varepsilon})$$

in the case that v is given by I and II, respectively. The very refined meshes of the Finite Element Methods were taken to ensure that these numbers only reflect the relative error introduced by the homogenization and not any numerical errors.

For Table 1, the various constants are given by the following values: $a_0 = 2$, $a_1 = 1$, $D = 0.5$, $b_0 = b_1 = 1$ and the right hand side f is the constant function = 1. The table clearly shows that for as smooth data as this there is only little to gain from the expression II.

TABLE 1

ε	Relative Energy Error with I in %	Relative Energy Error with II in %
0.5	6.77	7.84
* 0.45	9.26	5.09
* 0.4	11.92	10.02
* 0.35	6.51	5.04
* 0.3	7.70	4.48
0.25	5.01	4.97
0.2	4.31	4.15
* 0.15	4.04	2.12
0.1	2.48	2.25
0.05	1.34	1.19
0.04	1.07	0.93
* 0.03	0.89	0.45
0.02	0.55	0.45
0.01	0.28	0.20

Special attention should be called to the entries marked by an asterisk; these are the ones where ε does not divide 1, i.e., there is a chopped-off cell at the end of the interval [0,1]. In more than one dimension chopped-off cells will always occur near $\partial\Omega$ (unless the domain Ω is an interval and ε divides the respective lengths). As is clear from the table the decrease in error from I to II is most significant among the marked entries. This gives high hopes for the performance of our variational method in more than one dimension.

Table 2 contains the relative errors in energy for the case $a_0=100$, $a_1 = 1$, $D = 0.3$, $b_0 = b_1 = 0$ and $f \equiv 1$.

TABLE 2

	Relative Energy Error with I in %	Relative Energy Error with II in %
0.5	2119.9	92.4
* 0.45	4691.1	98.2
* 0.4	2925.9	88.4
* 0.35	1612.5	75.7
* 0.3	3162.5	95.6
0.25	1102.5	60.7
0.2	905.9	54.3
* 0.15	892.0	39.5
0.1	484.9	36.6
0.05	282.1	47.1
0.04	193.9	12.0
* 0.03	273.0	11.6
0.02	98.0	6.3
0.01	49.0	3.2

It is evident from the last four entries in Table 1 and Table 2 that all the energy errors converge like ε , exactly as predicted for this kind of smooth right hand side by Theorems 2.1 and 3.1, respectively.

It is quite strikingly clear however that the large jumps in the coefficient a make the expression II a much better approximation to u^ε than I. This is particularly clear for $\varepsilon \geq 0.03$, where I according to Table 2 consistently produces an error larger than 100%, meaning that even the constant = 0 is a better approximation. In terms of upper bounds for the energies this is reflected in the fact that I for $\varepsilon \geq 0.03$, in the example of Table 2, always gives a positive upper bound, which is not very informative as we well know that the exact energy is negative. II on the other hand consistently produces upper bounds that are strictly negative.

Let us for a moment imagine that one is only interested in finding a good approximate value for the energy, but is not particularly interested in good approximations to fluxes of u^ε or one-sided bounds. If u denotes the solution to (4) then one can form

(12) $\frac{1}{2} \iint_{\Omega} [\underline{\mathbf{a}} \nabla_x u \; \nabla_x u + \bar{b} u^2] \, dx - \iint_{\Omega} fu \, dx \; .$

According to [3] the difference between this and the exact energy behaves like $O(\varepsilon)$.

Table 3 lists the values of the exact energy of u^{ε} for the same case as before, namely $a_0 = 100$, $a_1 = 1$, $D = 0.3$, $b_0 = b_1 = 0$ and $f \equiv 1$.

TABLE 3

ε	Exact Energy
0.5	-0.02560
0.45	-0.01878
0.4	-0.02196
0.35	-0.02656
0.3	-0.02131
0.25	-0.02835
0.2	-0.02855
0.15	-0.02741
0.1	-0.02899
0.05	-0.02871
0.04	-0.02927
0.03	-0.02842
0.02	-0.02929
0.01	-0.02929

The energy computed directly according to (12) is -0.02929. From the table it is clear that -0.02929 is a very good approximation to the exact energy for most values of ε. (Here it appears as lower bound, but that is insignificant, there are examples where it is always above the energy and examples where it is neither.)

As a final example Table 4 contains the results for a singular load $f(x) = (1-x)^{-3/4}$ and $a_0 = 2$, $a_1 = 1$, $D = 0.5$, $b_0 = b_1 = 1$. Note that the expression I is well defined in $H^1([0,1])$ only because of the cut-off.

TABLE 4

	Relative Energy Error with I in %	Relative Energy Error with II in %
0.3	13.03	6.94
0.15	4.46	1.71
0.075	6.26	3.22
0.0375	1.66	0.55
0.01875	3.71	1.74

(The major contribution to the error in this example comes from near $x = 1$. This is the reason only to display entries with ε that do not divide 1. If ε divides 1 the two methods perform almost identical, even more so than in Table 1, but this is misleading).

Many more computations have been carried out with different coefficients a and b and other loads f, but the pattern that has emerged is exactly as illustrated by the previous numbers.

5. Conclusions

If either all the data is smooth or one is only interested in approximate values for integrated quantities, such as the energy, and convergence in norms not measuring derivatives, then "standard" homogenization seems to provide the kind of accuracy one would need for most practical applications. If however the data is unsmooth and it is essential to get good approximations also to the fluxes of the exact solution (in elasticity e.g. the stresses, in order to predict cracking), then "standard" homogenization leaves much to be desired. The same is true if one wants consistent upper (or lower) bounds for the energy. For these purposes our variationally based procedure (9) seems to be the right answer. (To obtain lower bounds for the energy one should of course develop a variationally based procedure from the dual variational formulation of (1).)

References

[1] Adams, R.A., Sobolev Spaces, Academic Press, 1975.

[2] Babuska, I., Homogenization approach in engineering. pp. 137-153, in Computing Methods in Applied Sciences and Engineering. Lecture Notes in Economics and Mathematical Systems, 134. Springer, 1976.

[3] Bensoussan, A., Lions, J.L., and Papanicolaou, G., Asymptotic Analysis for Periodic Structures. North-Holland, 1978.

[4] Ciarlet, P.G., The Finite Element Method for Elliptic Problems. North-Holland, 1978.

[5] Vogelius, M. and Papanicolaou, G., A projection method applied to diffusion in a periodic structure. To appear in SIAM Journal on Applied Mathematics.

EFFECTIVE MEDIUM APPROXIMATION FOR DIFFUSION ON RANDOM NETWORKS*

Itzhak Webman
Departments of Mathematics and Physics
Rutgers University
New Brunswick, N. J. 08903
and
Courant Institute of Mathematical Sciences
New York University
New York, N. Y. 10012

There is a rapidly growing interest in the problem of classical diffusion in random systems.[1-3] It is relevant to a number of physical processes in disordered media such as dispersive hopping transport in amorphous semiconductors[1,2,4] and the migrations of localized electronic excitations among guest molecules in a host.[5,6] The main current theoretical approach to these phenomena is based on the continuous time random walk theory.[1,2,4,7] Alternative methods were recently used to study the problem of one-dimensional systems where some aspects of it can be treated more rigorously.[3,8,9]

In this paper a new self consistent effective medium approximation is proposed for the related problem of a diffusion on a lattice characterized by random values of transfer rate $W_{n'n} = W_{nn'}$ between pairs of nearest neighbor sites. These values are assigned to the lattice bonds according to a given p.d.f. $\pi(W)$ in a random manner. The approach presented here is closely related to the effective medium theory for the macroscopic conductivity and dielectric properties of random inhomogeneous media (EMT),[10] and to the coherent potential approximation for the electronic properties of alloys (CPA).[11]

For a given realization of the random lattice the diffusion process is described by the following master equation for $P_n(t)$, the probability to be at site n at time t. Given that the diffusing quantity is at $n = 0$ at time $t = 0$.

$$(1)\qquad \frac{\partial P_n(t)}{\partial t} = \sum_{\substack{n' \\ \text{nearest neighbors} \\ \text{of } n}} \left(W_{nn'} P_{n'}(t) - W_{n'n} P_n(t)\right)$$

*Supported by AFOSR Grant No. 78-3522, and U.S. DOE Contract No. DE-AC02-79ER10353 and U.S. DOE Contract EY-76-C-02-3077.

with the boundary conditions $P_n(t = 0) = \delta_{n,0}$.

Consider the Laplace transform of Eq. (1),

$$\sum_{\substack{n' \\ \text{nearest neighbors} \\ \text{of } n}} \left[W_{nn'}\tilde{P}_{n'}(\omega) - W_{n'n}\tilde{P}_{n'}(\omega)\right] = \omega\tilde{P}_n(\omega) + \delta_{n,0}\tilde{P}_n(0) \tag{2}$$

Here

$$\tilde{P}_n(\omega) = \int e^{-\omega t} P_n(t)\, dt\ . \tag{3}$$

Eq. (2) can be recast as the following matrix equation,

$$A(\omega)\ \hat{P}(\omega) = S \tag{4}$$

where, using "bra-ket" notation

$$\hat{P}(\omega) \equiv \sum \tilde{P}_n(\omega)\ |n\rangle \tag{5a}$$

$$S = \sum_n \delta_{n,0}\ |n\rangle \tag{5b}$$

and

$$A(\omega) \equiv \sum_{k\ell} k\rangle\ [(\omega + \sum_i W_{ik})\delta_{k\ell} - W_{k\ell}]\ \langle\ell| \tag{6}$$

where the summation is over all pairs of nearest neighbor sites.

All the information concerning the diffusion process can be derived from $\{\langle P_n(\omega)\rangle\}$, where $\langle\ \rangle$ denotes an average over the ensemble of random lattices. For example, the mean square displacement of the diffusing quantity from the origin at time t is given by

$$\langle\vec{R}^2(t)\rangle = \mathcal{L}^{-1}\left[\sum_n \langle\tilde{P}_n(\omega)\rangle\ \vec{R}_n^2\right] \tag{7}$$

where R_n is the location of site n and $\mathcal{L}^{-1}$ denotes inverse Laplace transform. Accordingly what is needed is an approximation to $\langle A(\omega)^{-1}\rangle$.

$A(\omega)$ can be represented as a sum of a homogeneous term and a term which contains the random fluctuations:

$$A(\omega) = A_M(\omega) + \delta A(\omega) \tag{8a}$$

where

$$A_M(\omega) = \sum_{k\ell} |k\rangle\ [(\omega + zW_M)\delta_{k\ell} - W_M\Delta_{k\ell}]\ \langle\ell| \tag{8b}$$

$$\delta A(\omega) = \sum_{k,\ell} 2(W_{k\ell} - W_M)\hat{Q}_{k\ell} \tag{8c}$$

$$\hat{Q}_{k\ell} = \frac{1}{2}\ [|k\rangle - |\ell\rangle][\langle k| - \langle\ell|] \tag{8d}$$

and

$$(9) \qquad \Delta_{k\ell} = \begin{cases} 1 & \text{if } k, \text{ are nearest-neighbors} \\ 0 & \text{otherwise.} \end{cases}$$

A^{-1} can now be expressed as a t matrix expansion:[12,13]

$$(10) \qquad A^{-1} = A_M^{-1} + A_M^{-1} \, T \, A_M^{-1}$$

where

$$(11) \qquad T = \sum_{k\ell} t_{k\ell} + \sum_{k\ell \neq mn} t_{k\ell} A_M^{-1} t_{mn} + \sum_{\substack{k \neq mn \\ mn \neq pq}} t_k A_M^{-1} t_{mn} A_M^{-1} t_{pq}$$

and the t matrix for the bond $k\ell$ is:

$$(12) \qquad t_{k\ell} = Q_{k\ell} \frac{W_M - W_{k\ell}}{1 - \frac{1}{2} (\langle k| - \langle \ell|) A_M^{-1} (|k\rangle - |\ell\rangle)(W_M - W_{k\ell})}$$

An effective homogeneous lattice which represents the ensemble of random lattices will be characterized by a ω-dependent transfer rate $W_M(\omega)$ which solves

$$(13) \qquad \langle T(W_M(\omega)) \rangle = 0$$

such that

$$(14) \qquad \langle A^{-1} \rangle = A_M^{-1}(W_M(\omega)) \; .$$

Eq. (14) implies that the ensemble averages $\langle \tilde{P}_n(\omega) \rangle$ obey the following equation,[14]

$$\sum_{\substack{n' \\ \text{nearest neighbors} \\ \text{of } n}} W_M(\omega) (\langle \tilde{P}_{n'}(\omega) \rangle - \langle \tilde{P}_n(\omega) \rangle) = \omega \langle \tilde{P}_n(\omega) \rangle \; .$$

The effective medium approximation for $W_M(\omega)$ is obtained by setting

$$(15) \qquad T \cong \sum_{k\ell} t_{k\ell}$$

and solving

$$(16) \qquad \langle t \rangle = \int t(W', W_M(\omega)) \; \pi(W') \, dW' = 0 \; .$$

Using Eq. (12) one is led to the following equation for $W_M(\omega)$:

$$\text{(17a)}\quad \left\langle \frac{W_M(\omega) - W'}{W'(1-\varepsilon G_0(\varepsilon)) + [(\frac{z}{2} -1) + \varepsilon G_0(\varepsilon)]\, W_M(\omega)} \right\rangle = 0$$

where $G_0(\varepsilon) \equiv \langle 0|\ A_M^{-1}(W_M(\omega))\ |0\rangle$ is the diagonal element of the lattice Green function at $\varepsilon = \omega/W_M(\omega)$, and z is the coordination number of the lattice. For a cubic lattice in d dimensions $G(\varepsilon)$ is given by:

$$\text{(17b)}\qquad G_0(\varepsilon) = \int^{\infty} e^{-(\frac{z}{2} + \varepsilon)t} [I_0(t)]^d \, dt$$

where $I_0(t)$ is the modified Bessel function of order 0.

The mean square displacement $\langle R^2(t)\rangle$, and the frequency dependent conductivity $\sigma(\omega)$ for the hopping transport on the lattice are related to $W_M(\omega)$ by the following relations.[2]

$$\text{(18)}\qquad \begin{aligned} \langle R^2(t)\rangle &= \mathcal{L}^{-1}\left(\frac{z}{\omega^2} W_M(\omega)\right) \\ \sigma(\omega) &= \frac{z}{2} W_M(i\omega) \end{aligned}$$

The approximation presented here has the following features in common with the CPA and EMT.

(a) The condition $\langle t\rangle = 0$ leads to the vanishing of the averages of the next two terms in the expansion of T given by Eq. (11). This result is due to the absence of correlation between values of W' assigned to different bonds, together with the restrictions on the summations in Eq. (11). The first nonvanishing term in $\langle T\rangle_{EMA}$ is thus $O(t^n)$.

(b) Since this scheme is based on expansion in t matrices (rather than expansion in δA) it is not perturbative and thus not limited to weak disorder.

(c) The corrections to the effective inclusion approximation can be estimated by studying the nonvanishing term in $\langle T\rangle_{EMA}$.[16]

(d) It is amenable in principle to systematic improvement by including terms of higher order than t^3 in T and solving $\langle T(W_A(\varepsilon)\rangle = 0$.

For $\omega = 0$, Eq. (17) reduces to an equation equivalent to the EMT for the D.C. conductivity, and it yields a result for the diffusion coefficient $D = \frac{z}{2} W_M(0)$. The limits of validity of the EMT for the disordered resistor network have been studied by comparison with numerical results.[13,17] It was found to be a good approximation even for rather broad distributions of local transfer rates $\pi(W')$.

An interesting example for which one can obtain some analytical results from Eq. (17) is the case of a disordered lattice with the following distribution $\pi(W')$,

(19) $$\pi(W') = p\delta(W' - W_0) + (1-p)\delta(W')$$

i.e. the case where a fraction (1-p) of the bonds are characterized by a zero transfer rate.

Using the asymptotic expression for $G_0(\varepsilon)$ for $\varepsilon << 1$ for a S.C. lattice in Eq. (17) one obtains the following results for $W_M(\omega)$,

(20) $$W_M(p,\omega) = \frac{3}{2} W_0(p-p_c) \left[1 + \frac{2}{9} \frac{a\omega}{W_0(p-p_c)^2}\right], \qquad p > p_c$$

$$\frac{a\omega}{3|p-p_c|} \left[1 - \frac{1}{18} \frac{a\omega}{W_0(p-p_c)^2}\right], \qquad p < p_c$$

in the small ω limit, $\omega << W_0(p-p_c)^2$. Here $p_c = \frac{1}{3}$ and $a = G_0(0)$.

An analysis of the long time diffusive behavior based on Eq. (20) leads to the following results:

(a) In the limit of large t,

(21) $$\langle R^2(t)\rangle = D\,t, \quad D \propto W_0(p-p_c) \qquad \text{for } p > p_c$$

$$\lim_{t\to\infty} \langle R^2(t)\rangle \propto \frac{a}{p_c-p} \qquad \text{for } p < p_c$$

(b) For both $p > p_c$ and $p < p_c$ the above asymptotic behavior is obtained for $t >> \tau$ where τ diverges as $p \to p_c$.

(22) $$\tau \sim W_0^{-1}(p - p_c)^{-2}$$

These results reflect in a qualitative manner the properties of the clusters of bonds of one type on a lattice with a percolation threshold at $p = p_c$. The absence of infinite clusters of conducting bonds for $p < p_c$ leads to the vanishing of D at $p = p_c$ and to the absence of diffusion at $p < p_c$. The increasing tortuosity of the large clusters as $p \to p_c$ results in the divergence of the time of approach to asymptotic behavior at both $p > p_c$ and $p < p_c$.

At the percolation threshold $p = p_c$

(23) $$W_M(\omega) = \left(\frac{a\omega}{W_0}\right)^{1/2}$$

At long times one obtains anomalous diffusive behavior of the type

(24) $$\langle R^2(t)\rangle \propto (W_0 t)^{1/2}.$$

The expressions for $W_M(\omega)$ in Eq. (20) and Eq. (23) can be recast in the following scaling form:

$$W_M(p-p_c,\omega) = W_0(p-p_c)^{t_c} f(y) \tag{25}$$

$$y = a\omega W_0^{-1}|p-p_c|^{-\gamma} \quad .$$

A similar scaling form near p for an analogous problem has been suggested by Stephen.[18]

Eq. (25) leads to the following general time dependent diffusive behavior near the percolation threshold:

$$\langle R^2(t)\rangle = Dt \qquad p > p_c \tag{26}$$

$$D \propto W_0(p-p_c)^{t_c} \qquad t \gg \tau$$

$$\lim_{t\to\infty} \langle R^2(t)\rangle \propto a|p-p_c|^{-(\gamma-t_c)} \qquad p < p_c$$

$$\tau \sim W_0^{-1}|p-p_c|^{-\gamma} \qquad t \gg \tau$$

and at $p=p_c$

$$\langle R^2(t)\rangle \propto (W_0 t)^{(\gamma-t_c)/\gamma} \quad . \tag{27}$$

The EMA value for the percolation conductivity exponent is $t_c = 1$ while $\gamma_{EMA} = 2$. The numerical values for t_c are ~ 1.6 in $d=3$ and $1.1-1.3$ in $d=2$ [13,17]. A scaling law $\gamma = t_c + 2\nu - \beta$ where β is the percolation probability exponent has been proposed.[18] Accordingly $\gamma \cong 2.8$ in $d=3$. Thus, the EMA results follow the correct scaling behavior but yields incorrect values for the exponents. One can expect the EMA to be more accurate away from the critical region or for random systems with distributions $\pi(w')$ which do not lead to critical behavior.

Work is presently in progress on obtaining results for various distributions $\pi(W')$. Numerical studies intended to assess the range of validity of the effective medium approximation in various cases will also be carried out.

The author is grateful for stimulating discussions with M. H. Cohen, J. Klafter, J. L. Lebowitz and J. K. Percus.

NOTE: The author's present address is Exxon Research and Engineering, P.O. Box 45, Linden, N.J. 07306.

REFERENCES

1. H. Scher and M. Lax, Phys. Rev. B7 4491, 4502 (1973).
2. E.W. Montroll and B.J. West in "Fluctuation Phenomena," E.W. Montroll and J.L. Lebowitz, eds., North Holland, 1979; and references therein.
3. S. Alexander, J. Bernasconi, W.P. Schneider and R. Orbach, Rev. Mod. Phys. 53, 175 (1981); and references therein.
4. H. Scher and E.W. Montroll, Phys. Rev. B12, 2455 (1975).
5. S.W. Haan and R. Zwanzig, J. Chem. Phys. 68, 1877 (1977).
6. J. Klafter and R. Silbey, J. Chem. Phys. 72, 843 (1980).
7. E.W. Montroll and G.H. Weiss, J. Math. Phys. 6, 167 (1965).
8. J. Bernasconi, S. Alexander and R. Orbach, Phys. Rev. Lett. 41, 185 (1978).
9. T. Odagaki and M. Lax, Phys. Rev. Lett. 45, 847 (1980).
10. R. Landauer, Phys. Rev. 94, 1386 (1954).
11. R.J. Elliot, J.A. Krumhansl and P.L. Leath, Rev. Mod. Phys. 46, 465 (1974).
12. K.M. Watson, Phys. Rev. 103, 489 (1956); Phys. Rev. 105, 1388 (1957).
13. S. Kirkpatrick, Rev. Mod. Phys. 45, 574 (1973).
14. The time dependent averages $\{\langle P_n(t)\rangle\}$ will now be the solution of a Generalized Master Equation:

$$\frac{\partial\langle P_n(t)\rangle}{\partial t} = \sum_{n'} \int W_M(t-t')(\langle P_{n'}(t')\rangle)dt \quad .$$

This result agrees with the observation made by J. Klafter and R. Silbey, Phys. Rev. Lett. 64, 55, 1980.

15. G.F. Koster and J.L. Slater, Phys. Rev. 96, 1208 (1954).
16. I. Webman and M.H. Cohen (unpublished); J. Koplik (unpublished); D.J. Bergman and Y. Kantor (unpublished).
17. I. Webman, J. Jortner and M.H. Cohen, Phys. Rev. B11, 2885 (1975).
18. M.J. Stephen, Phys. Rev. B17, 4444 (1978).

LIST OF PARTICIPANTS*

G. Leigh Anderson
Exxon Production Research Co.
PO Box 2189
Houston, TX 77001

D. K. Babu
Dept. of Civil Engineering
Princeton University
Princeton, NJ 08544

Neil Berger
Dept. of Mathematics
Univ. of Illinois at
Chicago Circle
Box 4348
Chicago, Ill. 60680

Gregory Beylkin
Courant Institute

Richard Bourret
Dept. of Physics
University of Miami
Coral Gables, Fla.

Russel Caflisch
Dept. of Mathematics
Stanford University
Stanford, CA 94305

A. Callegari
Exxon Res. & Eng. Co.
Corporate Research Laboratories
PO Box 45
Linden, NJ 07036

I-Chung Chang
Courant Institute

Gary Chirlin
Dept. of Civil Engineering
Princeton University
Princeton, NJ 08544

Doina Cioranescu
Lab. Analyse Numérique
Paris VI France

Virginia A. Clark
Exxon Production Research Co.
PO Box 2189
Houston, TX 77001

Benoit Cushman-Roisin
Dept. of Oceanography, WB-10
University of Washington
Seattle, WA 98195

Robert Dautray
C.E.L. B.P. 27
94190 Villeneuve Saint-Georges
France
(French Atomic Energy Commission)

David Dellwo
U.S. Merchant Marine Academy

A. K. Didwania
Exxon Res. & Eng. Co.
Corporate Research Laboratories
PO Box 45
Linden, NJ 07036

Nat Fisch
Plasma Physics Lab.
Princeton University
Princeton, NJ 08544

Kenneth M. Golden
Courant Institute

Malcolm Goldman
Courant Institute

Lynne Ipiña
Courant Institute

Joel Koplik
Schlumberger-Doll Research Ctr.
PO Box 307
Ridgefield, Conn. 06877

Rolf Landauer
IBM Research Ctr.
PO Box 218
Yorktown Hts., NY 10598

Florian Lehner
Div. of Engineering
Brown University
Providence, R.I. 02912

G. Marshall
The Rockefeller University
1230 York Avenue
New York, NY 10021

---

*Excluding those participants who presented papers at the conference as their names and affiliations are included therein.

PARTICIPANTS continued

Jeff McFadden
Courant Institute

R. E. Meyer
Mathematics Research Center
University of Wisconsin
Madison, WI 53706

Satoshi Mochizuki
Exxon Production Research Co.
PO Box 2189
Houston, TX 77001

George Morikawa
Courant Institute

C. Morshedi
Courant Institute

C. Nicholson
New York University
Medical Center
Dept. of Physiology
New York, NY 10016

Mark Orman
Riverside Research Institute
80 West End Ave.
New York, NY 10023

Lowell Palecek
University of Minnesota

Bradley Plohr
Rockefeller University
1230 York Avenue
New York, NY 10021

Raghu Raghavan
Riverside Research Institute
80 West End Ave.
New York, NY 10023

Marty Reiman
Bell Laboratories
600 Mountain Ave.
Murray Hill, NJ 07974

Ed Rinehart
Exxon Production Research Co.
PO Box 2189
Houston, TX 77001

Joel C.W. Rogers
Applied Physics Laboratory
Johns Hopkins University
Laurel, Md. 20810

Vladimir Rokhlin
Exxon Production Research Co.
PO Box 2189
Houston, TX 77001

Rodolfo Rosales
Massachusetts Institute of Tech.
872 Massachusetts Ave. 507
Cambridge, Mass. 02139

P. Sarnak
Courant Institute

Len Schwartz
Exxon Res. & Eng. Co.
Corporate Research Laboratories
PO Box 45
Linden, NJ 07036

Tim Secomb
Bioengineering Institute
Columbia University
638 Mudd Bldg.
New York, NY 10027

Renato Spigler
Courant Institute
(Univ. of Padua (Italy))

George Stell
SUNY at Stony Brook
Dept. of Mech. Eng. & Chemistry
Stony Brook, NY 11794

D. Sulsky
Courant Institute

Charles Tier
Dept. of Mathematics
Univ. of Illinois at
Chicago Circle
Box 4348
Chicago, IL 60680

Aydin Tozeren
Dept. of Civil Engineering
Columbia University
New York, NY 10027
(Visiting Professor)

Eugene Trubowitz
Courant Institute

Hung-Sheng Tsao
Rockefeller University
500 E. 63 Street
New York, NY 10021

PARTICIPANTS continued

J. Watson
Dept. of Mathematics
Stanford University
Stanford, CA 94305

Abel Weinrib
Harvard University
Physics Dept.
Cambridge, MA 02138

J. Willemsen
Schlumberger-Doll Research Ctr.
PO Box 307
Ridgefield, CT 06877

Dennis Willen
Exxon Production Research Co.
PO Box 2189
Houston, TX 77001

Michael Williams
Dept. of Mathematics
Virginia Polytechnic Institute
and State University
Blacksburg, Va. 24061

Shao-Ping Wu
Courant Institute
(Hangchow University)

Erich Zauderer
Polytechnic Institute of N.Y.
333 Jay Street
Brooklyn, NY 11201

Xi-Chang Zhong
Courant Institute

Lecture Notes in Physics

Vol. 114: Stellar Turbule... D. F. Gray and J. L. Linsk...

Vol. 115: Modern Trends... Proceedings, 1979. Edit... IX, 597 pages. 1980.

Vol. 116: Mathematical... Proceedings, 1979. Ed... pages. 1980.

Vol. 117: Deep-Inelastic... Ions. Proceedings, 197... 394 pages. 1980.

Vol. 118: Quantum Chr... Edited by J. L. Alonso a...

Vol. 119: Nuclear Spectr... by G. F. Bertsch and D....

Vol. 120: Nonlinear Ev... Systems. Proceedings, ... nelli and G. Soliani. VI, 3...

Vol. 121: F. W. Wiegel, F... molecular Systems. V, 1...

Vol. 122: New Develop... Proceedings, 1979. Edite... 1980.

Vol. 123: D. H. Mayer, T... in Classical Statistical ...

Vol. 124: Gravitational ... Exact Solutions. Procee... VI, 487 pages. 1980.

Vol. 125: Nonradial and... ceedings, 1980. Edited b... VIII, 497 pages. 1980.

Vol. 126: Complex Analy... ativistic Quantum Theory... Iagolnitzer. VIII, 502 pag...

Vol. 127: E. Sanchez-Pa... and Vibration Theory. IX, ...

Vol. 128: Neutron Spin E... F. Mezei. VI, 253 pages. ...

Vol. 129: Geometrical an... Theories. Proceedings, ... S. Shnider. VIII, 155 page...

Vol. 130: Mathematical Methods and Applications of Scattering Theory. Proceedings, 1979. Edited by J. A. DeSanto, A. W. Sáenz and W. W. Zachary. XIII, 331 pages. 1980.

Vol. 131: H. C. Fogedby, Theoretical Aspects of Mainly Low Dimensional Magnetic Systems. XI, 163 pages. 1980.

Vol. 132: Systems Far from Equilibrium. Proceedings, 1980. Edited by L. Garrido. XV, 403 pages. 1980.

Vol. 133: Narrow Gap Semiconductors Physics and Applications. Proceedings, 1979. Edited by W. Zawadzki. X, 572 pages. 1980.

Vol. 134: γγ Collisions. Proceedings, 1980. Edited by G. Cochard and P. Kessler. XIII, 400 pages. 1980.

Vol. 135: Group Theoretical Methods in Physics. Proceedings, 1980. Edited by K. B. Wolf. XXVI, 629 pages. 1980.

...ctures in Modelling ...s 1980. Edited by

...rks in Nuclei. Edited ...I, 414 pages. 1981.

... Proceedings 1980. ...374 pages. 1981.

...ometric Methods in ...81. VII, 329 pages.

...metry of the Time- ...uantum Mechanics.

...ence on Numerical ...gs. Edited by W. C. ... 485 pages. 1981.

...ody Theories. Pro- ...de Llano, M. Fortes

... Quantum Electro- ...G. Gräff, E. Klempt

... A Comprehensive ...d by T.T.S. Kuo and

... A Comprehensive ...ceedings 1980/81. ...ng. VIII, 571-1.082

...spects of Nonlinear ... 341 pages. 1981.

...dent Thomas-Fermi

... Proceedings, 1980.

...alization. Proceed- ...Castro, and L. Peliti.

...ativitätstheorie und ...n. 1981.

...eory. Proceedings, 1981. Edited by J. Hietarinta and C. Montonen. V, 251 pages. 1982.

Vol. 152: Physics of Narrow Gap Semiconductors. Proceedings, 1981. Edited by E. Gornik, H. Heinrich and L. Palmetshofer. XIII, 485 pages. 1982.

Vol. 153: Mathematical Problems in Theoretical Physics. Proceedings, 1981. Edited by R. Schrader, R. Seiler, and D.A. Uhlenbrock. XII, 429 pages. 1982.

Vol. 154: Macroscopic Properties of Disordered Media. Proceedings, 1981. Edited by R. Burridge, S. Childress, and G. Papanicolaou. VII, 307 pages. 1982.

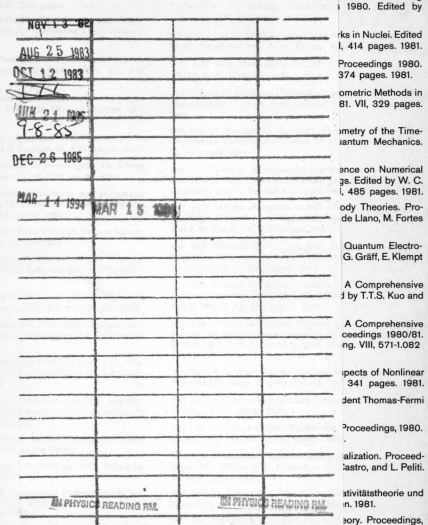